# GREEN INFRASTRUCTURE

With more than half of the world's population now living in urban areas, it is vitally important that towns and cities are healthy places in which to live. The principal aim of this book is to synthesise the disparate literature on the use of vegetation in the built environment and its multifunctional benefits to humans. The author reviews issues such as: contact with wildlife and its immediate and long-term effects on psychological and physical well-being; the role of vegetation in removing health-damaging pollutants from the air; green roofs and green walls, which provide insulation, reduce energy use and decrease the carbon footprint of buildings; and structural vegetation such as street trees, providing shading and air circulation whilst also helping to stop flash floods through surface drainage. Examples are used throughout to illustrate the practical use of vegetation to improve the urban environment and deliver ecosystem services.

Whilst the underlying theme is the value of biodiversity, the emphasis is less on existing high-value green spaces (such as nature reserves, parks and gardens), than on the sealed surfaces of urban areas (building surfaces, roads, car parks, plazas, etc.). The book shows how these, and the spaces they encapsulate, can be modified to meet current and future environmental challenges including climate change. The value of existing green space is also covered to provide a comprehensive textbook of international relevance.

**John W. Dover** is Professor of Ecology at Staffordshire University, Stoke-on-Trent, UK.

"This book comes at exactly the right time. The term 'Green Infrastructure' symbolizes new thinking in relation to planning and constructing cities. Vegetation offers many benefits to urban dwellers, and now is the time to integrate this knowledge into city planning procedures. Around the world, examples of GI now exist and national and international associations support and disseminate such ideas. This textbook highlights many such examples to introduce the concepts to a wider audience. I wish this book as many readers as possible; first of all students of related disciplines to take these ideas as seed into their future business lives, and also urban developers to integrate this new thinking into their daily business to spread out more green than grey infrastructure around the globe."

– *Manfred Koehler, Professor of Landscape Ecology, University of Applied Sciences Neubrandenburg, Germany, President of the World Green Infrastructure Network.*

"This is a very thoughtful, timely and comprehensive book. If our cities are to be considered as being truly sustainable then the integrated provision of green spaces will be a vital component of that sustainability. Green Infrastructure will help us to do just that."

– *Richard Sabin, Director, Living Green City, UK.*

# GREEN INFRASTRUCTURE

Incorporating plants and enhancing biodiversity in buildings and urban environments

*John W. Dover*

Routledge
Taylor & Francis Group

LONDON AND NEW YORK

earthscan
from Routledge

First published 2015
by Routledge
2 Park Square, Milton Park, Abingdon, Oxon OX14 4RN

and by Routledge
711 Third Avenue, New York, NY 10017

*Routledge is an imprint of the Taylor & Francis Group, an informa business*

© 2015 John W. Dover

*British Library Cataloguing-in-Publication Data*
A catalogue record for this book is available from the British Library

*Library of Congress Cataloging in Publication Data*
Dover, John W.
Green infrastructure : incorporating plants and enhancing biodiversity in buildings and urban environments / John W. Dover.
pages cm
Includes bibliographical references and index.
1. House plants in interior decoration. 2. Foliage plants. 3. Interior landscaping. I. Title.
SB419.D74 2015
635.9'65--dc23
2014048373

ISBN: 978-0-415-52123-9 (hbk)
ISBN: 978-0-415-52124-6 (pbk)
ISBN: 978-0-203-12199-3 (ebk)

Typeset in Bembo
by Fakenham Prepress Solutions, Fakenham, Norfolk NR21 8NN

# CONTENTS

# PREFACE

This book arose from an Urban Ecology module I developed for a Master's degree at Staffordshire University called 'Greening the Grey' which explored the value of biodiversity, mostly in the form of vegetation, to us as humans and also to wildlife. I became deeply interested in the subject and felt that some kind of summary was needed to help provide the background evidence to promote the greening of our cities. Writing took longer than expected, the subject matter is wide and took me down paths I had never before encountered and gave me diverse experiences. For example, I found beauty in the way that a geotextile membrane – unseen to us – helps degrade oil leaks from cars; I also found horror in the way that ultrafine particulates from pollution may enter our bodies. My personal view is that using biodiversity intelligently in our towns and cities is perhaps the only way we can confront some of the challenges of urban living facing us now, and in the future.

# ACKNOWLEDGEMENTS

I have a great debt of thanks to pay to the many people who have helped me during the process of writing this book. The greatest, of course, goes to my wife Christine and to my children Alice and Charlotte for their tolerance and encouragement. My postgraduate students on the Master's module 'Greening the Grey' at Staffordshire University and PhD student (now Dr) Caroline Chiquet have given me a great deal of pleasure and encouragement, and kept me on my toes over the years; without you I would not have started the work at all. Colleagues and friends from academia and the commercial world have been generous with their time and information, which I have truly valued. Non-academic friends and relatives (especially Shirley Dixon and Jerry West) also deserve a mention – for putting up with my constant talk on the subject and waiting patiently whilst I'd take a photograph of a particularly interesting wall, tree or roof (again!). I'd particularly like to thank Dr Caroline Chiquet, Christine Dover, Sean Farrell, Richard Sabin and Professor Philip James for reading and commenting on an earlier version of the manuscript and thus helping improve it; all remaining errors and inaccuracies are, of course, mine and mine alone. I'd also like to thank Tim Hardwick at Earthscan for agreeing to take on steering me through the process of the book's production, along with Ashley Wright, and not least for reminding me to get on with it! Thanks to Charlotte Dover for unearthing the information on the Holloway Sanatorium. Picture credits are given throughout the book, but I'd just like to thank, personally: Christine Dover and Caroline Chiquet, Staffordshire University, UK; Alan Goodkin, UK; Dr Alex Tan from the National University of Singapore; Sarah Fisk of Adnams plc, UK; Andrea Fürst from the Hundertwasser Archive, Vienna, Austria; Marc Grañén of Phytokinetic, Spain; Joel Nash of ANS Group Europe; Emily Hauth of the City of Portland, Oregon; Edmund Maurer of the Planning Department, Municipality of Linz, Austria; Dr Tom Smiley of the Bartlett Tree Research Lab, Charlotte, NC; Sean Farrell of Mobilane, UK; Richard Sabin of Biotecture, UK; Daniel Taylor and Liam Barrett of Bauder Ltd; David Luukas from Alumasc Exterior Building Products Ltd; Gary Grant of The Green Roof Consultancy; and Justin Hobson of Country Life. I am also exceedingly grateful to Rosie Duncan of Staffordshire University's Geography Department for redrawing several of the figures and to Christine Dover for indexing.

# 1

# WHAT IS GREEN INFRASTRUCTURE?

This chapter will:

- review some existing green infrastructure definitions
- present a modified definition of green infrastructure
- identify the reasoning behind the need to reduce the amount of sealed surface in urban areas.

## 1.1 Why green infrastructure?

I have been interested in green infrastructure, on and off, for a long time, since the early 1990s, even though it wasn't called green infrastructure then. In the dim and distant past, I worked for an environmental consultancy company, and one project I handled was to produce the evidence base to support the green corridors policy which Kirklees Metropolitan Borough Council (based in Huddersfield) were proposing for their Unitary Development Plan. It was clear to me then that green spaces and biodiversity in towns provided a number of benefits to the community (and I mean this in the broadest sense of people, public services and enterprises) and I recommended that they recognised that in their policy (Dover, 1992). Moving back into academia, I subsequently did some rather more esoteric work on wildlife corridors (Dover & Fry, 2001) and urban ecology (Dover, 2000), amongst other things, and in 2006 worked for the regeneration agency Renew North-Staffordshire on 'embedding green infrastructure in housing market renewal' (Dover, 2006a, 2006b) which 'renewed' my interest in using biodiversity to deliver important 'ecosystem' services to humans.

Ecosystem services are generally considered to be of four types:

- *Supporting services*, e.g. nutrient and water cycling, soil formation, primary production
- *Regulating services*, e.g. climate control and pollution removal
- *Provisioning services*, e.g. food, medicines, building materials
- *Cultural services*, e.g. societal appreciation of nature and environment (EASAC, 2009).

I passionately believe that green infrastructure is the only way to cost-effectively meet some of the most important environmental challenges we face in urban areas; challenges that we will have to meet continuously over the coming decades, and perhaps even over centuries to come. There are many good reasons for creating green infrastructure, but one which is of singular importance is coping with the impacts of climate change, and the associated intensification of other environmental problems such as air pollution.

## 1.2 Definitions of green infrastructure

Perhaps I should make it clear from the start: when I use the word 'green' I mean using plants, vegetation or even microbes (even those that are not green). So 'greening' buildings means, literally, growing plants on, around, up, or even in them. I will not be using the term for other more technological approaches to reducing the environmental impact of operating buildings such as using ground-source heat pumps (Cho & Choi, 2014) or in the sense of a 'green' political movement.

Green infrastructure is often seen as being about the value of 'open' green spaces, and mainly emphasises pre-existing ones. Definitions which exclude the built environment are unhelpful in that they do not give a policy steer to the greening (literally, with vegetation) of sealed surfaces (e.g. Benedict & McMahon, 2006; Goode, 2006; TCF, 2011; and NWGITT, 2007). In the UK the Town and Country Planning Association (TCPA) defined green infra-structure in this way:

> Green Infrastructure is the sub-regional network of protected sites, nature reserves, greenspaces, and greenway linkages. The linkages include river corridors and flood plains, migration routes and features of the landscape, which are of importance as wildlife corridors.
>
> Green Infrastructure should provide for multifunctional uses i.e., wildlife, recreational and cultural experience, as well as delivering ecological services, such as flood protection and microclimate control. It should also operate at all spatial scales from urban centres through to open countryside.
>
> *(TCPA, 2004)*

Yet ironically both the TCPA (2004) and Goode (2006) whilst using a definition that excludes building-related structures actually include them as examples within their publications (e.g. green roofs and sustainable urban drainage systems). Whilst many definitions persist in excluding built-environment components, some newer ones acknowledge the built-environment features explicitly, but only in respect of a limited set of attributes. For example, this one concentrates on sustainable urban drainage:

> Green infrastructure is management approaches and technologies that utilize, enhance and/or mimic the natural hydrologic cycle processes of infiltration, evapotranspiration and reuse. Green infrastructure approaches currently in use include green roofs, trees and tree boxes, rain gardens, vegetated swales, pocket wetlands, infiltration planters, porous and permeable pavements, vegetated median strips, reforestation/revegetation, and protection and enhancement of riparian buffers and floodplains.
>
> *(EPA, 2008)*

Others use a vague term that allows inclusion, but does not effectively promote built-environment features as such. For example:

> Green Infrastructure is a strategically planned and delivered network comprising the broadest range of high quality green spaces and other environmental features.
>
> *(NE, 2009a)*

The term 'other environmental features' was used in the UK government's White Paper on the natural environment (Defra, 2011b), although it is unclear exactly what these features are as they are not defined. Nevertheless, the document cites an example that includes green roofs and walls and refers to the Flood and Water Management Act 2010 that has provision for encouraging sustainable urban drainage.

So my preference would be for a definition such as (NE, 2009a) but which is more explicit. For example:

> Green Infrastructure is a strategically planned and delivered network comprising the broadest range of high quality green spaces and other environmental features *including surfaces such as pavements, car parks, driveways, roads and buildings that have been modified to incorporate biodiversity and promote ecosystem services.* (italics = my text)

The definition used by Tzoulas *et al.* (2007) is pretty good and seems to cover most eventualities and in very few words:

> all natural, semi-natural and artificial networks of multifunctional ecological systems within, around and between urban areas, at all spatial scales.

However, I have two remaining quibbles: 1) the last two definitions suggest a rather top-down approach and one which discounts individual features that have not been a priori included in some strategic planning and delivery vehicle – this seems perverse as surely green infrastructure is a sum of assets whether they fit into a strategic framework/network or not; 2) these definitions stop at the entrance to a building – I do not see why this should be so as ecosystem services can be delivered by biodiversity inside buildings as well as outside. So the definition which I will be using is:

> Green infrastructure is the sum of an area's environmental assets, including stand-alone elements and strategically planned and delivered networks of high-quality green spaces and other environmental features including surfaces such as pavements, car parks, driveways, roads and buildings (exterior and interior) that incorporate biodiversity and promote ecosystem services.

Now, having settled on a definition I am happy with, I am going to blithely ignore a huge chunk of it. Whilst I want to give an overview of the values of green infrastructure in the first part of the book, I want then to concentrate on the 'other environmental features including surfaces such as pavements, car parks, driveways, roads and buildings (exterior and interior) that incorporate biodiversity and promote ecosystem services'. Why? Principally because the 'strategic network' approach is well documented and a core component of spatial planning.

The subtitle of the excellent book *Green Infrastructure* by Benedict and McMahon (2006) is *Linking Landscapes and Communities*, and implies – incorrectly to my mind – that GI is the preserve of the spatial. I do not want to give the impression that spatial planning or the creation and promotion of networks of green spaces, linkages, etc. is not important: it clearly is. What I want to do is show how a green infrastructure approach that embraces the modification of our existing non-green areas, piecemeal or strategic, can improve our environment. I want to show how we can 'green the grey'.

## 1.3 Why should we want to 'green' buildings and reduce the amount of sealed surface in urban areas?

Whitford *et al.* (2001) identified four ways in which the ecology of cities is distinct from the surrounding countryside: 1) cities are warmer, 2) cities retain less water, 3) cities emit carbon dioxide, 4) cities have lower biodiversity; all these features are considered undesirable for one reason or another. These are generalisations, of course, and may not hold true in every case. Nevertheless, the point is that whilst cities are human habitats, humans are not very good at creating uniformly good ones for their inhabitants. The four features Whitford *et al.* (2001) identify are linked with poor environmental performance, but there are a whole clutch of others which impact directly and indirectly on human health which, for simplicity, can be bundled under the headings of 'pollution' and 'green access'. Human health, physical and mental, has been shown to be better where people have access to good quality green space (Croucher *et al.*, 2007; GSS, 2008) and especially that which is rich in biodiversity (Fuller *et al.*, 2007). Sadly, the poorest people in our urban communities are also those with least access to green space, and tend to have the poorest health. Poor health is correlated with income (Deaton, 2002; Smith, 1999), and if you do not have much income, you tend to live in densely populated areas and eat less well than if you are rich (Alaimo, 2001). The housing stock you live in will be poor too: old, leaky and probably also more expensive to heat (Burholt & Windle, 2006). The health problems associated with poverty are exacerbated by lack of access to green spaces (Mitchell & Popham, 2008) and this disparity in access to environmental goods comes under the general concept of environmental justice (Haughton, 1999). Deprivation is, sadly, nothing new (Newman *et al.*, 1990) and nor is the pollution that comes with urbanisation (Jacobson, 2002) and resultant health impacts (e.g. $PM_{2.5}$ related asthma (Kheirbek *et al.*, 2013)) which we know that plants can help to reduce (see Chapter 2 *et seq.*).

If you live in a town or city, especially a large one, you will recognise that it generates its own climate – it is warmer than the surrounding countryside (Changon, 1992). I like snow and always found it frustrating when I lived in cities: we had miserable rainy winters, but as soon as I drove out of town, the countryside was covered in beautiful snow – you might not agree with my preferences, but the reality is that urban climates are different. Back in 1982 a seminal book was published called *Urban Ecology* (Bornkamm *et al.*, 1982); it contained papers from the second European Ecological Symposium which was held in Berlin in 1980. I recall reading it some years later and being struck by the way that much work on vegetation was looking at how plants took up pollutants such as heavy metals (Greszta, 1982) and could be selected for tolerance to air pollution such as sulphur dioxide ($SO_2$) (Bell *et al.*, 1982). In the same volume Horbert *et al.* (1982) published a paper summarising their research which identified the relationship between the proportion of sealed surfaces in an urban area (buildings, roads, pavements, etc.) and the resulting local climate which was: warmer (on average 0.5–1.5°C higher, 2–6°C higher on clear days), wetter (5–10% more precipitation), with higher levels of runoff, more polluted (gaseous pollution 5–25 times greater) and had lower humidity (winter 2% lower, summer 8–10% lower) and air circulation (windspeed 10–20% lower) than surrounding areas with a higher density of vegetation. In the same paper, Horbert *et al.* (1982) identified the ameliorating effect of urban vegetation on temperature (the central part of the 212 ha Tiergarten park was cited as being as much as 7°C lower in comparison with adjacent built-up areas) and that it could have localised impacts on the surrounding areas (depending on the permeability of the built-up areas so as to permit air movement). These urban climate issues are exactly the same challenges that face us now, but with increased ferocity due to the challenges of climate change, and some of the solutions are exactly as Horbert *et al.* (1982) identified: using vegetation.

So, we humans have not only changed our local environments for the worse, through overcrowding and pollution, but we have also changed the global environment (IPCC, 2007, 2014; Solomon *et al.*, 2007). Very few serious scientists doubt that global climate change is real and that it is caused by human (anthropogenic) activity, principally through the burning of fossil fuels (IPCC, 2007). We know from modelling that one of the consequences of global climate change is that weather is going to become more unpredictable with extreme events such as heatwaves and severe storms becoming more frequent. Extreme events such as heatwaves kill people (Golden, 2004; Rey *et al.*, 2007), extreme events like storms cause flooding (Starke *et al.*, 2010). Towns and cities that have a lot of 'sealed surface' (buildings, roads, car parks) are less well equipped to deal with such, literally, lethal problems than those with a lot of green space (O'Neill *et al.*, 2009).

In April 2013 the UK government published its National Adaptation Programme aimed at reducing the impact of climate change (Defra, 2013b) which 'supports the use of green infrastructure'. So why is this book concerned largely with the 'built' part of the environment as opposed to large open spaces? It is partly to do with definitions and opportunities. Many green infrastructure definitions (see above) appear to exclude biodiversity on and in buildings and the benefits that it brings. This means that when policy is implemented and budgets set and spent based on such definitions, built-up areas are excluded. The reason for greening buildings and reducing the proportion of sealed surface is simple: if we don't, we are going to have increasingly uncomfortable urban areas with low aesthetic value, high levels of pollution, poor climate, excessive heat, and flash flooding. In combination, this will have an impact on physical and mental health.

It is also my perception that most organisations have departments that work in silos and rarely work together or even appreciate the value of each other's work. Green infrastructure, to many officers and political representatives in urban areas, means parks and nature reserves and the perception is therefore that it is really about aesthetics, recreation and wildlife. Of course managing such areas costs money, and viewed against the need for other essential services park management or nature reserve wardening appears a luxury and non-essential and, as a result, is often the first service to have its budget cut when times are hard. This view is, of course, to completely misunderstand what green spaces do, and to undervalue the wide-ranging benefits they provide to the whole community. Unfortunately, many current definitions (see above) have created new silo thinking: if green infrastructure is perceived as being only about green open spaces, then this is an important misconception. So this book aims to demonstrate: 1) the multifunctional values of green infrastructure generally – and how this cuts right across most of our concerns as citizens, and 2) to show how some aspects of the built environment can be modified to deliver many of these benefits in addition to green spaces. Why? The answer is very simple: few politicians are bold enough to suggest knocking down valuable buildings to create a new open space in the heart, or even periphery, of a town or city, but modifying built structures either by modifying existing building and curtilages (surrounding areas) by 'retrofitting' or incorporating green infrastructure elements in replacement or new builds is far easier and is really a matter of modifying planning policy and building regulations, and demonstrating economic and community value. It is also worth remembering that buildings are three-dimensional, whereas many urban green spaces are very flat and essentially two dimensional: because of this it means that you could, potentially, get more benefit from greening the surfaces of a building than you could from the same area of open space if the building was not there.

# 2

# BENEFITS OF GREEN INFRASTRUCTURE

This chapter will:

*   identify the general benefits of green infrastructure
*   review the different ecosystem services delivered by green infrastructure.

## 2.1 Introduction

Without biodiversity there is no green infrastructure. Unfortunately, as a term 'biodiversity' is really rather vague and can be considered at the level of genetic variability within a species or population of a species right up to some global measure such as the total number of species on the planet – which is a scientist's way of looking at it. Another, very anthropogenic (human-focused) way of viewing biodiversity is to consider what it can and has done for us. Well, biodiversity does rather a lot: it provides us with materials for food, clothing, construction, pharmaceuticals, aesthetic pleasure, oxygen production and carbon dioxide removal, and so much more – such values have been recognised for a long time (Spellerberg, 1992). What have become more, or perhaps simply more widely, recognised are the basic ecosystem services that biodiversity delivers: removing pollutants, acting as air conditioning, insulation, shading, removal and pre-treatment of rainwater, pollination services, increasing values of adjacent property, reducing healthcare costs, etc. Of course, in a kind of circular argument, enhancing green infrastructure can also increase and improve wildlife habitat, which improves the ecosystem services provided by biodiversity (Figure 2.1).

Vegetation also acts as a carbon store, and whilst the 'urban forest' (a term coined to collectively describe all the trees in urban, as opposed to rural, areas) appears to capture very little (e.g. less than 1% of the carbon emissions in the USA (Nowak, 1994b) or 6% for the West Midlands Region of the UK (Stewart *et al.*, undated-a)), this can add-up to a substantial amount. In addition, Nowak and Heisler (2010) in a report for the US National Recreation and Park Association estimated that the value of trees in urban parks was in the billions of dollars annually for air temperature reduction, $500 million per year for air pollution removal, an unknown but substantial value for shielding from UV radiation in preventing skin cancer

| Aspect | Attributes | Examples |
|---|---|---|
| Visual | Aesthetics, screening | Improved visual environment |
| Human Health | Exercise (walking, running, green gym), pollution, abatement, de-stressing, socialisation, recreation | Reduced costs to health providers through reduced admissions; improved mental health; faster recovery |
| Education | Study and experience of wildlife, schools, ranger services, volunteering | Formal and informal education, skills through volunteering, hobbies (photography, bird watching, etc.) |
| Food Production | Allotments, gardens, orchards, roofs, walls | Improved diet, community bonding, education, biodiversity, exercise |
| Transport | Alternative movement corridor for cyclists and pedestrians | Reduces road congestion, safer routes, reduced pollution exposure, more relaxed setting |
| Economics | Property prices, inward investment, tourism, improved business / shopping environment | Improved staff morale, reduced sick-leave, improved staff retention, attracts businesses; units let faster and fewer voids |
| Climate Control | Reduced heat-related mortality, heat island, and wind; improved air circulation and climate change mitigation | Provides shade against UV-related cancer, cardiovascular mortality, heatstroke, etc.; reduces air and surface temperature, freshens air |
| Sustainable Urban Drainage | Reduced runoff, flash flooding | Reduces risk, economic losses, trauma and distress, processing costs |
| Pollution Control | Water, light, noise, air pollution (particulates, gases and aerosols, odours) | Removes $PM_{10}$ and below from air; absorbs nitrogen oxides, ozone, volatile organic chemicals; acts as heavy metal sink |
| Energy Efficiency | Reduces air conditioning and heating costs | Insulates buildings, provides shelter against drafts, shades windows |
| Biodiversity | Wildlife habitat, wildlife corridors and stepping stones | Provides breeding habitat, food and other resources; promotes dispersal; reduces extinction risks |

FIGURE 2.1 Examples of the services provided to humans and wildlife by green infrastructure. Many of the aspects overlap; for example, human health is improved not just by exercise, but also by climate control, sustainable urban drainage and pollution control (modified from Dover, 2000).

and cataracts, \$1.6 billion a year for carbon storage and \$50 million a year for carbon removal; the value of the trees themselves was estimated at \$300 billion. Specific benefits will depend on the area under consideration – for the Piedmont region of the USA (from the south of New Jersey to eastern Texas), the value of trees to the community was dominated by reductions in stormwater runoff and elevated property prices followed by decreased energy use and lower levels of air pollutants and $CO_2$ (McPherson et al., 2006).

## 2.2 Biodiversity value

### 2.2.1 Wildlife and their habitats

In urban areas there are many green spaces, and urban woodland is known to be of value in European cities to vertebrates (birds, bats) and invertebrates (Tyrväinen et al., 2005). But not all green spaces will have much in the way of biodiversity value; for example, a closely cut park lawn. Although such areas have their place as amenity areas, and they will provide unsealed surfaces helping to reduce runoff following rain. Perhaps ironically, neglected 'waste' areas can be valuable for biodiversity; Sukopp (1990) estimated that abandoned urban sites could have four times as many species as city parks, although about half of the plant species would be non-native. Residential areas can host substantial communities of invertebrates as Owen (2010) demonstrated in her 35-year study of her suburban Leicestershire garden which had: 23 species of butterfly, 375 species of moth, 94 species of hoverfly, honeybees, 8 species of 'true' bumble bee, 5 species of 'cuckoo' bumble bee, 45 species of solitary bee, 7 species of social wasp, 55 species of solitary wasp, 87 species of sawfly, 183 species of bug (Hemiptera) and 422 species of beetle. Areas of high plant species richness and heterogeneity in vegetation structure (from patchy areas with bare ground to full tree canopy cover) are likely to deliver a wider range of ecosystem services and habitat for fauna than those of low richness and structure. However, this should not be interpreted as suggesting that every green space should try and pack everything into it, more as a plea to think more widely about green space design and functions. The character of a particular space will be reflected by its principal purpose: play areas, cycleways and walkways, churchyards, parks, sports fields, etc. But wherever possible it should incorporate native species, and be structurally variable (not all mown grass) because in this way society will get a bigger benefit *whatever the function* of the green space. For example, does a sports field all have to be low mown, hard-wearing, amenity grass species, or can the non-competitive areas be fringed by more naturalistic vegetation? Can golf courses become wildlife havens? Can churchyards be managed for wildlife? After all, Sjöman et al. (2012c) showed that Aarhus had high tree species richness in its cemeteries (Figure 2.2).

But what is wildlife habitat? There are very technical approaches to this question – for example, habitat has been described as an 'n-dimensional hypervolume' (Begon et al., 1996) – but we will take a more practical approach. Habitat is often thought of as a single physical space where a species lives, a place where all its requirements can be found and where the environmental characteristics are suitable. It might be a place with the right soil type, nutrient levels, aspect, slope, light and moisture for a particular plant. For some animals that definition may hold, but in the last decade a wider understanding has developed – that which highlights the resource needs of a species, and an acknowledgement that these resources may not be in the same place in space or time (Dennis et al., 2014) – something which is obvious if

FIGURE 2.2 Cemetery managed for wildlife in the village of Kirtlington, UK. It is important to let visitors know why grass is not cut to demonstrate that it is a deliberate management action for a specific, positive reason, otherwise it can be interpreted as merely a way to save money or because of neglect. High vegetative species richness generally leads to high faunal richness. © John Dover

**TABLE 2.1** Insects recorded as visiting 25 species of native or naturalised plant in the University of Cambridge's Botanic Garden 1996–1997

| Common name | Scientific name | Ants | Anthidium manicatum | Apis mellifera | Bombus hortorum | Bombus lapidarius | Bombus pratorum | Bombus pascuorum | Bombus terrestris | †Psithyrus sp. (Cuckoo species) | 'Solitary bee' | 'Solitary wasp' | Aglais urticae | Celastrina argiolus | Gonepteryx rhamni | Inachis io | Lycaena phlaeas | Maniola jurtina | Pieris brassicae | Pieris napi | Pieris rapae | Polyommatus icarus | Pieris tithonus | Unidentified Skipper | Vanessa atalanta | 'Moths' | 'Hoverflies' | Episyrphus balteatus | Eristalis intricarius | Eristalis tenax | Helophilus pendulus | Sphaerophora scripta | Platycheirus sp. | Syrphus sp. | Melanostoma sp. | Metasyrphus sp. |
|---|---|---|---|---|---|---|---|---|---|---|---|---|---|---|---|---|---|---|---|---|---|---|---|---|---|---|---|---|---|---|---|---|---|---|---|---|
| *Group* | → | *Ants* | *Hymenoptera (Ants, Bees, Wasps)* | | | | | | | | | | *Lepidoptera (Butterflies & Moths)* | | | | | | | | | | | | | | *Hoverflies* | | | | | | | | | |
| Cornflower | *Centaurea cyanus* |  |  | ★ |  |  |  |  |  |  |  |  |  |  |  |  |  |  |  |  |  | • |  |  |  |  |  |  |  |  |  |  |  |  |  |  |
| Greater knapweed | *Centaurea scabiosa* |  |  |  |  |  |  |  |  |  |  |  | • |  | • |  |  |  |  |  |  |  |  |  |  |  |  |  |  |  |  |  |  |  |  |  |
| Foxglove | *Digitalis purpurea* |  |  |  |  |  |  |  |  |  |  |  |  |  |  |  |  |  |  |  |  |  |  |  |  |  |  |  |  |  |  |  |  |  |  |  |
| Wild teasel | *Dipsacus fullonum* |  |  | • | • |  |  |  | • | • |  |  |  |  |  |  |  |  |  |  |  |  |  |  |  |  | • | • |  |  |  |  | • |  | • | • |
| Hemp agrimony | *Eupatorium cannabinum* |  |  | ★ |  |  |  |  |  |  |  |  | • |  |  |  | • |  | • | • | • |  |  |  | • |  |  |  |  |  |  |  |  |  |  |  |
| Field scabious | *Knautia arvensis* |  |  | ★ |  | ★ | • | ★ |  |  |  |  | ★ |  |  |  |  |  |  |  |  |  |  |  |  |  |  | • |  |  |  |  |  |  |  |  |
| White deadnettle | *Lamium album* |  |  |  | • | ★ |  | ★ |  |  |  |  |  |  |  |  |  |  |  |  |  |  |  |  |  |  |  |  |  |  |  |  |  |  |  |  |
| Toadflax | *Linaria vulgaris* |  |  |  |  |  |  |  |  |  |  |  |  |  |  |  |  |  |  |  |  | • |  |  |  |  |  |  |  |  |  |  |  |  |  |  |
| Bird's-foot trefoil | *Lotus corniculatus* |  |  | • |  | ★ |  | • |  |  |  |  |  |  |  |  |  |  |  |  |  |  |  |  |  |  |  |  |  |  |  |  |  |  |  |  |
| Ragged robin | *Lychnis flos-cuculi* |  |  | • |  | ★ |  | • |  |  |  |  |  |  |  |  |  |  |  |  |  |  |  |  |  | • | • |  |  |  |  |  |  |  |  |  |
| Purple loosetrife | *Lythrum salicaria* |  |  | ★ |  | ★ |  |  | • |  |  | • | ★ |  |  |  |  |  |  |  | • |  |  |  | • | • | • | • |  | • | • | • | • |  |  | • |
| Musk mallow | *Malva moschata* |  |  | ★ |  | ★ |  | ★ |  |  |  |  |  |  |  |  |  |  |  |  |  |  |  |  |  |  | • |  |  |  |  |  |  |  |  |  |
| Common mallow | *Malva sylvestris* |  |  | • |  | ★ |  |  |  |  |  |  |  |  |  |  |  |  |  |  |  |  |  |  |  |  |  |  |  |  |  |  | • |  |  |  |
| Marjoram | *Origanum vulgare* |  |  | ★ |  | ★ |  |  |  |  |  |  |  |  |  |  |  |  |  |  |  |  |  |  |  |  |  |  |  |  |  |  |  |  |  |  |
| Meadow clary | *Salvia pratensis* |  |  | ★ |  | ★ |  | ★ |  |  |  |  |  |  |  |  |  |  |  |  |  |  |  |  |  |  |  |  |  |  |  |  |  |  |  |  |
| Wild clary | *Salvia verbenaca* |  |  | ★ |  |  |  |  |  |  |  |  |  |  |  |  |  |  |  |  |  |  |  |  |  |  |  |  |  |  |  |  |  |  |  |  |
| Soapwort | *Saponaria officinalis* |  |  | ★ |  |  |  |  |  |  |  | • |  |  |  | • |  |  |  |  |  |  |  |  | • | • | • |  |  |  |  |  |  | • | • |  |
| Small scabious | *Scabiosa columbaria* |  |  | ★ | • | ★ | • | • |  |  |  |  | ★ |  | • |  |  |  |  |  |  |  |  |  |  |  | • |  |  |  |  |  | • | • |  |  |
| Red campion | *Silene dioica* |  |  |  | • |  |  |  |  |  |  |  |  |  |  |  |  |  |  |  |  |  |  |  |  |  |  |  |  |  |  |  |  |  |  |  |
| White campion | *Silene latifolia* |  | • |  | • |  |  |  |  |  |  |  |  |  |  |  |  |  |  |  |  |  |  |  |  |  |  |  |  |  |  |  |  |  |  |  |
| Marsh woundwort | *Stachys palustris* |  | • |  |  | ★ |  | ★ |  |  |  |  |  |  |  |  |  |  |  |  |  |  |  |  |  | • | • |  |  |  | • |  |  |  |  |  |
| Hedge woundwort | *Stachys sylvatica* |  | • | • |  | • |  | ★ |  |  |  |  |  |  |  |  |  |  |  |  |  |  |  |  |  |  | • |  |  |  |  |  |  |  |  |  |
| Red clover | *Trifolium pratense* |  |  | ★ |  | ★ |  | ★ |  |  |  |  |  |  |  |  |  |  |  |  |  |  |  |  |  |  |  |  |  |  |  |  |  |  |  |  |
| White clover | *Trifolium repens* |  |  | ★ |  | • |  | ★ |  |  |  |  |  |  |  |  |  |  |  | • |  |  |  |  |  |  |  |  |  |  |  |  |  |  |  | • |
| Common vetch | *Vicia sativa* |  |  |  |  | • |  | ★ |  |  |  |  |  |  |  |  |  |  |  |  |  |  |  |  |  |  | • |  |  |  |  |  |  | • |  |  |

Source: Comba *et al.* (1999b).

† Now *Bombus* sp; ★ = species most visited; • = species visited. Because of difficulties in identification, the bumblebees were grouped such that records of *B. lapidarius* may also have included some *B. ruderatus*; *B. terrestris* may include *B. lucorum*; *B. hortorum* may include *B. ruderarius*. Individuals not identified to species level are given either as Genus (e.g. *Syrphus* sp.) or more generally as 'hoverflies', 'moths', etc.

you think of the requirements of migratory species. So, resources may be in several different places for a butterfly: somewhere to lay eggs and food for caterpillars to eat, somewhere else to pupate, somewhere else to feed as an adult, somewhere else to roost or find a mate and the places that are used may be used at different times in the year. This concept is called complementation, that is, the provision of complementary resources (Dunning *et al.*, 1992). So even the tiniest flower bed, if it has the right plants, can provide 'habitat resources' and be vitally important as food for adult butterflies. But what are 'the right plants'? Different insects have different ways of accessing plants: some are generalist feeders and some are 'specialists' that have mouthparts that evolved to access nectar from plants with very specific flower structures (different species of bees have different mouthpart lengths, for example), so the 'right plant' may depend on a particular plant species' flower structure. However, in urban parks and gardens there can be a mismatch between plants that have been cultivated for their looks but which are of little value to wildlife. Comba *et al.* (1999a) compared a number of horticultural cultivars of plants that had been modified by breeding for 'looks' with the uncultivated 'wild type' version of the same species for nectar production and the ability of invertebrate species to get at the nectar. All the species they studied (nasturtium *Tropaeolum majus*, larkspur *Consolida* sp., snapdragon *Antirrhinum majus*, viola and pansy *Viola x wittrockiana*, French marigold *Tagetes patula* and hollyhock *Alcea rosa*) produced good levels of nectar, but the structural flower changes in the commercial varieties tended to modify the community of species visiting the flowers and reduce the availability of the nectar (although this was not always the case). Subsequently Comba *et al.* (1999b) screened 25 native or naturalised British plants for their value as nectar flowers for bees, butterflies and hoverflies; the work was carried out in Cambridge and reflected the species assemblage likely to visit flowers in urban areas. They identified a number of species that were popular with bees and which were visited by a wide range of species and could thus be of great value in urban plantings (Table 2.1).

Nectar for adult invertebrates like butterflies is one thing, but what about breeding 'habitat' or 'resources'? Some butterfly species may require areas of long grass to lay eggs on – a requirement that rules out many of our green spaces. For some species which do not fly very far, breeding habitat needs to be close to other resources so they are easily accessible but species that are very vagile (able to disperse widely), like the peacock (*Inachis io*), can move around urban areas pretty well to find what they need (in the case of the peacock: nettles

FIGURE 2.3 Left: peacock nectaring; right: female peacock ovipositing on nettle. The peacock is a highly mobile butterfly easily able to find nectar sources for food and nettles to lay its eggs on provided they are in the right situation. © John Dover

*Urtica dioica*) (Figure 2.3). Some species have very exacting requirements for egg-laying sites; I have seen the wall butterfly *Lassiommata megera* in a London park finickly selecting where to lay its eggs on the interface between bare ground and grass where turf had been removed around the base of an amenity tree to help it grow.

Native plant species are particularly valuable in the urban setting as they are more useful to our native fauna. For example, Perre *et al.* (2011) investigated the value of plants from the Family Asteraceae (daisies) in the urban environment of the campus of the State University of Campinas, São Paulo, Brazil, to insect herbivores living in flower heads. They found 30 species of Asteraceae on site and whilst only seven of the species were non-native, their abundance far exceeded that of the native species. Despite this dominance, they supported far fewer species of herbivore (4) than the native species (26), although there was no difference in abundance between natives and non-natives in herbivore abundance. However, being purist may miss opportunities; Angold *et al.* (2006) found that urban habitat patches developed a flora that contained both native and non-native species in 'recombinant communities' which provided habitat for a wide range of species.

### 2.2.2 Wildlife corridors

Wildlife corridors can be thought of as linkages helping species move through the urban landscape; discrete patches of habitat, known as stepping stones, provide a similar function (Diamond, 1975). It has recently become fashionable to suggest that green roofs may act as stepping stones or be part of wildlife corridor networks.

If the distance between populations of a species is greater than the distance the species is able to disperse, then populations are isolated and habitat is inaccessible. In the same way the urban form (buildings, roads) may act as barriers to movement. This can have serious consequences for species as, all things being equal, small populations on small habitat patches are more vulnerable to extinction than large populations on large habitat patches. The long-term viability of a species in a particular area may be dependent on many small habitat patches exchanging individuals in what is known as a 'metapopulation' or 'population of populations' (Forman, 1995; Levins, 1970). Habitat or resource patches, if they go extinct for some chance (stochastic) reason such as fire, disease, predation or parasitism, are unlikely to be recolonised if isolated by distance or barriers. Dawson (1994) identified five main reasons for employing corridors and stepping stones:

- To allow a species to escape from an extinction event, or to recolonise a vacant patch after extinction
- To allow movement between resource patches so that all habitat requirements are met
- To facilitate seasonal migrations
- To promote gene flow between populations; to reinforce flagging populations, etc.
- As an escape route from global warming.

In the context of a particular urban area, the global warming escape function is largely irrelevant as it is really a very large-scale attribute appropriate to the regional or national scale (Hill *et al.*, 1993).

Wildlife corridors and stepping stones were theoretical constructs devised at a time when spatial ecology was in its infancy and questions were being asked such as 'Is it better to have one large nature reserve or several small ones?' (the SLOSS debate: Single Large or Several Small) (Diamond, 1975; Game, 1980). Following on from this was the idea that if several habitat patches were available, but their overall area was too small to support a population, then perhaps linking them together with strips of habitat (corridors) or putting intervening small patches (stepping stones) to facilitate movement may effectively make a large habitat from several small ones. These concepts were enthusiastically taken up by planners, including those responsible for urban areas, but at the time the concept was not supported by much in the way of data and this caused some problems in the practical application of the concept and its defence in planning committees, at least in the UK (Dawson, 1994; Spellerberg & Gaywood, 1989). You only have to ask a few questions to see the problems:

- Which species benefit from corridors?
- What are the requirements of those species to get them to use corridors?
- What are the optimal dimensions for corridors for a particular species?
- Will corridors actually facilitate movement of disease, parasites and predators between populations (Dendy, 1987; Simberloff & Cox, 1987)?

Whilst general design advice on many aspects of spatial ecology has been available for some time (e.g. Dramstad *et al.,* 1996), planners, without advice from ecologists, may not know what they are dealing with, and there is still too little data on which species use corridors, but at least the concept is now much more widely accepted. We do know that some species are strongly affected by even the smallest interruption in habitat – for example, gaps in hedges affect dormouse *Muscardinus avellanarius* dispersal (Bright, 1998), but you don't tend to get dormice in urban centres. Fortunately studies are appearing that support the use of corridors in urban areas such as those carried out in the region of Paris, France, by Vergnes *et al.* (2012) who reported the positive effect of corridors on spiders, staphylinids and carabids and by Vergnes *et al.* (2013) who worked on shrews.

In designating urban corridors the default, in the absence of evidence, has often been to simply look for contiguous linear open spaces with some greenery and label them 'wildlife corridors' and assume that they are. Within urban areas the banks of rivers, streams and canals are often labelled corridors as are networks of green spaces even if they are not all the same kind of biotope (biotope = a generalist description of a general vegetation type e.g. grassland, woodland). There are problems in this approach as 'connectedness', as we see it, is not necessarily the same as 'ecological connectivity' for a given species (Baudry & Merriam, 1988). In other words, we may see something as connected such as two woods linked by a stretch of grass, but for a given species there may be some attribute missing in the grassland, or some subtle barrier attribute present, such that it cannot be used as a conduit.

## 2.3 Direct value to humans

The environment has been credited with an incredible range of values: providing food, building materials, natural drugs and pesticides (which may also act as starting points for development of more effective synthetic alternatives), inspiration for works of art (poetry, paintings, music), social engagement (e.g. volunteering), spiritual value, horticultural products (including plants), a repository of human history (cultural landscapes, artefacts, archaeological finds), ecosystem services (local and global), promoting physical and psychological well-being, study of natural history, acting as an educational resource, and much more (Lees & Evans, 2003). Not all of these aspects, by any means, will be covered here but I will highlight some provided by urban green infrastructure; coverage of any issue is not intended to be comprehensive.

### 2.3.1 Visual amenity

The incorporation of vegetation into urban areas has an immediate and striking effect: it softens the impact of the hard landscape. In purely visual terms it can screen ugly buildings and structures, enhance beautiful ones, reduce glare, and introduce colour (Figure 2.4). In a photomontage-based study of street vegetation in Sapporo, Japan, Todorova et al. (2004) found that trees along the edge of pavements were the most favoured elements of vegetation in streetscapes but that other elements enhanced the visual amenity. Responses to photomontages manipulating the type of vegetation below trees suggested that neat arrangements of low-growing bright flowers in single species stands were highly favoured, whereas tall flowering stands with an irregular structure were not. Moving outside the streetscape, larger areas such as parks, commons, nature reserves, woodland, can be highly structured and 'horticultural' or be managed more for wildlife with a more 'unkempt' appearance. These different management approaches, with different values, may attract different segments of the population depending on their interests and perceptions of beauty. A recent trend in the UK has been to use wildflower plantings on wide verges and on roundabouts. The aesthetic improvement is substantial (Figure 2.5), and may have some biodiversity value as well for invertebrates using them as pollen and nectar sources. But smaller areas can have substantial aesthetic benefits including balconies, street planters, doorstep plantings and window-boxes (Figure 2.6).

Access to green space can clearly contribute to the ability of urban dwellers to take exercise with both physical and mental benefits, but there are significant benefits in terms of stress reduction which derive purely from the sensory environment. Pretty et al. (2005a) compiled the evidence base to promote the countryside for health and well-being in the UK; the review is useful in an urban context as it draws together a substantial body of data on a wide range of topics. Various ideas have been put forward to try to explain why humans respond positively to nature, the most famous is probably the 'Biophilia' hypothesis which suggests that we have a built-in response to natural settings and the lack of them creates a discord with the environment in which we evolved (de Groot & van den Born, 2003; Grinde & Patil, 2009; Wilson, 1984). Ulrich (1983) proposed a 'psychophysiological stress reduction framework' whereby nature promotes recovery from stress and other psychological disorders (Croucher et al., 2007; Morris, 2003). The 'attention restoration theory' followed, as promulgated by Kaplan and Kaplan (1989), and centres on the restorative effects of nature on the results of prolonged concentration and effort (Croucher et al., 2007). Reid and Hunter (2011) explored different measures of well-being and the philosophy between different camps of how well-being happens: the 'hedonic' approach being the absence of negative factors (experiences, emotions, feelings) and the 'eudaimonic' view whereby well-being is possible despite the presence of negative factors. Groenewegen et al. (2006) proposed a study on the effects of green space on health, well-being and social safety in the Netherlands, at a range of spatial scales; the document is simply a study protocol, but I like the term they used as a

**FIGURE 2.4** Top left: climbers on the railings of a building in Oxford, UK, in April bring a dash of welcome colour to the streetscape; top right: English ivy (*Hedera helix*) growing over ugly, rust-splattered corrugated iron fencing in Stoke-on-Trent; bottom left: concrete fencing being removed from the Staffordshire University campus in Stoke-on-Trent; and bottom right: immediately after replacement with ivy green screens (extension panels are provided for the ivy to grow up). © John Dover

**FIGURE 2.5** Wildflower plantings on wide verges and roundabouts in Warrington, UK.
© John Dover

catchy 'sound-bite' to sum up the value of green space: 'Vitamin G'. The following sections explore aspects of how green infrastructure affects our health and well-being.

### 2.3.2 Human health

The idea that green spaces have health benefits is not a 21st-century concept. A report by Natural England 'Our Natural Health Service' (NE, 2009b) quoted the Registrar for Births, Marriages and Deaths who wrote in 1839: 'A park in the East End of London would probably diminish the annual deaths by several thousands … and add several years to the lives of the entire population'; Victoria Park, named after Queen Victoria, resulted in 1850. There are many excellent reports that explore the relationship between the environment and health, including those by interest groups such as the Royal Society for the Protection of Birds (Bird, 2004), government agencies including forestry (Henwood, 2001), nature conservation (Lees & Evans, 2003), health (GSS, 2008), recreation (Pretty *et al.*, 2005a), landscape (Croucher *et al.*, 2007) and academic reviews such as Morris (2003) and Tzoulas *et al.* (2007), although not all of these reports relate specifically to the urban environment.

Green infrastructure can impact on human health in a number of ways, the most obvious are: by increasing the amount of shade available on hot days, through promoting air circulation, cleaning the air and water of pollutants, helping people to 'de-stress' and relax tension, improving recovery from illness or operations, providing food, community integration

**FIGURE 2.6** Even small-scale plantings such as balconies, street planters, doorstep plantings and window-boxes can have a substantial impact on an area's aesthetic. © John Dover

and opportunities for physical exercise (see section 2.7 for more on heat effects). The link between city greening and its potential to reduce healthcare costs was already being made in the late 1970s (Doernach, 1979). Bird (2004) suggests a single park of 8–20 ha could save the UK economy annually between £1.6 million and £8.7 million depending on the city; this represents savings of between £0.3 million and £1.6 million to the National Health Service (NHS), again depending on the population density of the settlement.

Gerlach-Spriggs *et al.* (1998) report that the earliest hospitals were in monastic institutions where gardens were part of the healing process. In the design of the Holloway Sanatorium (lunatic asylum) for women in the UK (Anonymous, 1872), Thomas Holloway explicitly included gardens and countryside views for their therapeutic effect and the retention of trees for shade. Green spaces have been shown to have positive effects on human health in a number of primary studies (de Vries *et al.*, 2003; Maas *et al.*, 2006; Mitchell & Popham, 2007; Takano *et al.*, 2002; White *et al.*, 2013) and reports by influential organisations, such as the Royal Society for the Protection of Birds (RSPB) (Bird, 2004), Natural England (Lees & Evans, 2003) and Greenspace Scotland (Croucher *et al.*, 2007). Grahn and Stigsdottir (2003) in Sweden showed that the more often a person visits green spaces, the less likely they are to report symptoms of stress. Proximity of green space to home was vitally important, as respondents to their questionnaire survey claimed that they did not actively seek out green space to visit elsewhere if there was none nearby. There is a well-established link between health in different income groups, with those on lower income typically experiencing poorer environments and thus poorer health, as Chaudhuri (1998) graphically described for child poverty in Canada. Mitchell and Popham (2008) showed that proximity to green space helped reduce health inequalities in lower income groups. Fuller *et al.* (2007) in Sheffield, UK, showed that the psychological benefit of green space to people increases with increasing levels of biodiversity. White *et al.* (2013) found lower levels of mental distress and higher well-being were correlated with higher levels of urban green space. Luck *et al.* (2011) examined the impact of vegetation cover, species richness and abundance of birds and density of plants on well-being and feelings of connectedness to nature in a study of 1,000 individuals from 36 communities in south-eastern Australia. They found that 'personal well-being' was linked to the extent of vegetation cover of an area, but nature connectedness was only weakly related to the other variables. Once they had controlled for demographic characteristics of their subjects (e.g. age, level of activity), they found positive relationships between 'neighbourhood well-being', a measure of satisfaction with the locality, and species richness and abundance of birds and vegetation cover. Van Dillen *et al.* (2012) found both quantity and quality was important for higher levels of (self-reported) health; the findings held for both green spaces and streetscape vegetation. Even images of 'restorative environments' have been shown to relieve mental fatigue (Berto, 2005). Of particular interest from the study by White *et al.* (2013), of 10,000 people from the British Household Panel Survey, was that they highlighted the importance of small improvements of mental well-being at the individual level which, when aggregated, were amplified at the community level. The studies of de Vries *et al.* (2003), Maas *et al.* (2006) and Takano *et al.* (2002) indicated that the effect of urban greening was strongest in groups most likely to be relatively tied to the locality, that is, housewives, the old, the young and those with poor educational backgrounds. Taylor *et al.* (2002) found that self-discipline was higher in inner-city girls whose homes had more natural views; the relationship was not found for boys, possibly because boys spent less time playing in and around the home. Kuo and Taylor (2004), in the USA, claimed that outdoor activities in green spaces reduced attention deficit hyperactivity disorder (ADHD) symptoms significantly more than activities held outdoors but in a built environment or indoors. Mitchell and Popham (2007) using census and economic data for the whole of England noted that though a general effect can be found between green space and health, the relationship appeared to depend on income. The expected positive relationship between good health and green space was evident in urban high-income and low-income groups as well as rural low-income groups, but there was no relationship in suburban high-income areas or rural high-income areas; for suburban low-income areas, however, there was a negative association with green space and health. Mitchell and Popham (2007) suggested that this may indicate a disjunct relationship between quality, quantity and accessibility in poor suburban areas. On a much broader scale, Donovan *et al.* (2013) correlated tree mortality (100 million trees lost in 15 American states as a result of the emerald ash borer beetle *Agrilus planipennis*) with human mortality between 1990 and 2007 from cardiovascular disease and lower respiratory tract illness – they estimated there were some 21,193 excess deaths resulting from the reduced public health benefits afforded by trees.

### 2.3.3 Exercise

Exercise is well known to have a wide range of health benefits, and Bird (2004) indicated that a 3 km footpath could contribute, annually, between £0.1 and £1 million to the economy depending on the population density of the area, and would save the local National Health Services (NHS) in the UK between £21,000 and £210,000. One estimate by the UK Department for Culture, Media and Sport in 2002 (DCMS, 2002) indicated that increasing the physical activity of adults by 10% would save 6,000 lives and be worth £500 million annually to the economy. The UK National Institute for Clinical Excellence went further and suggested that lack of exercise costs the UK £8.2 billion in healthcare costs and absence from work (NICE, 2008a). The Institute issued guidance on how to promote physical activity in built as well as natural environments; they also identified what kinds of organisations were responsible for implementation and made links to policy documents (NICE, 2008b).

For older people, the presence of parks and tree-lined streets has the effect of encouraging walking and has been shown to positively influence longevity in city dwellers (Takano *et al.*, 2002). Bell *et al.* (2008) reported that the body mass index of children was lower in neighbourhoods where there was more vegetation and suggested that this was probably because more time was spent outdoors in physical activity. Coombes *et al.* (2010) in a study of 6,821 adults in the City of Bristol, UK, found that reduced accessibility (increasing distance) to green space reduced frequency of visits, and proximity of a park to home decreased the probability of individuals being obese. The concept of the informal 'green gym' has been popular for some time, but doctors may even refer patients to organised sessions whereby exercise programmes are carried out in green spaces (Lees & Evans, 2003). However, the relationship between physical activity and the perception of better health associated with green areas is probably not simple. For example, Maas *et al.* (2008), using a sample of 4,899 Dutch people, examined the relationship between physical activity (walking and cycling), the amount of green space around the home (within 1 km and 3 km radii of the home post code) and how it related to feelings of health. Their findings were not straightforward; it appears that living in a 'green environment' resulted in less time walking or cycling for leisure – although they spent more time gardening (in itself an activity known to promote good health and reduce stress (Milligan *et al.*, 2004; Van Den Berg & Custers, 2011; van den Berg *et al.*, 2010)). Fewer commuting trips were made with increasing amounts of green space; but if commuting was undertaken by cycle, trips were longer if the commuter had more green space around the home. There did not seem to be a link between the additional time spent gardening and cycling to work and the relationship between green space and health. Part of the difficulty with identifying links between health perception and green space was probably because of confounding factors; for example, a comparison of urban with non-urban areas. Maas *et al.* (2008) noted that, in their study, subjects with the most green space around their homes lived in agricultural areas and hence had bigger gardens (requiring more gardening time). As a result, they were further from major infrastructure (such as shops) which would require the use of a car to access (compared with those living in urban areas). Additionally, if they did commute to work, the journey would of necessity take longer.

Pretty *et al.* (2005b) examined the effect of natural views on physical and mental health in a laboratory setting by projecting a series of 'green' and urban images onto walls whilst 100 volunteers (55 female and 45 male, ranging from 18 to 60 years of age) from the University of Essex in Colchester, UK, exercised on a treadmill. The images comprised 'rural pleasant' such as sylvan grassland, 'rural unpleasant' such as groups of abandoned rusting cars, 'urban pleasant' such as tree-lined riverside, 'urban unpleasant' such as derelict buildings with smashed windows, or a control where no image was projected. When analysed by 'scene' type (20 volunteers per category), only those viewing rural pleasant scenes were shown to have significantly reduced blood pressure measures. When analysed by scene group, self-esteem was still better in every group following exercise compared with before exercise, but the response was strongest for the pleasant rural and urban scenes. Mood changes were complex with Anger-Hostility being significantly lower in urban unpleasant and control groups, but significantly higher in the rural unpleasant group. Confusion-Bewilderment was significantly reduced only in the urban pleasant and urban unpleasant groups; Depression-Dejection was significantly reduced in the urban pleasant and control groups; Tension-Anxiety was significantly reduced in the control group but the response was stronger in the rural and urban pleasant groups; Vigour-Activity responses were significant both in the urban and rural pleasant groups and in the control group. No significant responses were detected for the Fatigue-Inertia mood measure.

## 2.3.4 Recreation, relaxation and socialisation

### Socialisation

Green spaces are places where communities and individuals can come together to meet, celebrate or remember. They may be the only places that community events can be held, or they may be the place of choice by nature of their setting and the lack of special permissions that may be needed for road closures and the consequent need to pay for extensive policing. Events can vary from the purely commercial, such as visiting fairgrounds, to non-commercial, such as arts festivals or produce shows. Such events allow community groups to have stands to advertise their services and give demonstrations, combining entertainment with recruitment and fundraising (Figure 2.7).

Kweon *et al.* (1998) demonstrated that for older people in Chicago, USA, outdoor green spaces were important for maintaining social ties engendering a sense of community

FIGURE 2.7 Green spaces are useful for community activities. Top left: morris dancing in a public park (Portsmouth, Hampshire, UK); top right: nature conservation organisations promoting their activities (Pershore, Worcestershire, UK); bottom left: community event in a wood (Pershore); bottom right: a 'family and friends' visit to a pond (Fleet, Hampshire, UK). © Christine Dover, except bottom right © John Dover

and general social integration. These features are considered to be correlated with various measures of longevity and well-being (see references in Kweon *et al.*, 1998). Open spaces can generate both positive and negative concerns about personal safety, which may be quite distinct from the actual risk. The design of such spaces needs to have regard to the perceptions of danger that some landscape configurations and management practices might evoke as they may reduce the willingness to use such spaces (Jansson *et al.*, 2013). Kuo (2003) reviewed research carried out in Chicago, including that with photomontages, which showed that residents preferred images of their area with, rather than without, trees; and the more trees the better. The photomontage studies were confirmed by residents' actual use of areas with and without trees. Children also prefer to use green spaces for play, and more creative play was found associated with green spaces. Green cover also increased social interactions between adults and between adults and children and decreased graffiti, social disorder and crime.

## Natural history study

A number of organisations are now beginning to use the enthusiasm of the public for nature as a way of engaging them more directly and purposefully by creating participatory initiatives

**FIGURE 2.8** A university ecology practical in bird surveying using a green space on the edge of a housing estate in Staffordshire, UK. © John Dover

that provide data for large-scale projects, typically for monitoring purposes. These are collectively termed 'Citizen Science' projects and can be local initiatives or national ones (Pocock *et al.*, 2014a, 2014b). In some cases they are branded as such, but often they are projects that are linked to membership organisations and are ongoing. The city of Glasgow, in Scotland, had the 'BIG' project: 'Biodiversity in Glasgow' (BTO, 2009) which involved the public in recording birds and butterflies. A regional initiative in the UK was a hedgerow survey run by the Cheshire Landscape Trust (CLT, 2013). National projects include the 'Nature's Calendar Phenology Network' run by the Woodland Trust (WT, 2014), which records the first appearance of a range of wildlife each year and helps track the influence of climate change, and the Breeding Bird Survey (BTO, 2014) (formerly the Common Birds Census run by the British Trust for Ornithology) (Lees & Evans, 2003). Not all natural history study in urban areas needs to be formal, or organised, or part of a major project; probably the majority of 'study' is likely to be something along the lines of watching birds on a pond or lake, or idly watching squirrels leap from branch to branch – and is none the worse for that. Green spaces are, however, the obvious places for ecological training to take place (Figure 2.8), though green space has the potential to be used for almost any part of a curriculum given a sufficiently inventive teacher. Local spaces could be used for teaching history, mathematics, biology, chemistry, philosophy, sports, physics, etc.

## 2.4 Food production/urban agriculture

Most of the food consumed in urban areas is grown elsewhere. In developed countries much of it is sold by a small number of large supermarket companies; in small shops on housing estates, fresh fruit and vegetables, which are essential to improve health, are not always available whilst cheap processed food is (Garnett, 2000). Encouraging local food production can reduce 'food miles', improve food security, provide jobs and have substantial health benefits including having a better balanced diet, exercise and socialisation, and is a global phenomenon (Ackerman et al., 2014; Brebbia et al., 2002; Cockrall-King, 2012; Rich, 2012; Specht et al., 2014; Viljoen & Bohn, 2014) (Figure 2.9). Indeed, in 1996 urban agriculture was estimated to be practised by some 800 million people (Smit et al., 1996 cited by Lee-Smith, 2010). The majority of this activity is located in the developing world where many are dependent on it, but interest in urban agriculture has recently been increasing in the developed world too, primarily in the USA but now also in Europe and elsewhere (Lee-Smith, 2010; Mitchell et al., 2014). Urban agriculture is not necessarily a ground-level activity, and Despommier (2011) explores the possibilities of growing upwards with his book *The Vertical Farm*. Urban agriculture is not restricted to the vegetable domain; chicken

FIGURE 2.9 Allotment gardens provide opportunities for local food production and in doing so provide a wide range of benefits including opportunities for exercise, quiet contemplation and socialisation; biodiversity should benefit too. This example is from the village of Beer, Devon, UK. © John Dover

keeping, primarily for eggs, has recently become very popular in the UK (Rustin, 2012), but small animal production including goats, rabbits, guinea pigs, snakes, fish farming and beekeeping are all practised in urban areas (Iaquinta & Drescher, 2002). Local laws may not encourage livestock, however, as LaBadie and Scully (2012) found with chicken keeping in Iowa. Fleury (2002) identifies other features of society enhanced by the introduction of urban agriculture, including a sense of identity, the strengthening of local heritage values and the creation of a distinct local landscape. De Lange (2011) found food production was often not the primary aim of urban agriculture in Amsterdam: it was frequently used as a vehicle for education and promoting social cohesion.

Food production typically takes place in private gardens, allotments, city farms and community gardens, school gardens, community orchards, some parks, and even on balconies and roofs. Edmondson *et al.* (2014) found soils in gardens and allotments in the City of Leicester, UK, to be superior to that of farmland surrounding the city which was subject to intensive, conventional agriculture. Generally speaking, produce grown in gardens has been shown to have levels of contaminants, such as lead, which are within acceptable levels, but some have exceeded such ranges and, in a study of vegetables grown in urban gardens by McBride *et al.* (2014), this was associated with particles adhering to leaves or from aerosol deposition. In a test of contaminant levels in the soil of 54 New York City community gardens, 78% of the 564 samples taken were shown to be within safe limits set for all ten metal contaminants studied. However, this meant that 70% of the sites had some contaminated samples (principally barium: 46% of gardens, and lead: 44%) (Mitchell *et al.*, 2014). Contaminants were identified less frequently in raised beds, where clean soil and compost had been imported onto the site, and may be considered a viable 'risk reduction' method along with other practices including careful washing and preparation of food (see references in Mitchell *et al.*, 2014) and growing fruit rather than vegetables (McBride *et al.*, 2014). Clearly caution should be exercised when growing food on abandoned land due to the risk of contamination from old industrial processes, through using contaminated water, and the inappropriate use of pesticides (Garnett, 2000; Madaleno, 2002; Ranasinghe, 2009; van den Berg, 2002).

One lovely example of how creative thinking can produce food and have wider benefits was that reported by Lees and Evans (2003): many allotments become run-down and are sometimes threatened by closure, due to the small number of active plot holders (EFF, 2006). A 4.85 ha (12 acre) site in Cowley, Oxfordshire, UK, instead of being closed was regenerated by planting a small woodland, an orchard and a wildflower meadow. To encourage families with children to take up plots, the development also included play areas. The changes made to the site have had a much wider community impact, including being used by local schools and being part of a visual arts festival. In terms of biodiversity, birds, at least, do benefit from allotments, though Müller (2007, cited by Strohbach *et al.*, 2009) in his study of bird diversity in Leipzig's (Germany) allotments only found those species that are generally associated with human settlements rather than any rarities.

Whilst urban parks are typically maintained by local authority staff and with visual amenity in mind, there is a trend to allow food production to take place in defined areas (Garnett, 2000) and some parks have had fruit and vegetable growing for some time. The National Urban Park in Stockholm, Sweden, which is biodiversity-rich, has six allotment areas in Söderbrunn, Kvarnvreten, Frescati, Jakobsdal, Bergshamra and Stora Skuggan. The first allotment in Stockholm was dug in 1904 at Djurgården and the others were put in place over the following 50 years. In area terms, the allotments take up a very small part of the park, but are still considered to add value in ecosystem services (pollination, seed-dispersal, pest control) (Barthel *et al.*, 2005). Even sophisticated Paris has its urban agriculture; for example, Montmartre has communal vines, as do some public parks (Fleury, 2002).

Urban orchards are also undergoing something of a renaissance with the interest in growing local food. Several reasons for this interest are evident in addition to food production: their heritage and biodiversity value are perhaps the most obvious additional benefits, but social and economic capital has also been identified as accruing from community orchards (Countryman, 2009). Traditional orchards, that is, those not under commercial production and which use older fruit varieties, are known to be valuable for saproxylic invertebrates – those that live on, or in, dead and decaying wood. Species particularly associated with old orchards in the UK include the beetle the noble chafer *Gnorimus nobilis* and the group of specialist species that live on mistletoe *Viscum album* (which is itself strongly, though not exclusively, associated with fruit trees) (Biggs, 2003; Williams, 2000). Old orchards are mainly found in the countryside, but some may have become encapsulated by urban development. One such is the La Sainte Union School Orchard in Kentish Town, London. The school moved to the site in 1864 and still has the original convent orchard with apple *Malus domestica*, pear *Pyrus communis* and quince *Cydonia oblonga* trees (Cruikshank, 2001). Urban 'orchards' may not resemble traditional orchards: the University of East London has an urban orchard which consists of six fruit trees (two apple, two pear, one plum *Prunus* sp., one cherry *Prunus* sp.) planted in half-barrels at its Stratford campus (UEL, 2011). Another example of a small orchard is a collection of 12 crab apple trees *Malus* sp. in St John's Gardens, Lower Byrom Street, Manchester, UK. The trees, along with the rest of the floristically rich garden, started out as a winning demonstration display, designed by Daniela Coray, at the Royal Horticultural Society's flower show at Tatton Park, Knutsford – after the show it was transferred to St John's Gardens and has been adopted by local residents who manage it (CityCo, 2011). In Chicago the 'Chicago Rarities Orchard Project' is trying to encourage the planting up of unused open spaces using rare varieties, using the plantings for educational and general open space use. Interestingly, historical research has shown one of the targeted areas on Logan Square was once planted up with trees and one orchard produced some 600 bushels of cherries in 1871 (CROP, 2009, 2012)!

## 2.5 Transport (walking/cycling)

Green spaces themselves, and traffic-free links between them, provide movement corridors for wildlife, but also act as 'zero-carbon' transport corridors for people either as pedestrians or cyclists (Chapman, 2007) (Figure 2.10). Such links are expected to promote an active lifestyle with social benefits and improved physical and mental well-being as well as helping to reduce traffic congestion (Goode, 2006; Humphreys *et al.*, 2013; Martin *et al.*, in press; Mell, 2007; Ogilvie *et al.*, 2011; Pucher *et al.*, 2010; Wen & Rissel, 2008). A cost-benefit analysis carried out by Saelensminde (2004) showed that investment in cycle networks can return benefits of four to five times the cost of implementation. Green infrastructure can provide more pleasant and safer routes than roads, factors that have been shown, or suspected, to be important in encouraging commuters to walk or cycle rather than use cars (Panter *et al.*, 2014; Pucher & Dijkstra, 2000; Sonkin *et al.*, 2006). In the UK, the Connect2 initiative developed 79 projects

FIGURE 2.10 Zero-carbon transport. Green spaces and the links between them provide opportunities for pedestrians and cyclists to travel in pleasant surroundings away from polluting vehicle traffic. © John Dover

across the UK to encourage walking and cycling (Ogilvie *et al.*, 2011) and in Southwark, London, links between parks have been developed with planting improvements in streets (LS, undated). The provision of new routes, such as those carried out under Connect2, does appear to have increased walking and cycling after two years (Brand *et al.*, 2014; Goodman *et al.*, 2014).

Chapman (2007) reviewed the literature on transport and climate change and suggested that walking and cycling were viable approaches to reducing cars. However, an analysis by Brand *et al.* (2014) suggested that the increased walking and cycling activity resulting from the Connect2 programme had not yet been translated into a substantial reduction in $CO_2$ emissions from transport. Instead, they suggested that the increase in cycling and walking they recorded was probably not due to more people engaging in the activities, and hence reducing the number of car journeys, but rather a result of those already using walking and cycling as a mode of transport changing their patterns of use in response to the new routes. Brand *et al.* (2014) considered that provision of opportunities for walking and cycling may need to be coupled with disincentives relating to car use (e.g. higher parking charges, fuel taxation, road tolls, etc. (Chapman, 2007)), incentives and other policy measures. Worryingly, Steinbach *et al.* (2011), using London, UK, as a case study, found that cycling as a commuting activity had an ethnic, gender and class bias towards professional white males, and this is clearly not a desireable situation. Various schemes have been tried to improve the use of walking and cycling over the use of cars. The evidence from previous work, reviewed by Ogilvie *et al.* (2004), is that general schemes such as publicity campaigns have little impact whereas those targeted specifically at behaviour change of an already motivated group generally work, though with differing levels of effectiveness. Some approaches including engineering solutions, financial incentives, and alternative approaches, such as telecommuting, have had mixed fortunes, with some having the opposite effect to that desired. Green infrastructure is potentially an excellent way of improving population health and reducing fossil fuel consumption by encouraging walking and cycling, but the simple provision of GI without other support programmes may mean that its value is not fully realised; there is clearly a data lack in this area.

## 2.6 Economics/enhanced building values

Doernach (1979) predicted that greening urban structures (green roofs, green walls) would generate new jobs; whether the scale of his prediction of 25,000 in Germany alone has proved to be accurate is open to question, but new companies have certainly sprung up to service this new industry.

Does biodiversity make a difference in terms of economic benefits to an area? Benefits can be both tangible and intangible in financial terms as Abdullah *et al.* (2002) found in their study of the building and operation of a new large park in the centre of Kuala Lumpur, Malaysia. The park was built in 1998 as part of a large mixed development including a large shopping centre complex, hotel and associated high-rise buildings (the largest being 88-storey twin towers) in an area which was rather low on green space. The park itself is 20 ha in size, designed by the Brazilian landscape designer R.B. Marx to be biodiversity rich, especially for birds. It has about 1,900 indigenous trees and palms from 74 species, some of which were transplanted from the old Selangor Turf Club, and also provides a range of recreational amenities (Abdullah *et al.*, 2002; SKLCC, 2014). Park use is 10,000–20,000 on weekdays rising to 60,000–70,000 at weekends. Interviews and questionnaires revealed that the shopping centre was used by 42% of visitors before visiting the park and 21% afterwards. The park is used as a marketing feature, along with the shopping centre and high-rise twin towers, to attract businesses into the area. The shopping centre's success was strongly associated with its location (next to the park) both by the management of the centre and by visitors who cited the park as being important. Just over half of park visitors gave it as their main reason for visiting the area, with others in the park primarily intending to visit the adjacent buildings. The Mandarin Oriental Hotel and local cafés reported high demand for outdoor seats overlooking the park, preferring them to indoor seats. The park is considered to have reduced congestion, attracted visitors and contributed to making Kuala Lumpur 'a comfortable city'; it is also a low crime and vandalism area. Maintenance costs of such a large park are equally high and are paid for by the 22 'lot holders'; given that these are commercial organisations and payments come out of their profits, continued support may be vulnerable to economic slumps.

Studies in the USA have shown that vegetation can enhance the business prospects of an area (Wolf, 2005). McPherson (1988) indicated that vegetation near residential areas could increase house prices and also increase feelings of privacy. A report for the UK Commission for Architecture and the Built Environment (CABE) called 'Does Money Grow on Trees' (CABE, 2005) concluded that:

- there is a positive relationship between domestic property prices and proximity to a high-quality park;
- the enhanced value is affected by local factors including the type of park, the type of property, the layout of buildings around the park, the location and the local population;
- a park's impact is enhanced when it is linked with tree planting along adjacent streets, and where the park's vegetation can be seen down streets or over rooftops;
- a property that looks onto a park will command a higher premium than one that backs onto it;
- housing near parks without associations with crime will generally command a premium compared with those where there is.

Because of the multiplicity of factors and the wide range of park types considered in the study, there was a wide range in estimated uplift from 0 to 34% for living in the vicinity of a park and 3 to 34% for overlooking a park. In the latter situation most properties had a 5 to 7% uplift in value. These findings seem to hold generally: Kong *et al.* (2007) demonstrated a housing price uplift due to proximity to green space in their case study of Jinan City in China; Tyrväinen and Miettinen (2000) in the district of Sale, Finland; Melichar and Kaprova (2013) in Prague, Czech Republic; Pandit *et al.* (2013) in relation to street trees in Perth, Western Australia. Mansfield *et al.* (2005) gives the uplift estimates from a number of studies, and probes the impact of different definitions of urban forest in relation to the housing market (e.g. trees on private land compared to large forest parcels and their ecological value) whilst Gatrell and Jensen (2002) examine how different urban forest greening approaches fulfil economic local development goals.

Gore *et al.* (2013) reviewed the literature relating to green infrastructure's impact on the economic growth of a locality and concluded that:

- having high-quality parks in an area increases inward investment and local property prices;
- an area's attractiveness and the quality of parks has an effect on visitor numbers and their spending;
- green infrastructure provides ecosystem services such as pollution control, sustainable urban drainage and temperature moderation, and at a reduced cost compared to more technological approaches. Reduced costs associated with damage (e.g. from flooding) allow greater local investment;
- access to green space reduces mental health/stress issues and promotes exercise resulting in increased productivity;
- local food production makes a (very) small contribution to the local economy;
- creation and servicing of green infrastructure creates employment opportunities.

Whilst ecosystem service valuation has started to become influential in policy terms (ten Brink *et al.*, 2009), valuation of specific services in specific urban areas (other than for pollution – see below) are infrequent. Zhang *et al.* (2012) calculated the reduced runoff in Beijing, China, due to its 61,695 ha of green space to be 154 million $m^3.y^{-1}$ (or 2,494 $m^3.ha^{-1}$), which equated to an economic value of RMB 1.34 billion in 2009 in runoff reduction. As might be expected, the reduced runoff was uneven across Beijing, varying with locality due to different proportions of green space.

## 2.7 Climate control/climate change proofing

### 2.7.1 Health effects of heat

Too much heat can be lethal, with the old being particularly susceptible; other vulnerable groups include children, those in hot-environment occupations, the physically or mentally ill, alcoholics, tranquiliser users and those living in the upper stories of tall buildings (Hoffmann *et al.*, 2008; Ishigami *et al.*, 2008; Kilbourne *et al.*, 1982; McMichael *et al.*, 2006). However, Rey *et al.* (2007), in their study of heatwaves and 'ordinary' summer temperatures over the period 1971–2003 in France, concluded that no sector of the population was immune to heat effects. Mortality is typically highest in the most urbanised and deprived areas (Rey *et al.*, 2009) as Johnson and Wilson (2009) found for heat-related deaths resulting from the 1993 heatwave in Philadelphia. The heatwave of 2003 is thought to have caused 2,000 excess deaths in the UK (Defra, 2013b) and 15,000 in northern France (Defra, 2013a). In Paris alone the 2003 heatwave resulted in over 2,600 extra emergency hospital visits, 1,900 extra admissions and 475 excess deaths (Dhainaut *et al.*, 2004). Reducing the impact of the heat island effect (see below), which will be exacerbated by climate change and the increased frequency of predicted extreme events, is clearly of importance for human health (Golden, 2004). The Heatwave Plan for England was initially developed in 2004 and explicitly mentions green infrastructure as a way of reducing heat-related mortality (Defra, 2013a) and O'Neill *et al.* (2009) include trees and vegetation as ways of 'preventing heat-related morbidity and mortality' (see also Kilbourne *et al.*, 1982). Climate change has potentially considerably more impact on health than mortality directly caused by heat. Predicted impacts include an increase in: cold-related illness and deaths (as a result of extreme events); flooding, resulting in drowning and sewer overflows (leading to disease and psychological trauma); algal blooms releasing toxins into reservoirs; incidence of food poisoning and water-borne disease; infections resulting from invertebrate vectors such as ticks and mosquitoes; skin cancer; pollution-related health impacts (Kovats, 2008; McMichael *et al.*, 2006).

### 2.7.2 The urban heat island effect

The city climate differs from that of the countryside around it (Oke *et al.*, 1999; Stülpnagel *et al.*, 1990; Taha, 1997; Villiger, 1986), with northern hemisphere towns and cities having lower relative humidity, less sun, more cloud, more snowfall and more rain than rural areas (Kusaka, 2008; Shepherd *et al.*, 2002; Taha, 1997). Perhaps the most well-known facet of city climates is the 'urban heat island effect' whereby cities are warmer, by several degrees Celsius, than the surrounding countryside, and the larger the city the larger the temperature difference (Kusaka, 2008; Oke, 1973; Sieghardt *et al.*, 2005). Whilst a warmer urban area might be considered a benefit during winter (McPherson, 1994) ('the winter heat island' (Whitford *et al.*, 2001)), as experienced in Tokyo (Ichinose *et al.*, 1999), it is most definitely a negative impact in the summer, making urban areas unpleasantly warm. The impact is felt most at night, and especially when the weather is calm and skies are clear, potentially resulting in 'heat stress, circulation problems and insomnia', especially if high temperatures last for some days (Stülpnagel *et al.*, 1990). The heat island also exacerbates other urban problems such as increasing levels of pollution (Rosenthal *et al.*, 2008). The general pattern of temperature difference between built areas and vegetation becomes pronounced mid-afternoon and continues building during the day, eventually becoming greatest at night as buildings, roads, etc. release their stored heat energy more slowly than vegetation (Gedzelman *et al.*, 2003) and heat release is impeded by the reduced 'sky view' available as a result of tall buildings (Kusaka, 2008; Oke, 1981). This 'sky view factor' is important as it is the fraction of the sky to which the urban surface can radiate heat to, and this is lower in densely built-up areas with many tall buildings than (say) in parks and the countryside (Graves *et al.*, 2001). The heat island then dissipates rapidly towards daybreak and, because built-up areas warm more slowly than vegetation, pockets of cool air can develop in the early morning soon after dawn (Gedzelman *et al.*, 2003; McPherson, 1994). The development of a heat island and its dissipation also needs to be seen in the context of urban versus rural interactions. At night, as heat rises above the urban areas, cool air from the rural perimeter replaces it, and a system known as the Urban Heat Island Circulation can develop (Haeger-Eugensson & Holmer, 1999). Oke (1979) made a distinction between the 'canopy-layer heat island' (CLHI) from

**TABLE 2.2** Factors influencing the development of the urban heat island (Nitis *et al.*, 2005)

| *Radiation balance* |
|---|
| ↑ anthropogenic heating |
| ↑ absorption of short-wave radiation |
| ↓ outgoing long-wave radiation |
| ↓ long-wave radiation loss (reduced sky) |
| ↓ evapotranspiration |
| ↓ heat transport |
| |
| *Physical factors* |
| topography |
| relief |
| urban structure |
| building density |
| building materials |
| |
| *Urban–rural differences in:* |
| albedo moisture |
| roughness length |
| thermal capacities |

about roof height to the ground and the 'boundary layer heat island' (BLHI) above building levels. The CLHI is most affected by local urban processes, whilst the BLHI is more affected by larger-scale processes (Golden, 2004). However, there is clearly an interaction between the boundary-layer heat island and mesoscale processes, as the urban heat island has been demonstrated to influence the wider (mesoscale) climate, with increased rainfall downwind of cities (Shepherd et al., 2002). Nitis et al. (2005) document the factors which contribute to the urban heat island effect (Table 2.2).

The heat island phenomenon is not new, Golden (2004) tracks the earliest publication on it back to Howard (1833). New York may well have had an urban heat island since the late 1800s, increasing in intensity as the city has grown (Rosenthal et al., 2008). Recent estimates give New York's heat island effect as approximately 4°C in summer and autumn, with a winter heat island (including spring) of about 3°C (Gedzelman et al., 2003). London's heat island was estimated to be 3°C higher than average peak temperatures in the surrounding countryside in the summer of 1999 (Graves et al., 2001). A number of studies have identified long-term trends in increasing urban temperatures (Brazel et al., 2000; Wilby, 2003). The additional heat comes from darker surfaces (buildings, roads, pavements) absorbing more solar heat, energy use for industrial processes, transport, air conditioning and reduced amounts of vegetation resulting in less shade (e.g. from trees) and evapotranspirative cooling (Akbari et al., 2001) and, because rainwater is diverted to drains, there is also less water for evapotranspiration (Taha, 1997).

## 2.7.3 Reducing the heat island effect with vegetation

Buyantuyev and Wu (2010) considered that vegetation was most efficient at reducing the heat island effect during the day as most plants close their stomata at night reducing transpiration. Stülpnagel et al. (1990) noted that on calm nights there could be a 9°C difference in temperature between the centre of Berlin and the surrounding countryside. Buyantuyev and Wu (2010) found the same difference when comparing nearby vegetated and non-vegetated parts of Phoenix, Arizona, and Gedzelman et al. (2003) estimated New York's at 8°C. The heat island is subject to modification according to local, seasonal and periodic weather conditions, with winds typically reducing it or shifting it downwind from its 'normal' location (Gedzelman et al., 2003). In his German thesis, Stülpnagel (1987) (given in English in Stülpnagel et al., 1990) concluded that the impact of a vegetated area on the local climate:

- related to area (the bigger the green area the better)
- extended further downwind than upwind
- extended further at higher windspeeds
- extended further in urban areas that are more open and have additional green spaces
- showed vegetated areas that are situated in a dip, or are enclosed (by walls, tall vegetation) have less impact
- showed internal fragmentation of green areas by sealed surfaces (e.g. roads) reduces their value.

Increased summer urban temperatures result in increased electricity consumption for air conditioning (Akbari et al., 2001) and hence increased $CO_2$ emissions that contribute to global warming (Akbari, 2002). As a result, the carbon reduction value of a shading tree in urban Los Angeles was considered by Akbari (2002) to have a greater value than an equivalent forest tree by three to five times in terms of carbon reduction (urban tree saves 18 kg carbon, of which 4.5–11 kg is direct sequestration in tree tissues and the rest from reduced electricity use).

Most vegetation is green because plant leaves contain pigments that absorb red and blue visible light; carotenoids take in blue light and chlorophyll takes in red and blue light, other pigments (flavonoids) absorb ultra-violet (Yoshimura *et al.*, 2010). Contrast the relatively soothing colour of plants with that of the two main urban ground surfaces: a typical concrete plaza or walkway/pavement which is pretty close to white and asphalt/tarmac roadways which are black and age to grey. Light surfaces (i.e. those with a high albedo (Taha, 1997)) reflect light (and heat) and black surfaces do not (Akbari *et al.*, 2009). Light surfaces, in reflecting heat, have their place in reducing the urban heat island and light-coloured paving and roofing is recommended for that reason (Akbari *et al.*, 1992, 2009). However, on sunny days many people walking in paved areas with white or light surfacing (such as concrete) would find that they would need sunglasses to avoid squinting or risk UV damage to their eyes (Heisler & Grant, 2000), whereas light surfacing on roofs would not impact on human vision. At the opposite end of the spectrum are tarmac road surfaces which, being black, reduce the need for pedestrians, drivers, etc. to wear sunglasses but, by absorbing heat and later re-radiating it, create uncomfortably warm conditions and contribute to the urban heat island (Pomerantz *et al.*, 2000). Asphalt surfaces do lighten with age though: when freshly laid they have a very low albedo (0.04–0.05) after five years this increases to 0.12±0.03 until replacement after about 10–15 years (in California) (Pomerantz *et al.*, 2000). Pomerantz *et al.* (2000) have suggested that asphalt surfaces could practicably be given higher albedos by surfacing with lighter coloured chippings achieving (say) an albedo of 0.35 by using chippings with an albedo of 0.5. Of course albedo is subject to seasonal variation especially with snowfall, loss of leaves from deciduous trees, etc. and urban structure (Brest, 1987) (Table 2.3). Even under snow conditions albedo is variable, with flatter surfaces having higher albedo than those with substantial vertical structure like trees; urban areas retain even snow cover (and thus high albedos) for a shorter time than rural areas as roads are cleared for traffic, pavements for pedestrians, and heat from poorly insulated buildings causes snow melt on roofs (Brest, 1987).

Shade cast by tall vegetation, such as trees, can substantially reduce the surface temperature of sealed surfaces (Fintikakis *et al.*, 2011) (Figure 2.11). Plants also, by transpiring, cool the ground (on a warm day try touching a road or path surface then some grass). But, taller vegetation is cooler than shorter vegetation by some considerable margin. Jeremy Thomas, whilst trying to unravel the ecology of the Adonis blue butterfly (*Lysandra bellargus*), showed that there was an 8°C difference between the ground temperature when grass was 1 cm high (24°C) compared with grass 10 cm or more high (16°C) (Thomas, 1983). Studies carried out in municipalities such as Stuttgart and Munich in Germany show that large parks, and especially those with woodland and trees, are some of the coolest parts of the urban environment being 2–3°C cooler than surrounding areas (Tyrväinen *et al.*, 2005). Peters and McFadden (2010) demonstrated that soil and surface temperatures were negatively correlated with increasing number of trees and leaf area index. In a study of Greater Manchester, UK, 28 'urban morphology' types were recognised and their proportion of sealed surface:green space estimated (Handley & Gill, 2009). Maximum surface temperatures were clearly positively correlated with proportion of sealed surface. Highest temperatures (>30°C) were

TABLE 2.3 Albedo of different land-uses derived from Landsat imagery of Hartford, Connecticut, on snow-free days (Brest, 1987)

| Location | Albedo (%) |
| --- | --- |
| City centre 'downtown' | 8–12 |
| High density residential | 8–12 |
| City outlying | 11–15 |
| Suburban | 11–17 |
| Trees | 7–19 |
| Wetland | 5–20 |
| Agriculture | 11–19 |
| Range | 11–19 |
| Non-forest park | 17–23 |
| Lake | 0.2–5 |

found in the town centre with about 20% vegetation; the coolest (<20°C) were categories with over 90% of their surface vegetated (or, as they say, evaporative) such as informal open space, woodland, or rivers and canals (Handley & Gill, 2009). Studies in a Manchester park showed air temperatures to be, on average, 0.8°C cooler in the park compared with the surroundings. Grass was shown to be substantially cooler than sealed surfaces and that tree shade had a cooling effect on surface and air temperatures (Armson *et al.*, 2012). Grass and trees (because of shade and evaporative cooling) were considered by Armson *et al.* (2012) to be better in combination than alone, and capable of producing surface temperatures between 4 and 7°C below that of the air. The urban forest in Beijing was estimated to cover 16.4% of the city area and reduce air temperature by 1.6°C which, because of a reduced load on air conditioning systems, also reduced electricity use by 0.238 GWh (Yang *et al.*, 2005). Parks in Singapore have been shown to have lower temperatures than built-up areas with the effect extending into the adjacent areas (Chen & Wong, 2006).

The cooling effect of urban vegetation extending into the surrounding areas is clearly an area that excites interest (e.g. Chen & Wong, 2006) but the distance within which the effect is evident is relatively short, only 100–500 m at best (Chen & Wong, 2006; Tyrväinen *et al.*, 2005; Wilmers, 1988, 1991), suggesting that urban areas need an extensive network of green spaces and street vegetation for their effect to be maximised. Interestingly, Natural England's minimum set of standards for accessibility to green space (ANGSt: Accessible Natural Green space in Towns and Cities) includes one criterion which is: 'that no person should live more than 300m from their nearest area of natural green space of at least 2ha in size' (Handley *et al.*, 2003), a figure that dovetails rather well with the estimate of how far a cooling effect extends from green space! Buyantuyev and Wu (2010) contributed an increased understanding of the heterogeneity of urban heat islands in their study of Phoenix, Arizona; they concluded that heat islands should not be considered a uniform entity, but a mix of hot and cold spots with temperature differences of such areas sometimes exceeding the difference between the urban centre and the surrounding countryside. They coined the term 'urban heat archipelago' to describe the scatter of hot and cold patches

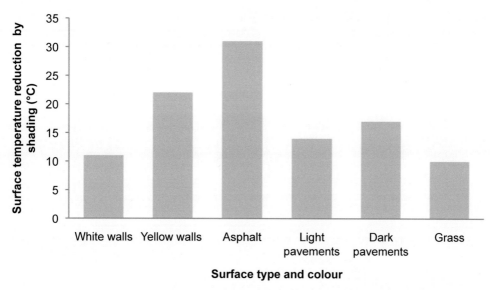

FIGURE 2.11 Reduction in the surface temperature (°C) of various urban surfaces in the city of Tirana, Albania, when shaded. Data from a microclimatic survey carried out in 2008, source Fintikakis *et al.* (2011).

within urban areas (Buyantuyev & Wu, 2010). Giannopoulou *et al.* (2011) used a network of 25 fixed locations to examine the urban heat island in five zones of Athens, Greece, and demonstrated substantial differences (5°C) within the city with the highest temperatures in the city centre and in the western industrialised zone. City parks are now being considered as 'cooling islands' and sophisticated approaches are being used to model their impact on the urban heat island (Declet-Barreto *et al.*, 2013; Vidrih & Medved, 2013). Arnfield (2003) pointed out that urban climate should be considered at different scales from the individual components to the whole city (Table 2.4). Recent work by Connors *et al.* (2013) suggests that very-fine-grain modifications to spatial patterns, such as complexity of grass patch shape, use of high albedo roof surfaces, shade cast by vegetation and tall buildings (Lindberg *et al.*, 2003), etc., may have an impact on local temperature. Rosenthal *et al.* (2008) have produced a report on mitigation approaches for New York's heat island which includes a range of measures including low albedo surfacing, green roofs and increased tree planting. Onishi *et al.* (2010) examined the potential of tree planting in car parks as a way of mitigating the urban heat island in Nagoya, Japan. By greening a car park with 30% tree cover and 70% grass their simulations indicated that summer temperatures could be reduced by 7.26°C (Onishi *et al.*, 2010); whilst not perhaps being practically achievable in area greened, the study certainly suggests that greening car parks could reduce small hot spots (Figure 2.12). For Atlanta, modelling studies indicated that a 100% increase in vegetation cover or a tripling of its albedo could reduce the UHI; on a practical basis increasing the amount of vegetation was considered to be the more feasible option (Zhou & Shepherd, 2010).

**TABLE 2.4** Hierarchical nature of urban settlements based on Arnfield (2003)

| Scale | Type | Components |
|---|---|---|
| 0 | Individual surfaces | Walls, roofs, gardens, lawns, parks, road, etc. |
| 1 | Building | Aggregation of walls + roof |
| 1 | Ground-level | Mix of irrigated and non-irrigated green space components (including gardens) and sealed surfaces |
| 2 | Urban canyon | Buildings and space between (e.g. roads) |
| 3 | City blocks | Urban canyons + building roofs |
| 4 | Neighbourhoods | Collection of city blocks |
| 5 | Land-use zones | Urban centre, industrial, residential, etc. |
| 6 | Entire city | Aggregation of land-use zones |

**FIGURE 2.12** Trees shading car parking places in France help keep local temperatures down and potentially reduce fuel evaporation. © John Dover

## 2.7.4  Plants and wind

Vegetation that has a lot of structure, e.g. trees, shrubs and hedges compared with cut lawns and flower borders, will also have the beneficial effect of reducing windspeed. Farmers have always known this, planting lines of trees (windbreaks) or tall hedges to slow it down around sensitive vegetation, or shorter hedges to protect animals (Pollard *et al.*, 1974). If it is very windy and warm, plants can wilt as they try to combat the heat through opening their stomata and having the water evaporated off faster than it can be replenished via the roots; plants can also be killed by wind-chill on cold days as the wind increases the exposure factor. It always surprises me how many architects and urban designers still create draughty urban plazas and roads without any apparent thought of the need to stop the wind tunnel effect (Figure 2.13). Klaus Tham designed the Bo01 development in the Western Harbour area of Malmö with the need to create shelter in mind – tall buildings were placed along the water's edge to shield lower-level buildings behind them; the street layout was more sinusoidal than straight and the streets narrow (Kruuse, 2011). A rough rule of thumb is that the lee-side shelter footprint of an object (hedge, tree, building) extends to 12 times its height (Lewis, 1965); so a hedge 2 m high would throw a shelter footprint of 24 m. Of course the shelter effect is greatest near the structure, within 2 times and 3 times of its height (Lewis, 1965), and the effect is greatest with windbreaks that have a dense structure (i.e. those that do not let a lot of wind through) (Lewis & Stephenson, 1966). However,

**FIGURE 2.13** On a summer's day this walkway in Liverpool, UK, is cooled by a strip of mini-fountains, though the surface reflectance (albedo) is high and sunglasses are needed. In the winter it is likely to be a cold wind-tunnel. © John Dover

the optimum permeability is considered to be 40% (Pollard *et al.*, 1974) with very dense structures resulting in vortices on the lee side and a rapid return to the windspeeds found on the upwind side. Heisler (1990) compared windspeeds at 2 m above ground at a local airport with those in residential areas with no trees and those with the same building density and different levels of tree cover (24–77%). Results indicated that the buildings reduced windspeed by 22%, but in the area with 77% tree cover average windspeeds could be reduced by 70% in summer and 65% in winter.

## 2.7.5 Climate change mitigation

Towns and cities are wetter than surrounding areas because the hot air from sealed surfaces and heat-sources rises promoting cloud formation (and hence rain) (Stülpnagel *et al.*, 1990). Climate change predictions indicate drier summers and wetter winters with more extreme events in both (Solomon *et al.*, 2007). Green infrastructure can help cope with increased storm events by reducing the likelihood of flooding that would result (see section 2.8 Sustainable urban drainage below) and with heatwaves by evapotranspiration and shading. Climate change is likely to exacerbate urban heat island effects (Wilby, 2003). The importance of green infrastructure was underlined by climate change modelling of the Greater Manchester data (Handley & Gill, 2009) described above. This work suggested that increasing the area of vegetation in the town centre by 10% would keep temperatures in the range experienced between 1961 and 1990 under both high and low emission scenarios for 2020 and the low scenario for 2080; for the high 2080s scenario the 1961–1990 range would just be exceeded using 10% vegetation (Handley & Gill, 2009). Unfortunately, to increase vegetative cover requires unused open space if conventional ground-level planting is used. Solecki *et al.* (2005) note, from their study of the urban heat island effect in New Jersey, that the more deprived, high-density residential areas of inner cities are less likely to have capacity for such solutions as tree planting than more affluent lower density areas. Greening solutions in such situations will require modifications to building surfaces (green roofs, green walls) to achieve thermal benefits; an alternative is to use high albedo roofing materials (Solecki *et al.*, 2005).

Climate scenarios that predict more extreme temperatures are likely to impact on all sectors of society, but most severely on the elderly and frail (Ishigami *et al.*, 2008). With elevated temperatures, people may become effectively trapped inside their homes and offices, reducing socialisation and opportunities for exercise. In such circumstances it is not unlikely that air conditioning will be increasingly relied upon to moderate internal building temperatures: increasing costs and contributing to greater emission levels of greenhouse gases (which then feed into increased climate change) – of course if there are also extreme lower temperatures in winter heating costs will increase as well. If the likely impacts of climate change and the role of vegetation in mitigating those effects are clearly understood by local authorities, businesses, and industry, relatively low-cost action can be taken now to meet the threat in future years when it will cost more to combat using more technological approaches. For example, it costs less to plant a small tree than a large tree (and survival and growth may well be superior). Consider the case of a care home for the older members of society where residents have access to a patio or grassed area for exercise, or simply to take the air: planting with suitable shade trees will allow the area to continue to be used in periods with elevated temperatures whilst having relatively low start-up and maintenance costs. The more ambulatory residents of the care home may have access to local shops, libraries or green spaces, but if the streets do not provide sufficient shade, they may be effectively prevented from accessing them: the planting of street trees and the use of hedges may create conditions that facilitate access and the strategic siting of seating underneath trees or in strip parks may actually improve, as opposed to simply maintain, quality of life. The correct siting and species use/mix is important to ensure that shading in summer and wind-shielding in autumn/winter does not mean that solar heating is compromised in winter (e.g. deciduous trees shade in summer but allow most sunlight through in winter). Whilst the scenario described is for a care home, it is equally applicable to domestic residences, businesses and schools, and for all age groups. Vegetation can also be used to shade the windows of conservatories, windows or verandas to moderate indoor temperatures.

A number of authors have demonstrated the energy-saving value of trees (Akbari, 2002; Akbari *et al.*, 1992, 2001; Heisler, 1986a, 1986b; Huang *et al.*, 1987; McPherson & Rowntree, 1993; McPherson & Simpson, 2003; Parker, 1983; Simpson, 1998). For example, McPherson and Simpson (2003) estimated that the 177 million trees in California, at the time of their study, saved 2.5% (6,408 GWh) on air-conditioning costs ($486 million at wholesale prices) – to customers this was collectively worth about $970 million. There can be a downside to shading as it can increase the need for winter heating (even after taking into account the windspeed-reducing effects on drafts); nevertheless, the net economic benefit of trees taking into account cooling and heating was $458 million at wholesale prices. The value of trees in energy saving was greatest during times of peak demand – peak load reductions were estimated at 10%. Even trees that do not cast direct shade on building surfaces can have benefits (estimated at 25–50% of shade trees). McPherson and Simpson (2003) considered that there were 242 million unused planting sites, and if 50 million were planted up to shade the east and west walls of homes, savings would increase by 1.1%.

The value of trees in reducing summer air-conditioning costs varies due to aspect (McPherson & Simpson, 2003). In a detailed study of 460 single-family homes in Sacramento, California, Donovan and Butry (2009) demonstrated that trees on the east-facing side of a building appeared to have no effect on electricity use – this was probably because trees on the east cast morning shade before air temperatures become uncomfortable. South-facing trees saved energy on air conditioning, but only if they were within 12.2 m of the house – primarily because the sun is higher in the sky when they would be effective, and trees further from the house would not cast long shadows and hence effective shade. West-facing trees cast shade later in the day, and at the time of peak temperatures and with lower sun-angles; as a result they could be further away than south-facing trees and still have an effect on energy consumption. Trees on the north-side did not cast shade on the study houses, yet curiously trees planted close to the north side of a house (within 6.1 m) actually increased energy consumption; factors suggested to explain this included close proximity reducing the cooling effect of wind, promoting the retention of heat at night, or simply because electric lights might be required more frequently because of light-blocking. Donovan and Butry (2009) considered that, combined, south- and west-facing trees saved 5.2% electricity and north-facing ones increased it by 1.5%.

## 2.8 Sustainable urban drainage systems (SUDS)

### 2.8.1 Introduction

Increasing urbanisation results in increasing areas of sealed surface and the requirement for large capacity stormwater systems to rapidly route water away to prevent flooding; the higher the proportion of sealed surface, the higher the volume of water to be dealt with (Tyrväinen et al., 2005). Where the sewer system combines both rainwater collection and sewage, heavy rainfall events can result in sewer overflows, creating flash flooding contaminated by faecal matter. Such events are an issue globally; for example, in San Francisco, USA, (Bandy, 2003) and in the UK as the floods in 2007 (Coulthard & Frostick, 2010; Pitt, 2008) and February 2014 demonstrated (Helm, 2014; Helm & Doward, 2014). Higher rainwater volumes are something that it looks like we are inevitably going to have to contend with because of climate change and the implications, as predicted by modelling, are frightening. In Europe heavy rain is predicted to increase in the central and northern areas and decrease in the south (Beniston et al., 2007). Whitehead et al. (2009) reviewed the impacts on aquatic systems caused by a combination of increased temperatures and extreme events. Whilst there may be droughts, there will also be more extreme rainfall events and heavy flushing after drought may have negative consequences in terms of mobilisation of concentrated physical, chemical and also microbiological entities (Bandy, 2003; Whitehead et al., 2009). Estimated effects on streams, rivers and lakes include:

- changes in river morphology as a result of erosive power of increased flows and sediment loads (and increased siltation of lakes);
- changes in mobility and dilution of contaminants (low flows increasing concentrations);
- decreased water quality leading to reduced habitat quality resulting from:
  - increased frequency of flash floods
  - increased stormwater runoff from urban areas
  - increased nutrient loads
  - changes in pH levels
  - increased temperatures increasing chemical reaction speed
  - toxic algal blooms
  - reduced dissolved oxygen levels;
- increased colonisation and movement of alien species (Whitehead et al., 2009).

Extreme rainfall events are more important than overall averages because they result in substantial economic losses (including damage to property) and have public health implications (including direct mortality) (Easterling et al., 2000; Katz & Brown, 1992; Sillmann & Roeckner, 2008). Effects are primarily due to the overloading of sewer systems. In the EU the Water Framework Directive (2000/60/EC) (WFD) (EU, 2000) was devised to (amongst other things):

- protect and enhance aquatic ecosystems (and terrestrial systems and wetlands that depend on the aquatic ecosystems);
- promote sustainable water use and protect water resources;
- enhance protection and improvement of the aquatic environment;

- progressively reduce pollution of groundwater;
- contribute to mitigating the effects of floods and droughts.

In the UK, responses to the WFD and flooding evident in 2007 (Pitt, 2008) included: the Flood and Water Management Act 2010 (Defra, 2010) for England and Wales, the Water Environment and Water Services (Scotland) Act 2003 (SG, 2003) and the Flood Risk Management (Scotland) Act 2009 (SG, 2009).

The central purpose of sustainable urban drainage systems is: to reduce, or slow down, the flow of (rain)water to sewers. As such, the use of SUDS is clearly relevant to the implementation of several of the primary WFD aims (EU, 2000).

But what is a sustainable urban drainage system? There is not a simple answer as there are so many components that can be considered as being part of SUD; perhaps the easiest approach is to start by breaking SUDS down into their basic functions and then thinking about what can be used to supply those functions later (Figure 2.14).

While rain is falling two things can immediately happen depending on the SUD component: 1) rainfall can be intercepted before it reaches the ground and is slowed down, or 2), on reaching the ground, it is immediately infiltrated ('soaked-up'), so that it does not runoff to the sewer system. If the rain has been intercepted, it then follows three routes depending on circumstances: it can continue to the ground to be infiltrated, if not soaked up it can be transferred to a containment area or, if it has stopped raining, it can be evaporated off the interception surface. If the rain was infiltrated, there are a number of possible subsequent fates: it can be re-evaporated from the infiltration surface, it can be taken up by plant roots

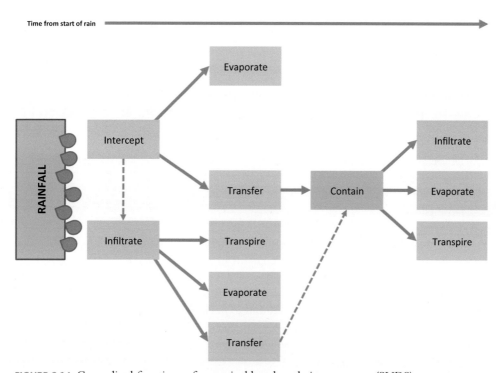

FIGURE 2.14 Generalised functions of a sustainable urban drainage system (SUDS).

and transpired, it can be transferred to a containment area, or it can remain in the infiltration medium and be transferred to the surrounding soil. The transfer to the containment area can be via conventional pipes or, preferably, via 'swales' (a nice name for a shallow vegetated ditch system). The containment area could be a pond, soakaway or 'rain-garden' from where the water can infiltrate, evaporate or be transpired off by marginal or aquatic plants (this can also happen during the transfer process). The containment area may also act as a temporary processing area (to remove pollutants, suspended solids, etc.) prior to disgorging into a stream or river. This means that planning for SUDS needs to be considered at a strategic level, and at a number of scales from the very local to regional (EPA, 2008). In Malaysia, SUDS are seen as a key way to prevent flash flooding, improve a low water supply and treat water pollution; implementation is via swales for transport and subsurface detention and various kinds of pondage for biofiltration and water storage (Zakaria *et al.*, 2003).

In the UK a major barrier to the introduction and use of SUDS was the wordage in legislation: SUDS simply were not covered by the definition of a sewer. As EU directives have to be translated into the laws of member states, the need for legislation to incorporate the WFD was been taken as an opportunity to modify those definitions, as in Scotland (SG, 2003; SW, 2011). The Water Environment (Controlled Activities) (Scotland) Regulations 2005 (SG, 2005) required runoff from new developments to be treated using SUDS (Stewart & Hytiris, 2008). In 2007 the various organisations concerned with environmental protection in the USA joined together to collaborate on the use of green infrastructure (GI) as a means of reducing stormwater runoff and published an Action Strategy (EPA, 2008) composed of the following elements:

- *Research* – on the performance and effectiveness of GI.
- *Outreach and communication* – dissemination of information including the need to examine and modify the way things are done – including the administrative realm.
- *Tools* – the development and provision of predictive and design models to aid in implementation.
- *Regulatory support* – identification of the provisions that already exist for use of GI in water management but that may not be recognised as such due to custom and practice.
- *Economic viability and funding* – assisting in making the financial case for use of GI including highlighting financial support mechanisms.
- *Demonstration and recognition* – promoting take-up by provision of case studies and existing projects, etc.
- *Partnerships* – expanding the number of organisations that signed up to the inital collaborative effort.

Following the publication of the Action Strategy, the US Environmental Protection Agency published a points-based 'Water Quality Scorecard' in 2009 and noted that regulatory changes were, after all, necessary to facilitate the use of SUDS, but that these were in what might have been considered ancillary activities such as landscaping, car park and street design (Hall *et al.*, 2009). The aim of the Scorecard was thus to help identify where changes needed to be made and to provide information on policy approaches and materials, and to give examples of successful implementation and how SUDS incorporation could be incentivised (Hall *et al.*, 2009). An example of the latter being a stormwater tax discount the 'Clean River Rewards Program' as operated in Portland, Oregon (CoP, 2014). In completing the Scorecard, a multi-agency or department approach was urged to maximise opportunities – poor communication within and between large bodies is an obvious problem when multidimensional problems need addressing. The Scorecard was divided into five sections which reflect some of the major responsibilities found in different municipal departments (or agencies):

1. Protect natural resources (including trees) and open space.
2. Promote efficient, compact development patterns and infill.
3. Design complete, smart streets that reduce overall imperviousness.
4. Encourage efficient provision of parking.
5. Adopt green infrastructure stormwater management provisions (Hall *et al.*, 2009).

The Scorecard asks basic questions linked directly to stormwater management outcomes and as such has a wider applicability than in the purely US context.

SUDS components have a range of different functions and most incorporate vegetation as an important part of the solution rather than as an incidental inclusion. Pratt (2004) reviewed the literature on the following SUDS components:

- Constructed wetlands
- Detention basins
- Detention ponds
- Filter drains
- Infiltration pits and trenches
- Pervious surfaces
- Retention/infiltration basins
- Swales and filter strips.

A nice set of examples is given in SUDSWP (2010):

- *Collection* – infiltration drains and ditches and permeable materials may collect water directly from roadsides (sometimes with pre-treatment by sediment-removing gullies). Typically these are filled with some kind of graded stone media which allows runoff to fill air gaps between the stones before running into a perforated drainage pipe at the base of the trench for transport to a storage area. Another approach is to use shallow, wide vegetated ditches (swales) to transport water. Swales may also have gullies catching sediment before runoff enters them. They may also have small 'check dams' incorporated so that they can store water in the deeper part of their profile. So as well as transporting water, swales can store and also infiltrate – reducing the amount of water reaching the storage ponds or basins.
- *Storage* – typically in shallow basins and ponds where water is captured, held and slowly released. Such structures can be dry most of the time or have a deeper, permanently water-filled pool incorporated. This 'attenuation storage' captures runoff and prevents the stormwater surges that overwhelm sewers. The water may have a (narrow) outflow to traditional sewer systems, or run to infiltration areas. Some designs may include a sediment trap by having a 'forebay' which is divided off from the rest of the pond area by a permeable material and typically takes up about a third of the pond length. Water entering the forebay slows and deposits its silt, with the permeable material

also removing suspended solids. Such structures offer pre-treatment and attenuation storage.

*   *Infiltration* – infiltration basins may have outflows as for storage basins, but their primary role is to hold water so that it can slowly infiltrate into the ground. This will be determined both by the nature of the ground (clay-based soils would not be suitable) and by the level of the water table.

*   *Wetlands* – these combine the concept of ponding with marshes and have a combination of deeper and shallower areas. They have inflows and outflows at either end of the wetland and are intended to allow sedimentation (by slowing water down) and to remove pollutants and nutrients through uptake by plants.

All the above use plants in their operation, but there are opportunities with most of them to increase the biodiversity component at the design stage. However, in concentrating on engineered solutions (whether or not they include plants), it is possible to overlook the obvious: that plants *on their own* are SUDS components, something that San Francisco has understood. Bandy (2003) explored options to decrease rainwater-caused sewer overflows in San Francisco and examined three options: permeable pavements, harvesting of rainwater (water can be captured in tanks for direct reuse, e.g. in toilets, car washing, garden watering, and not simply for attenuation storage (see EA (2008)) and trees. In Portland, Oregon, a stormwater tax discount is available via the 'Clean River Rewards Program' if properties have four or more trees over 4.57 m (15 feet) high – or can demonstrate a canopy cover of 18.58 m$^2$ (200 square feet) (CoP, 2014). Schmidt (2009), in terms of mitigating the urban heat island effect, considers rainwater harvesting for cooling and irrigation to be of higher priority than simple trough and trench infiltration, or harvesting for toilet flushing etc., but lower than the use of unpaved green areas (including street trees), green roofs and green façades, artificial lakes and open waters.

Using a SUDS approach therefore uses (plant) biodiversity to achieve specific storm drainage aims; they can also increase faunal biodiversity. Jackson and Boutle (2008) reported findings from the SUDS system incorporated in a new development at Upton in Northampton, UK. They found that swales and retention ponds were colonised by 34 aquatic macroinvertebrate families from the River Nene Valley as well as newts, frogs and toads. Birds typical of wetland habitats had also been observed using the SUDS elements, including the reed bunting *Emberiza schoeniculus*. In New Jersey, USA, McCarthy and Lathrop (2011) found that 92% of stormwater basins with continuous water (i.e. they never run dry) contained fish and thus successful reproduction in such water bodies depended on the juvenile stages of anurans (frogs and toads) being resistant to fish predation. Basins that dry out in late summer, however, make potentially good anuran habitat (Brand & Snodgrass, 2010). In Melbourne, Australia, Kazemi *et al.* (2011) compared the above-ground invertebrate fauna of roadside 'bioretention' swales with two types of roadside verge composed of either grass with occasional trees or a more ornamental verge which the authors termed 'gardenbed-type' – typically small patches of vegetation below a tree with the patches separated by bare ground. The swales were designed to hold water which, after infiltration through a filter medium, was collected by perforated pipes and transferred to a storage area. Invertebrates were sampled by sweep netting vegetation and the study showed that species richness and abundance were both higher in the swale vegetation. Swales were probably better because of the improved food and shelter provided by the increased structural vegetation cover, abundance of flowering plants, lower pH, and sloping profile of the swales compared with conventional verges.

## 2.8.2 Plants and rain

If we look at some of the fundamentals of higher plants, we see that they have roots, a stem and (usually) green leaves. Features will vary between species with, for example, some plants appearing to have no, or very short, stems whilst others have huge ones (e.g. carline thistle *Carlina vulgaris* compared with giant redwood trees *Sequoiadendron giganteum*). These plants also share an internal water transport system with roots taking up water from the soil, transporting it round the plant in tubes of 'woody' xylem, and the leaves letting it out through holes called stomata. Releasing water at the leaf surface can be thought of as being similar to humans sweating, and has a similar end result: it cools leaves down. Of course if it gets too hot, the plant can lose too much water and wilt, so stomata can be opened and closed by the plant to increase or slow down the amount of water that flows through it. Plants that live in very hot conditions have fewer stomata, to reduce water loss, and have various other adaptations to their environment such as waxy cuticles, leaves reduced to spines, and swollen stems to act as water storage organs. This process of giving-off water at the leaves is called transpiration.

So why is the structure of plants and their water transport mechanism important in green infrastructure? If we first start with rain; it is pretty obvious that anything underneath it is going to get wet. It is also obvious that, as long as the rain is not part of an electrical storm (thunder and lightning), standing under a tree with leaves on is a good idea as it keeps you dry, or rather dryer. Eventually water seeps through the canopy until you start to get wet, when it stops raining it is a good idea to get out from under the tree because it takes some considerable time for it to stop dripping water. This is the first reason why trees, and vegetation in general, are useful in green infrastructure: they slow rain down. One of the predictions of climate change is increased variability in weather (Murphy *et al.*, 2009), including heavier bouts of rainfall. Whether or not the flooding events that have happened in 2007 in the UK and elsewhere (Coulthard & Frostick, 2010) are specifically climate change events (Whitehead *et al.*, 2009), one of the things that can help reduce flash floods is green infrastructure because vegetation slows rainfall down, reduces overall rain volumes that reach the ground, and in consequence results in lower peak flows (the largest volume of water flowing to the drains at any given time). It is obvious, given the above, how structural vegetation slows rainfall down – but how does it reduce the amount that reaches the ground? Well hopefully the sun starts to shine after the rain, and the water still on the surface of the leaves evaporates and so never gets to the ground. 'Steam' rising from trees after rainfall is always a delight to see! The second way plants help to reduce flash floods is simply because they take water up from the soil and 'transpire' it off, reducing the water held in the ground so that new rain has somewhere to go. Having green areas allows the rain to percolate into the soil instead of running over pavements, roads, plazas and parking lots to the drains (Bolund & Hunhammar, 1999; Guo *et al.*, 2000; Niemelä *et al.*, 2010). Now, it is also pretty obvious that at different times of the year different kinds of plants will have different values: deciduous species are going to be better at slowing down rain when they have leaves, for instance (i.e. in spring, summer and early autumn), whilst evergreens have leaves all year round. But even evergreen species are not going to do much transpiring in the winter – though they will be better at catching rain and snow.

### 2.8.3 Implementing SUDS

It is clear that plants are central to the operation of sustainable urban drainage, and green roofs, green walls, grassland, street trees, hedges, shrubs, allotments, etc. can all be considered elements of SUDS. However, it does not have to be green to be part of a SUDS – permeable pavements (where water drains through permeable concrete, asphalt, gravel gaps between paving blocks, etc.) often do not have any vegetation, though some incorporate spaces for grass to grow. Nevertheless, even in such situations where there is no obvious surface component, at the subsurface level there is likely to be a microbial community that is actively degrading pollutants such as oil (see section 4.3.1). Ponds could be considered 'blue' infrastructure. Whatever the individual components of a particular implementation in a particular place, the general guiding principles are the same, but the multiplicity of elements which can be used as SUDS components means they are extremely flexible and can be adapted to almost any situation.

The City of Portland, Oregon, USA has been very active in introducing SUDS (Liptan, 2006) and there are two particularly nice examples. The first comes from NE Siskiyou Green Street where a residential road was narrowed at one end through having two SUDS installed. The net result was to improve the visual amenity, control traffic and, most importantly from the initial design viewpoint, to reduce and slow down water flow into the drains (Figure 2.15). In a test using a fire hydrant to provide water for a 25-year storm event, the two elements reduced the peak flow by 81%, delayed the peak flow by 16 minutes and reduced the volume going to the drains by 82% (Liptan, 2006).

FIGURE 2.15 Before (left) and after (right) installation of two SUDS elements at the end of NE Siskiyou Green Street, Portland, Oregon, USA (the SUDS were installed on both sides of the road). © City of Portland (see Liptan, 2006).

FIGURE 2.16 The installation of stormwater planters in SW 12th Avenue Green Street, Portland, Oregon, USA. Top left: before installation, right: old plantings removed and soil excavated; middle left: wooden shuttering in place for concrete, right: after planting up; bottom: the planters during rain showing collection of kerbside runoff. © City of Portland (see Liptan, 2006).

Another project was the conversion of conventional surface plantings along SW 12th Avenue Green Street to SUDS. In this situation, existing rather routine planting beds were excavated, SUDS planters were concreted in in situ, using shuttering, and finally planted up with a mixture of trees and plants tolerant to periodic inundation (Figure 2.16). The result is visually more pleasing than the initial plantings and reduces runoff.

## 2.9 Pollution control

Pollution comes in many forms: as light, noise, air, gases and other vapours, soil and water contamination, smells and particulates; vegetation has a role in reducing all these.

### 2.9.1 Water

The implications for removing soil and water contaminants from the human environment are obvious. Apart from the primary reason for preventing flash floods, SUDS (see section 2.8) are particularly useful in preventing 'non-point source pollution' from being discharged into streams and rivers as part of those flood events. This type of pollution has no single obvious source that can be controlled; for example, oils and particulates containing toxic chemicals that accumulate on roads (Bandy, 2003). Wetland plants such as reeds (*Phragmites australis*) and cattails (reed mace, or bulrushes) (*Typha latifolia*) often found in SUDS are known for their ability to sequester heavy metals (Grisey *et al.*, 2011). Cunningham *et al.* (2010) looked at the conductivity (primarily from de-icing salt) and total inorganic nitrogen loading along the 18 km-long Casperkill stream in New York state, USA and also measured the quality of the stream using a biotic index (of aquatic macroin-vertebrates). They showed convincingly that nitrogen loads in the water decreased when the stream went through open green space (presumably because runoff from such areas had lower nitrogen levels) with a consequent improvement in the biotic index. Conductivity levels did not follow this pattern, however, and they suggested that management of this pollutant needed to be tackled at the watershed level. Even some of the most potent chemicals can be taken in by plants and detoxified. Whilst only a laboratory demonstration, it is impressive that some Chinese plant species (Chinese elder *Sambucus chinensis*, upright hedge-parsley *Torilis japonica*, hybrid willow *Salix matssudana*, water lily *Nymphea teragona* and poplar *Populus deltoides*) were all shown to be capable of taking in cyanide from an aqueous solution and presumably converting it, via a well-known metabolic pathway, into the amino acid asparagine. Of the species tested *S. chinensis* was the most effective (Yu *et al.*, 2005).

### 2.9.2 Light

Vegetation's ability to improve the 'look' of an area by acting as a visual screen (McPherson, 1988) extends to helping reduce unwanted light from industrial buildings, shops, sports pitches, roads, etc. Light pollution is implicated in impacts on human health by affecting production of the hormone melatonin by the pineal gland and hence disrupting operation of the natural sleep cycle (circadian rhythm, body clock) with potential impacts on heart disease, diabetes and obesity, and reducing the protection it affords in reducing cancer (Falchi *et al.*, 2011). For protection year-round, clearly using some form of evergreen shrub, tree or climber, perhaps coupled with an earth bund to increase the height of the structure, will be important (rather than using deciduous species). Whilst the use of vegetation to reduce obvious nuisance light pollution is hardly a novel concept, there appears to be little research on the positive or negative impact of light pollution on nocturnal ecosystem services in the scientific literature and Lyytimäki (2013) attributes this to shifting baseline syndrome. This syndrome is typically seen as a generational issue whereby older generations (e.g. grandparents) have seen more change over their lifetimes and (typically) deprecate the changes they have seen, whilst the youngest generation, having seen no change, accepts the current situation as normal (Leather, 2010; Papworth *et al.*, 2009) (see RCEP (2009) and Longcore and Rich (2004) for impacts of light pollution on wildlife).

### 2.9.3 Noise

Apart from its nuisance value, noise has human health implications including cardiovascular disease (Lercher *et al.*, 2013) and diabetes (Sorensen *et al.*, 2013); its role in such diseases probably results from promoting stress and especially sleep disturbance (Sorensen *et al.*, 2013). Diabetes has also been linked with nitrogen dioxide levels in the environment (Raaschou-Nielsen *et al.*, 2013b) but Sorensen *et al.* (2013) were able to control for $NO_2$ levels and were able to demonstrate a plausible cause-and-effect model based on noise-induced stress increasing cortisol levels which in turn depresses insulin production by pancreatic cells. Noise also affects wildlife, including birds (Fontana *et al.*, 2011; Herrera-Montes & Aide, 2011).

Vegetation can act as a sound barrier, reducing noise from adjacent areas, especially if the vegetation is on top of an earth bund (Cook & Van Haverbeke, 1974, 1977; Herrington, 1974; McPherson, 1988). Cook and Van Haverbeke (1977) demonstrated that one to two rows of dense shrubs in front of one to two rows of trees (either the shrubs or trees, or both, should be evergreen) should reduce noise from cars and light commercial vehicles travelling at 35 mph (56 kph) to acceptable levels in suburban areas. However, this assumed that the building being studied was 25 m from the road centreline; vehicles closer to the road edge would probably require some other acoustic shielding. Trees have been recommended as noise barriers in towns (Leonard & Parr, 1970) and evaluated for their resistance to air pollution for the purpose (Pathak *et al.*, 2011). In Uttar Pradesh, India, the species that Pathak *et al.* (2011) found to be most valuable as noise barriers (graded as 'excellent'), and also due to their tolerance of air pollution as air cleansers, from a test list of 35 species were: the white-fruited wavy fig tree *Ficus infectoria*, mango *Mangifera indica*, and the sacred fig *Ficus religiosa*. Three other species graded as 'very good' were: the Indian banyan *Ficus benghalensis*, Indian rosewood *Dalbergia sissoo* and the false ashoka or Buddha tree *Polyanthia longifolia*. Fang and Ling (2003) examined the characteristics of 35 evergreen tree plantations in Taiwan for their ability to reduce traffic noise. The experimental approach was ingenious in that urban traffic noise was recorded and then used on-site to produce noise conditions at 48 + 2dB A (this is a measure weighted towards the medium-range frequencies that the human ear is most sensitive to). The noise was then sampled in transects through the plantations at 5 m intervals until the rear of the plantation under study was reached. Along with the usual parameters that one might expect to be recorded (e.g. height of trees, length and width of plantation), a measure of the density of the plantation 'visibility' was used – this is the distance into the plantation that an observer standing outside the wood can see an object before it disappears from view. In this study the 'object' was one of the researchers; although estimates were made in three of the plantations as their visibility exceeded their width. To determine the noise reduction effect of the trees, the same sound measured at known intervals over open ground had the sound measured at the same distance through the plantations subtracted to give the 'relative noise attenuation'. The strongest relationship identified was a negative logarithmic relationship between visibility and relative noise attenuation (i.e. the lower the visibility the greater the noise reduction). Relative attenuation was also shown to have a positive logarithmic relationship with the width, length and height of the plantations. Clearly the plantations and arrangements of trees within plantations will be related to geographic region, growing conditions and management practices, but the results of Fang and Ling (2003) show that 'visibility' appears to be a good surrogate for estimation of noise reduction. The other parameters, especially width, were also important, but given that sound does not propagate along a single transect line but effectively in a cone (from a point source) the length and height of a shelter belt can also be influential in noise attenuation. Ten of the 35 plantations were shown to give effective relative attenuation values in excess of 6 dB A; such tree belts were composed of large shrubs with less than 5 m visibility; 14 plantations gave moderate reductions of 3–5.9 dB A and had visibilities between 6 and 19 m. Both groups had lateral branches at the height of the sound receiver, the best had substantial foliage and much forking of the branches; the poorer performing plantations (<2.9 dB A reduction) were composed of species with branches above the height of the receiver. The clear message is that an understory of shrubs is likely to increase the performance of tree belts planted for noise reduction. Subsequently, Fang and Ling (2005) developed a map of noise reduction by trees to help in belt design using three dimensionless parameters. The three components were: $h'$ – receiver and noise source height/tree height; $d'$ – distance between noise source and receiver/tree height; and $m'$ – belt width/visibility.

The perception of noise, and the ability of vegetation to reduce it, does not seem to be straightforward, but appears to be a visual and auditory interaction (Herrington, 1974) influenced by the landscape setting. The perception of traffic noise by residents in their homes (in Hong Kong) was reduced more by views of wetland and garden parks than by views of grassy hills (Li et al., 2010).

### 2.9.4 Air pollution

'Air pollution' groups together the effects of a wide range of gases and aerosols – including very fine particulate matter. Some have human health impacts such a nitrogen dioxide and particulates on asthma (Anderson et al., 2013; Gehring et al., 2010; Kheirbek et al., 2013) and increased cancer risk from particulates (Raaschou-Nielsen et al., 2013a); others have impacts on ecosystems such as sulphur and nitrogen oxides (Bignal et al., 2004; Lovett, 1994); some have global effects such as carbon dioxide (climate change) and chlorofluorocarbons (CFCs) (ozone hole formation) (Solomon, 2004). Substantial mortality from air pollution in 1952, the 'London Smog', lead to the UK Clean Air Act (1956) (Harrison, 1992) and air quality has improved since with successively revised and new legislation. In the UK, between 1990 and 2001 improved air quality was estimated to have saved 4,200 lives and 3,500 hospital admissions/year (Defra, 2007a). In the USA, Fann and Risley (2013) estimate that reductions in $PM_{2.5}$ and ozone over a seven-year period (2000–2007) have reduced premature deaths by between 22,000 and 60,000 for the former and between 880 and 4,100 for the latter. Whilst the estimates have rather broad ranges, as might be expected working with such a large country and with constantly varying air quality and weather data, the message is clear: improving air quality saves lives.

The 1995 UK Environment Act introduced a requirement for the development of an Air Quality Strategy: the first was produced in 1997, revised in 2000, and the current one (at the time of writing) introduced in 2007 (Defra, 2007a). The strategy contains the current air quality objectives and the EU directive limit and target values for pollutants. The executive summary makes a bald and sobering statement: 'Air pollution is currently estimated to reduce the life expectancy of every person in the UK by an average of 7–8 months.' Of course, seven to eight months is an average – with some people (those exposed to the highest levels of pollution, or the most susceptible (O'Neill *et al.*, 2012)) having considerably shorter life expectancy than someone living in an unpolluted area (all other things being equal). In Estonia, Orru *et al.* (2011) estimated that reduced life expectancy could be in excess of 12 months in polluted city centres. The new UK Air Quality Strategy aims to lower reduced life expectancy to five months by 2020 (Defra, 2007a), though many air pollutants are proving difficult to control (particulates, nitrogen dioxide, ozone (Defra, 2007b)). There are places in this world with far worse air pollution than the UK.

Vegetation can help reduce a range of air pollutants either by direct deposition of particulates to surfaces, with lipid-soluble volatile organic chemicals being taken in directly through the cuticle (e.g. PCBs – polychlorinated biphenyls) or, as with gases, by the stomata (Barber *et al.*, 2002) as are some ultrafine (<0.1 μm) particles (Fowler, 2002). For some compounds, such as PCBs, the compound may be, at least in part, subsequently re-released into the air whilst other fractions are held in the leaf epidermis or mesophyll (Barber *et al.*, 2003; Wild *et al.*, 2006). In terms of their pollutant removal capacity, different types of vegetation, species, ecotypes, and varieties can all have different attributes and values and it is becoming increasingly clear that deposition to vegetation is not a simple process (Lovett, 1994). Inevitably, vegetation size is a factor: large trees remove more air pollutants than small trees; Nowak (1994a) estimated that trees with a diameter at breast height (dbh) of 76 cm removed 70 times the amount of a tree of less than 8 cm dbh (1.4 kg vs 0.02 kg). Trees are also better than grassland at removing air pollutants. Jonas *et al.* (1985) compared the capture efficiency of a range of tree species with grass in removing particulates from the air (size range 0.4–17 μm) and found trees to be substantially better. Trees that were well spaced from other trees captured more particulates, on a tree-for-tree basis, than those in close groups because of improved air circulation (Figure 2.17), but groups of trees were very effective overall (Stülpnagel *et al.*, 1990). The soil under trees has also been shown to have higher levels of lead, for example, than grassland (Fowler *et al.*, 2004). In 1991 Nowak (1994a) estimated urban trees removed 591 tonnes (t) of air pollution in Chicago (which had 11% tree cover) = 15 t CO carbon monoxide, 84 t $SO_2$ sulphur dioxide, 89 t $NO_2$ nitrogen dioxide, 191 t $O_3$ ozone and 212 t of particulates $PM_{10}$ and below. A recent project to evaluate the ecosystem services provided by Torbay's urban forest in the UK showed that 11.8% of the ground was covered in 818,000 trees and that they removed 50 t of air pollutants per year (0.0005 t CO, 7.9 t $NO_2$, 22.0 t $O_3$, 18 t $PM_{10}$, 1.3 t $SO_2$) equivalent in value to £1.5 million in externality costs and social damage (Treeconomics, 2011). The trees have

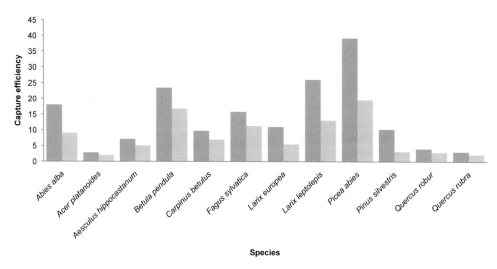

**FIGURE 2.17** Relative particulate capture efficiency of trees compared with grass (Deposition on trees/Deposition on grass). Dark green columns: data for single tree well spaced from other trees; light green columns: data for trees with reduced air circulation (e.g. planted in groups). Data from Jonas *et al.* (1985).

a replacement value of £280 million and store about 98,000 t of carbon a year. Luley and Bond (2002) produced a report aimed at integrating the management of urban trees into the air quality management plan of New York. Across the USA removal of ozone, $PM_{10}$, $NO_2$, $SO_2$ and CO by trees and shrubs was estimated at 711,000 tonnes per year worth $3.8 billion by Nowak *et al.* (2006). In Beijing, Yang *et al.* (2005) estimated that there were 2.4 million trees in the metropolitan area which removed about 772 tonnes of $PM_{10}$ per year and stored 0.2 million tonnes of $CO_2$. In addition to the 'normal' gaseous pollutants such as nitrogen dioxide, a wide range of volatile chemicals have also been shown to be removed from the air by plants (Fries & Morrow, 1981; Ockenden *et al.*, 1998; Smith & Jones, 2000; Topp *et al.*, 1986) including PCBs (polychlorinated biphenyls).

### 2.9.5 Particulates

Air pollution in urban centres is a worldwide problem (e.g. in China (Jim & Chen, 2008), Europe (Raaschou-Nielsen *et al.*, 2013a), India (Deshmukh *et al.*, 2013; Rai & Kulshreshtha, 2006), Thailand (Phoothiwut & Junyapoon, 2013), Greece (Vlachokostas *et al.*, 2012)) and particulates are an important component of it. Plants filter out dust (Rentao *et al.*, 2008) and small particles from the air (Beckett *et al.*, 1998; Dochinger, 1980; Hosker & Lindberg, 1982; Smith & Jones, 2000; Stülpnagel *et al.*, 1990; Varshney & Mitra, 1993) and deposition rates are higher than for buildings and sealed ground surfaces (Pugh *et al.*, 2012). Vegetation is of particular value to urban dwellers; low air circulation in cities results in high concentrations of small particles given off by motor vehicles, especially those fuelled by diesel, in their exhaust (Chaloulakou *et al.*, 2003; Paoletti *et al.*, 1989). In addition to carbon particles from diesel exhaust, road dust also contains other particles that originate from engine wear and tear:

typically composed of iron, chromium and nickel. Espinosa *et al.* (2001) found cadmium, calcium, cobalt, copper, iron, magnesium, manganese, lead, nickel, titanium and vanadium in urban aerosols in Seville, Spain. In countries, such as Finland, where winter snow cover is inevitable and snow-tyres with studs are used, particles with a high proportion of tungsten are also likely to be found (Bućko *et al.*, 2011). Airborne particles are removed by three main processes: sedimentation (via gravity), deposition (via precipitation), and, particularly important with respect to the role of vegetation, impaction via wind – the smaller the particle the more effective is the process of impaction (Smith, 1977).

The particle ranges most often quoted in the literature are those known as $PM_{10}$ and $PM_{2.5}$. PM stands for 'Particulate Matter' and the number (e.g. 10) is the diameter of the particle in 'microns' or 'micrometres' ($\mu$m); a micrometre is $1 \times 10^{-6}$ (or 0.000001) of a metre. The distinction between the two size ranges is because particles of 10 $\mu$m or below can get into the lower respiratory system (and are called 'thoracic' particulates) whilst particles of 2.5 $\mu$m and below can penetrate further, into the narrower spaces in the lungs, and are therefore called 'respirable' particles (Brunkeef & Holgate, 2002). These particles are easily breathed in and have serious health effects such as increased likelihood of death from respiratory and cardiovascular disease (Brunkeef & Holgate, 2002; Pope III *et al.*, 2011; Sunyer *et al.*, 2000), atherosclerosis (Araujo, 2011), lung cancer (Raaschou-Nielsen *et al.*, 2013a), and increased incidence of asthma (Anderson *et al.*, 2013). Ultrafine particles which are $\leq 0.1$ $\mu$m in diameter can cross into the body through the lungs and other parts of the respiratory tract and can thus be transported throughout the body; they have been found in the liver, spleen and kidneys and can move rapidly (<1 h) from the nose, via the olfactory nerves, to the brain (Solomon *et al.*, 2012). The level of mortality caused by particulates is not trivial, with one study estimating an annual death toll of 2,100 deaths per year in the Netherlands alone (Brunkeef & Holgate, 2002). Particulates are not simply inert carbon particles but can carry some quite nasty organic chemicals including cancer-causing polyaromatic hydrocarbons (PAHs), the highest concentrations of which are carried on the respirable $PM_{0.35-1.2}$ size class (Burkart *et al.*, 2013; Phoothiwut & Junyapoon, 2013). Sram *et al.* (2011) in a study of 950 policemen and bus drivers estimated that non-smokers exposed to more than 1 ng/m$^{-3}$ of benzo [a]pyrene bound to particulates could cause damage to DNA. Stewart *et al.* (undated-a) indicate that for every 10 $\mu$g.m$^{-3}$ increase in $PM_{10}$ concentration in the air there is an associated increase in human mortality of 1%. This was based on a 24-hour average of $PM_{10}$ concentrations and the authors claimed that doubling the number of trees in the West Midlands area of the UK could save 140 lives per year.

Taking the UK as an example, the European Union Directive 2008/50/EC on ambient air quality and cleaner air for Europe (EU, 2008) laid down particulate thresholds for human health. This has resulted in a set of UK air quality objectives for $PM_{10}$ such that from 2010:

- The daily threshold concentration must not exceed 50 $\mu$g.m$^{-3}$ on more than 35 occasions/year.
- The annual average must not exceed 40 $\mu$g.m$^{-3}$.

Scotland, with its devolved administration, has tighter thresholds because its background level of pollution is lower than in the rest of the UK:

- Daily: 50 $\mu$g.m$^{-3}$ on seven occasions.
- Annually: 18 $\mu$g.m$^{-3}$.

$PM_{2.5}$ annual mean values were set at 25 $\mu$g.m$^{-3}$ for the whole of the UK (excluding Scotland); for 2020 more stringent limits (12 $\mu$g.m$^{-3}$) have been set for Scotland. For urban areas, specifically, $PM_{2.5}$ reduction targets between 2010 and 2020 were set at 15% reduction compared to the urban background (Defra, 2007a). Local authorities in the UK have also been given duties under the Environment Act (1995) and the Environment (Northern Ireland) Order 2002 to establish local air quality management systems, including the introduction of 'air quality management areas' (AQMAs) for $PM_{10}$s and six other air pollutants if areas are unlikely to achieve compliance with national standards (Defra, 2007a). In 2012, 98 AQMAs for particulates had been declared by the UK's 405 local authorities (LA) (Defra, 2012) – although it is worth noting that an individual LA may have more than one action plan in force within its area. Defra (2012) indicates good compliance with particulate pollution standards in 2012 (compared with when monitoring started in 1992) with only London exceeding the daily $PM_{10}$ limit, and all areas complying with the annual limits for $PM_{10}$ and $PM_{2.5}$. Compliance with standards does not mean the absence of pollution, as comparison of the standards for Scotland and the rest of the UK demonstrate – and the standards also allow the subtraction of any natural particulates in the area before testing against threshold limits – so they are not about absolute pollution levels, but those only due to anthropogenic sources. It is also worth remembering that a 24-hour mean could equate to quite high levels of pollution during the period of maximum commuting during the day (and especially in the morning), when an urban area's population is most likely to be exposed, compared with perhaps half that peak level at night when far fewer people are about (Muir, 1998). Whilst Freer-Smith et al. (2005) did not present any data, they noted two peaks in particulate concentration (Monday–Friday) during the day (8–9am and 3–6pm) in their study of particulate deposition on trees. Pedestrians are exposed to higher levels of particulates than those driving a car and at much higher levels than static monitoring devices (which are the basis for air quality monitoring) indicate (70% higher for pedestrians and 25% for people in cars) (Briggs et al., 2008; Gulliver & Briggs, 2007). However, Kaur et al. (2007), reviewing personal exposure studies, considered that pedestrians and cyclists experienced lower particulate and carbon monoxide concentrations than those inside vehicles. It is also interesting that $PM_{10}$ and $PM_{2.5}$ levels were forecast to rise in the UK after 2015 (Defra, 2007b). Baxter et al. (2013) in a study of $PM_{2.5}$ and 'respiratory and other non-accidental mortality' in 27 US communities found an unexpected negative relationship with distance travelled in a vehicle; in other words, the longer your commute, the less likely you are to have an air-pollution-related health impact. This rather counterintuitive finding was suggested to arise as a result of those living in polluted urban settlements having shorter travel distances and thus being exposed to (in this case) $PM_{2.5}$ for longer than those living in suburban or out-of-town areas with cleaner air. An alternative interpretation offered was that the data was identifying the difference between the younger, working-age population with long commutes compared with the older/retired segment of the population with short or no commutes – the group which would be more likely to suffer air-pollution-mediated illnesses. Kaur et al. (2007) recommended that pedestrians use 'back street' routes that take them away from major traffic areas, and hence pollutant levels, as a way of reducing exposure.

Zhu *et al.* (2002b) examined the deposition of ultrafine particulates (in the 6–220 nanometre range; 1 nm = 1 x $10^{-3}$ μm) perpendicular to Freeway 405 along Constitution Avenue in Los Angeles's National Cemetery and demonstrated that loads declined exponentially with distance from the road; loads were heaviest within the first 30 m and were half that at about 100 m, background levels were achieved at 300 m from the road primarily because of coagulation (smaller particles clumping together) and atmospheric dilution. Zhu *et al.* (2002b) concluded that people living, working, and commuting 0–100 m from traffic will have enhanced exposure to $PM_{10}$ and below. See also Zhu *et al.* (2002a) for results from Interstate 710, which had more diesel vehicles than Freeway 405.

Options for reduction include removing particulates at source (including more efficient engines, fuels and catalytic particulate filters), removing them from the air, or moving particulate emission sources from areas where high concentrations build up. In the latter case, this is primarily by trying to influence road travel by having 'green travel plans', routing traffic away from hotspots, establishing low emissions zones, congestion charging and workplace charging for parking (Defra, 2007a). Given that much particulate pollution comes from motor vehicle exhaust (Ayala *et al.*, 2012), it is a tricky task to tackle without efficient diesel particulate filters that work well in both urban and long-haul situations and this remains a problem (Chilton, 2011; Fino, 2007; Fino *et al.*, 2003; van Setten *et al.*, 2001). Encouraging use of public transport is also often mooted as a way of reducing emissions by reducing car use; however, those waiting at bus stops in polluted areas may be subject to increased risk due to their proximity to traffic (Godoi *et al.*, 2013) and Kaur *et al.* (2005) found pedestrians in London were exposed to greater levels of particulates if they walked next to the kerb (road edge) compared with next to buildings (furthest from the kerb).

The ability of vegetation to remove particulates has been known for some time (e.g. Zulfacar, 1975) and especially in a rural context by the use of windbreaks to reduce pesticide drift (e.g. by 70–90% in the 3 m zone downwind; see references in Hewitt, 2001). Dochinger (1980) studied the entrapment of particulates by trees in urban areas of Steubenville, Ohio, USA from March to October using simple dustfall traps. He demonstrated that the canopies of deciduous trees (black locusts *Robinia pseudoacacia*, sugar maples *Acer saccharum* and red oaks *Quercus rubra*) could reduce dustfall by 27% and conifers (red pines *Pinus resinosa* and blue spruce *Picea pungens*) by 38% compared with mown grassy areas. The deciduous trees also had an understory of elms (*Ulmus* sp.), sassafras (*Sassafras albidum*) and dogwood (*Cornus florida*). In the absence of leaves (March, April and October) the deciduous species did not differ from the open grassy areas in the level of dust, although some weather conditions may improve capture rates (e.g. with snow covering bare branches) (Hosker & Lindberg, 1982). Over the five-month period when deciduous species had leaves (May–September) they were very similar to the coniferous species with particulate extraction of 37 and 38% respectively; there were monthly variations in collection efficiency. Dochinger (1980) also used a second method to sample particulates, in this case not dustfall, but the number of suspended particles in the air. The method involved sucking air at a constant rate through a sampling device with the particles being collected on filter paper. Using this method, over the eight months, deciduous species removed 9% of the suspended particles and conifers 13%; bare deciduous trees did not remove suspended particulates. Hagler *et al.* (2012), working in North Carolina, studied the effect of trees as particulate barriers alongside roads and found particulate loads varied, sometimes being

**TABLE 2.5** Methods of dry deposition of aerosols to vegetation

| *Deposition method* | |
| --- | --- |
| **Brownian diffusion** | Random motion of small particles; a 1 μm particle can knock into others at the rate of about $10^{16}$ a second. Deposition is limited by the ability of the particle to penetrate the boundary (air) layer surrounding the surface |
| **Interception** | Affects particles with low inertia which follow air currents and encounter a surface and are retained – typically by surface roughness or hairs |
| **Impaction** | Affects particles of high inertia which cannot follow the air currents round a surface and hit it |
| **Sedimentation** | Settling out of particles from the air due to gravity |
| **Rebound** | Mainly affects particles of 5 μm and above: the kinetic energy of the particle causes it to rebound from the surface it hits – but affected by the nature of the surface adhesion. |

Sources: UoC (UoC, undated); Petroff *et al.* (2008)

more, sometimes less than in an area without tree cover. They suspected that the variability related to the nature of their study system, which was highly porous with rather thin, gappy tree stands with modest leaf area.

Smith and Jones (2000) noted that particulates deposited on plants are generally smaller than 100 μm and described the various ways that they were trapped by vegetation: wet or dry deposition; if dry, by interception, impaction or sedimentation (although there are other processes that operate i.e. Brownian diffusion and rebound (Lovett, 1994; Petroff *et al.*, 2008)) (Table 2.5). Wet deposition is more important for particles present at great heights, whilst dry deposition is more important for particles in the air nearer the ground (the atmospheric surface layer) (Sehmel, 1980). Another 'wet' deposition method is via occult deposition – from 'wind-driven cloud water' (Unsworth & Wilshaw, 1989). Hosker and Lindberg (1982) indicate that wet and dry deposition remove roughly equal levels of pollutants, with wet deposition removing them by two methods: 1) within-cloud scavenging (rainout) and 2) below-cloud scavenging (washout). Smith and Jones (2000) highlighted uncertainties in the literature relating to the capture of particulates by vegetation; for example, grass appeared to have higher particulate loads than more complex vegetation such as herbs, shrubs and trees, but they pointed out that this may be because grasses have longer to accumulate particulates through retaining dead leaves from the previous growing season. The implications of this observation are substantial, as vegetation management and seasonality may strongly impact the efficiency of particulate trapping. Hosker and Lindberg (1982) list different vegetated and inert surfaces and the atmospheric features that influence deposition.

Despite the uncertainty in the literature, the findings of some studies reviewed by Smith and Jones (2000) may be taken as generalities: plants which have smooth, water-repellent leaves being less likely to collect or retain particulate matter compared with those that are sticky, hairy or have rugose surfaces; smaller-sized particles (1-5 μm) are likely to be trapped more efficiently by 'hairy' plants. Wang *et al.* (2011) examined 14 species commonly used in urban greening in Xi'an, China for their particulate-capturing ability. The species were predominantly trees and shrubs, but also included white clover (*Trifolium repens*) – the results showed substantial differences with *T. repens* capturing only 0.23 g of particulates m$^{-2}$ compared with 4.51g.m$^{-2}$ by the evergreen shrub Japanese mock orange *Pittosporium tobira*. Scanning electron microscopy revealed that cell surfaces of the most poorly performing species were covered in a dense layer of wax tubules, and that the species which performed best in particulate capture had rough surfaces with some having hairs and pits. The leaves of evergreens are exposed to air pollutants for longer than deciduous species, and as a result the surfaces of such species (e.g. Chinese red pine *Pinus tabulaeformis*) appeared much modified, and were considered to have enhanced capture abilities partially due to secretions from leaves and microbial colonisation. Sehmel (1980) gives a comprehensive run-down of the factors that influence the deposition of particulates, including vegetative characteristics. Particulates that are collected may remain on plant parts, be re-mobilised (e.g. by wind), or be washed off onto other plants below them, onto the soil surface, or transferred via vegetated swales to ponds. Whether in situ, or following immobilisation or transfer, some particulate contaminants such as nitrogen and phosphorus may be utilised directly by plants (Abe & Ozaki, 1999) or bound to organic material and immobilised or move through the soil horizon (Kalbitz *et al.*, 1997; Smith & Jones, 2000). As leaves age, and ultimately die, their collection efficiencies will change – in some cases more particulates will be collected, in others less – and leaf fall will ultimately bring collected particulates to the soil surface (though they may be remobilised in the process – via strong winds). More dense vegetation is likely to remove more particulates, whilst grazing may reduce collection efficiency (by removing plant material). This latter is more likely to have a substantial effect in rural rather than urban areas, although grass cutting and hedge trimming is likely to have an equivalent effect. It is also obvious that weather conditions will have a substantial impact on the amount of particulates that are removed by vegetation (Sehmel, 1980).

Whilst plants are capable of capturing particulates and removing them from the air, they are not immune to damage from them, with particulates clogging stomata, affecting photosynthetic efficiency, promoting disease attack, etc. (Prajapati & Tripathi, 2008; Prusty *et al.*, 2005; Rai & Kulshreshtha, 2006). Particulates can contain materials that are toxic to some plants, and even the physical action of particles impacting on leaves can result in abrasive damage (see Beckett *et al.*, 1998).

McDonald *et al.* (2007) modelled the impact of tree planting on PM$_{10}$ concentrations in the West Midlands and Glasgow, UK, and considered that tree planting, on its own, could make measurable improvements to PM$_{10}$ levels. Increasing tree density from 3.7% to a realistic target of 16.5% in the West Midlands was estimated to result in a reduction in PM$_{10}$ by 10% (equivalent to 110 tonnes/year), for Glasgow an increase in tree densities from 3.6 to 8.0% would result in a 2% reduction in PM$_{10}$ (= 4 tonnes/year). These modest 'global' levels may mask considerably higher particulate removal rates at the local level. Pugh *et al.* (2012) estimated reductions in urban canyons (streets with buildings either side) could be as high as 60%! Deshmukh *et al.* (2013) studied particulates in the industrial city of Raipur, India, and concentrations exceeded the annual average air quality standards of 60 μg.m$^{-3}$ for PM$_{10}$ and 40 μg.m$^{-3}$ for PM$_{2.5}$ by about 4 times and 3 times respectively; the 24-hour standards (100 μg.m$^{-3}$ for PM$_{10}$ and 60 μg.m$^{-3}$ for PM$_{2.5}$) were exceeded all the time for PM$_{10}$ and almost all the time (94%) for PM$_{2.5}$. The air quality was rated as 'moderate to hazardous' (Deshmukh *et al.*, 2013). As a comparison, Chaturvedi *et al.* (2013) attribute the low air pollution in Nagpur (24 h PM$_{10}$ concentration = 53 μg.m$^{-3}$), to it being one of the greenest cities in the country with 18% tree cover (including parks) and 17% in cultivation.

Convincing local authorities of the merits of vegetation planting for particulate control is difficult and especially so in terms of quantifying expected health benefits especially given the differing capture efficiency of different types of vegetation (grasses, trees, shrubs) and their different morphological characteristics (e.g. leaf size, shape, surface characteristics) (Tiwary et al., 2009). Modelling approaches have therefore attempted to bridge the gap. An interesting example was based on a 10 x 10 km section of the Boroughs of Newham and Greenwich in East London; different planting scenarios were modelled for the 5.5% of the area planned as green space (proposed and existing) using two tree species – sycamore (*Acer pseudoplatanus*), which has relatively poor $PM_{10}$ capture efficiency (similar to grassland), and Douglas fir (*Pseudotsuga menziesii*), which is much better – and grassland. Although several modelling scenarios were used from all grassland to all conifer, the 'realistic' scenario of green space planting with 75% grass, 20% sycamore and 5% fir was estimated to remove 0.17 t of $PM_{10}$ per ha.y$^{-1}$. Modelling of the health benefits suggested that modifying this 10 x 10 km section of London would reduce mortality by two deaths and two hospital admissions per year (Tiwary et al., 2009). Health benefits appear modest, but over ten years in a 10 x 10 km area the 'savings' are 20 premature deaths. The 'savings' are also likely to be an underestimate, as the study did not include street trees or other vegetation in the analysis. The tree composition could also be tweaked for better capture efficiency – it is unlikely, for example, that sycamore would be the only deciduous tree planted.

## 2.9.6 Gases and aerosols

Burnett et al. (1998) examined the impact of gaseous air pollutants on mortality in 11 Canadian cities from 1980 to 1991 and estimated that nitrogen dioxide pollution resulted in a 4.1% increased risk of death, ozone 1.8%, sulphur dioxide 1.4% and carbon monoxide 0.9%. As this class of compounds is wide, this section will be limited to briefly touching on the global effects of carbon dioxide, a global pollution issue due to its climate change impacts, but will mainly concentrate on the major gases implicated in human health impacts and the nuisance value of odours. Volatile organic chemicals as indoor pollutants are covered in Chapter 3.

### Carbon dioxide

Whilst carbon dioxide is a major polluter (resulting from anthropogenic activity such as clearing land, burning biomass and fossil fuels) and climate change gas, it is also a natural product of respiration by organic life. In terms of climate change, the additional $CO_2$ in the atmosphere reduces long-wave radiation (heat) emissions from our planet, resulting in global warming and resultant climate change as weather systems respond to an altered planetary heat balance (IPCC, 2014). Vegetation (and, come to that, animal life) can act as a carbon sink, incorporating it into biomass, some of which will, in turn, be incorporated into soils and some retained in longer-term structures such as trees and ultimately timber (though that will probably end up being burned or decay releasing the stored $CO_2$). As a result carbon sequestration by vegetation has become a hot topic with such actions as tree planting and biomass burning frequently being promoted (on the assumption that the latter is carbon neutral or at least better than fossil fuel burning).

Increasing vegetation density in urban areas will obviously make some contribution to reducing atmospheric carbon through sequestration, although it is likely to be slight in global terms and reducing carbon emissions in the first place is likely to be more effective. Nevertheless, collectively, trees can act as substantial carbon offsets. Nowak and Crane (2002) using data from ten cities and national data on tree cover estimated that urban trees in the USA stored about 700 million tonnes of carbon with an annual increment of 22.8 million tonnes of carbon a year. More recently, Nowak *et al.* (2013) expanded this data set to cover 28 cities and reported that each $m^2$ of tree cover stored 7.69 kg of carbon (C); the annual sequestration rate was estimated at 0.28 $kgC.m^{-2}$. In total they estimated that urban trees across the USA stored 643 million tonnes of carbon (data for 2005) with annual sequestration of 25.6 million tonnes, which makes their earlier estimates pretty good. Peper *et al.* (2008) calculated that in 2005 Indianapolis's street trees collectively absorbed 5,489 kg of $CO_2$ per year – the equivalent of that produced by 2,338 vehicles.

The age structure of trees in urban areas also affects the carbon storage and uptake figures; Nowak (1994b) indicated that annually a small tree (<8 cm diameter at breast height (dbh)) only sequesters the amount of carbon emitted by driving one car 16 km whereas a large tree (>76 dbh) takes-up the carbon emitted by driving a car 1,460 km. Growth rates also impact on carbon storage and Stewart *et al.* (undated-a) ranked 30 species growing in the UK by their growth rates – the fastest growing being Lawson cypress *Chamaecyparis lawsoniana*, Leyland cypress *Cupressocyparis leylandii*, larch *Larix decidua*, poplar (*Populus* sp.), silver birch *Betula pendula*, and willows (*Salix* spp.). However, slow growing, long-lived trees will tend to store more carbon; Stewart *et al.* (undated-a) cite English oak *Quercus robur* as storing the most carbon of any tree in the West Midlands for this reason. In Chicago, Nowak (1994b) estimated that the urban trees stored about 0.9 million tonnes of carbon and sequester some 40,100 tonnes/year.

The value of planting additional urban vegetation in relation to climate change is likely to be rather more immediately felt in the area of mitigating its effects, such as through reducing flooding risk and mortality from heatwaves, etc. However, urban vegetation also has a role to play in reducing primary carbon emissions by reducing energy consumption in buildings; for example, by reducing the need for air conditioning by shading buildings from direct sun (deciduous trees can shade windows in the summer but let light in in the winter) and also by reducing winter heating costs (e.g. evergreen wind shields). Also, paths and cycleways that are pleasant visually are more likely to be used (reducing fossil fuel consumption in the form of petrol and diesel); the aesthetics of paths and cycleways should be an important component during their design and in subsequent maintenance programmes. Getting people out of cars (and even buses) is important not only to promote exercise, but also because air quality inside vehicles may not always be optimal (Kadiyala & Kumar, 2012, 2013). However, as many (but not all) of the contaminants in vehicles arise from outside, getting people onto paths/cycleways as a positive benefit in air-quality terms does assume that the outdoor air quality is better (see above, under 'particulates').

## Ozone

Ironically, whilst ozone in the stratosphere (the upper layer of our atmosphere) protects our planet from dangerous UV radiation and is not a pollutant, low level ozone produced in the 'boundary layer' where we live (below the troposphere) (Clarke, 1992) is dangerous to humans and vegetation (Davidson & Barnes, 1998). Ozone is known as a secondary pollutant as it is a result of reactions between other 'primary pollutants' (Fowler, 2002). Volatile organic compounds in the air combine with nitrogen oxides (collectively known as $NO_x$), the most important of which are nitrogen monoxide NO and nitrogen dioxide $NO_2$, in the presence of sunlight (a photochemical reaction) to form ozone. Impacts on humans are as varied as cardiovascular and respiratory diseases, including increased mortality rates, rashes, dermatitis/eczema, urticaria, conjunctivitis and cellulitis (a bacterial inflammation of the skin exacerbated by ozone damage) (Anderson et al., 2012; Berman et al., 2012; Hunova et al., 2013; Szyszkowicz et al., 2012a; Szyszkowicz et al., 2010; Szyszkowicz et al., 2012b). Chen and Schwartz (2009) in a study of 1,764 US adults showed that increased ozone levels had measureable, negative impacts on cognitive performance. For every 10 ppb increase in annual ozone exposure, the decrease in coding ability (measured by a symbol-digit substitution test) was equivalent to ageing 3.5 years, and attention and short-term memory by 5.3 years. The EU have a number of threshold reporting values for ozone: the 'Long-term objective for the protection of human health' (LTO) is a maximum daily eight-hour mean of 120 $\mu g.m^{-3}$; the 'information threshold' is a 1 h mean concentration of 180 $\mu g.m^{-3}$; the 'alert threshold' is a 1 h mean concentration of 240 $\mu g.m^{-3}$. The LTO is used to assess whether a member state has met the 'target value for the protection of human health' (TV). The TV (as in EU Directive 2008/50/EC) is achieved if the LTO is not exceeded on more than 25 days/year (based on a three-year mean). In 2012 the TV was exceeded in 17 member states (Černikovský et al., 2013).

Ozone ($O_3$), along with components of vehicle exhaust, is removed by vegetation (Manes et al., 2012). However, ozone may also damage the plant (Chauhan, 2010a; Davidson & Barnes, 1998; Joshi & Bora, 2011). Paoletti et al. (2009) suggest that such damage can be turned to good use by using particularly susceptible species, such as the rose mallow *Hibiscus syriacus*, which is used as an ornamental hedge in Italy, as ozone bioindicators. Ozone produced characteristic damage to the rose mallow foliage, with extent (of damage) related to the $O_3$ concentration. Chauhan (2010b) also suggested that trees could be used as bioindicators of vehicle pollution and Ockenden et al. (1998) suggested pines (*Pinus sylvestris*) and lichen (*Hypogymnia physodes*) could be used similarly to monitor PCBs.

## Nitrogen oxides

Nitrogen dioxide ($NO_2$) is a common urban air pollutant and can have impacts on human health including asthma (Anderson et al., 2013) and also by acting as a neurotoxin affecting cognitive development in children (Freire et al., 2010). This latter study, carried out in the Granada region of southern Spain, used nitrogen dioxide as a proxy for air pollution from traffic in general. Children exposed to the highest $NO_2$ levels showed decreases in general cognitive scores, quantitative and working memory, and gross motor areas as measured by the McCarthy Scales of Children's Abilities. Whilst only the gross motor area scores were shown to be significantly lower in high-exposure children, the results were suggestive of a general effect of traffic pollution on children's neurodevelopment. Air pollution is also well known to have impacts on biodiversity with lichens being particularly sensitive and used as biomonitors (Jovan, 2008). Gombert *et al.* (2003) investigated the nitrogen concentration of two epiphytic lichens: *Physcia adscendens* a known nitrophyte (species reacting positively to nitrogen in the environment) and *H. physodes* an acidophyte (responding positively to a more acidic environment) with the traffic density in the cities of Meylan and Saint Martin d'Hères in Grenoble, France. They found a relationship between traffic density, proximity to roads and nitrogen levels in *P. adscendens*, but not *H. physodes*. Honour *et al.* (2009) demonstrated, in an experimental 'solardome' experiment, that nitrogen oxides at realistic concentrations for the urban street environment ($NO_x$ concentrations of 77–98 nl.l$^{-1}$ and $NO:NO_2$ ratios of 1.4–2.2) reduced growth, produced early senescence, delayed flowering and induced changes in the leaf wax characteristics of 12 species of ground flora (broadleaved plantain *Plantago major,* black knapweed *Centaurea nigra,* common sorrel *Rumex acetosa,* autumn hawkbit *Leontodon autumnalis,* bird's-foot trefoil *Lotus corniculatus,* white clover, sowthistle *Sonchus oleraceus,* fat hen *Chenopodium album,* Oxford ragwort *Senecio squalidus* and the grasses perennial rye-grass *Lolium perenne,* annual meadow grass *Poa annua* and timothy *Phleum pratense*). Responses were, however, species-specific. Further work by Bell *et al.* (2011) confirmed that diesel exhaust, and specifically the nitrogen oxides emitted, affected urban plant performance – stimulating some aspects and reducing others – and called for air-quality laws to reflect impacts on urban vegetation.

Trees are known to be nitrogen dioxide sinks, with Scott *et al.* (1998) estimating that Sacramento's urban trees removed 1–2% of the area's daily emissions, an annual equivalent of 13.9 kg/ha. Pugh *et al.* (2012) suggest that greening urban canyons could reduce $NO_2$ levels by up to 40%.

## Sulphur dioxide

As with other air pollutants, sulphur dioxide exposure is implicated in an increased mortality risk as Katsouyanni *et al.* (1997) found in a study of 12 European cities. An increase in 50 $\mu g.m^{-3}$ of sulphur dioxide was associated with a 3% increase in daily mortality in five western European cities (with 4% increase in cardiovascular and 5% increase in respiratory mortality). For settlements in central eastern Europe, the increase in daily mortality was 0.8%. Impacts were greatest in the summer months. As with nitrogen oxides, $SO_2$ can be taken up by plants (Bell *et al.*, 1982; Peper *et al.*, 2008; Tan & Sia, 2005).

## Odour

Some animal husbandry and waste treatment processes generate unpleasant odours, and it is largely from this body of literature which the value of outdoor vegetation to outdoor odour removal can be seen. Whilst the examples are rural, the principles are likely to be similar in urban situations where industry and local authority facilities generate odours. Agriculturally generated odours can be particularly troublesome, and research on poultry units and on pig production units has demonstrated that vegetated barriers such as shelter-belts can help control odours (Parker *et al.*, 2012; Tyndall & Colletti, 2007; Tyndall & Larsen, undated). The vegetation seems to act primarily by removing particulates that contain the odiferous volatile organic chemicals, and by promoting dilution through turbulent air mixing (Parker *et al.*, 2012). Chapter 3 covers the value of indoor vegetation in removing odours.

## 2.10 Energy efficiency

Dunnett and Kingsbury (2004) suggest that reducing the internal temperature of a building by 0.5°C can reduce energy use of air conditioning by as much as 8%; Bass *et al.* (2003) state that above a threshold temperature (not elaborated in their report) every 1°C increase results in 5% increased electricity use. Peck *et al.* (1999) noted that about 33% of the heating requirement of a building is caused by wind-chill and that wind protection by vegetation could reduce the effect by 75%, equivalent to a reduced heating requirement of 25%. Pérez *et al.* (2011a) summarise the mechanisms by which vegetation helps save energy either from avoiding or reducing the use of cooling or heating systems:

- shade cast by vegetation intercepting sunlight
- thermal insulation provided by the plants and the growing medium (and associated structures)
- cooling effect of evapotranspiration
- moderation of windspeed.

These features will naturally be affected by the general climate of the area, the season, specific weather on any given day, the type of plants used, their seasonal nature (evergreen/deciduous); the structure of the vegetation (i.e. just a single layer or several different layers of plants of different height), density and depth of the plant layer, the proportion of the wall covered, the nature of the growing medium and its water retention capacity, and any additional infrastructure such as drainage layers, waterproof membrane, additional insulation, as well as the type of building material and wall orientation (Kontoleon & Eumorfopoulou, 2010; Perez *et al.*, 2011a, 2011b).

In general terms, the energy-saving aspects of vegetation through tree shading require trees to be in the right place and in leaf at the right time. So providing shade in the summer to sunlit surfaces will save energy used in air conditioning, but the same trees will not be helpful if they are in leaf in the winter as they would stop sunlight warming the house. So deciduous trees would be a good choice for aspects of the house that need shade in summer but sun in winter; house aspects that need insulation from cold weather, especially wind, in the winter (the north in the northern hemisphere) would be better off with evergreen species (McPherson *et al.*, 2006). Trees are most effective in shading, and reducing air-conditioning costs, on the west-facing side of walls (in the northern hemisphere) and windows; and, for shade, the density of tree crowns is less important than shape (McPherson, 1994) (see section 2.7.4 for trees and wind and 2.7.5 for value in reducing climate change impacts including through improving energy efficiency). Chapter 5 on green walls and Chapter 6 on green roofs include sections on how cladding buildings with vegetation can improve insulation and hence reduce heating and air-conditioning costs.

## 2.11 Summary

Vegetation in urban areas has multifunctional value.

- Human health and comfort are directly enhanced by vegetation (e.g. by pollution control, shade).
- Mental health is improved by exposure to vegetation.
- Green spaces provide opportunities for exercise, recreation, socialisation, education and food production.
- Climate change and urban heat island impacts can be reduced by sustainable urban drainage, shading, and the climate moderation effects of vegetation.
- Areas can experience economic benefits from intelligent use of vegetation.
- Energy efficiency of buildings can be improved by the use of vegetation.
- Wildlife habitat can be created and enhanced by a green infrastructure approach which in turn can improve the quality of life for humans.

# 3
# INDOORS

This chapter examines the value of vegetation inside buildings including:

- improvement of the work environment
- improved physical and mental health
- reduction of pollutants including volatile organic chemicals.

## 3.1 Why vegetation indoors?

Green infrastructure is not just about the outside of buildings (Figure 3.1). This chapter will demonstrate the value that incorporating green elements inside buildings, and even views of green spaces from windows, has for mood and mental and physical health. Vegetation can also moderate the microclimate of buildings and remove air pollutants. Examples will include a range of situations such as the home, computer rooms, hospitals, offices and department stores. Green elements in such situations can vary from potted plants (Burchett *et al.*, 2010) through to internal green walls adjacent to stairways, walkways and foyers (Butkovich *et al.*, 2008; Cosgrave, 2009; Knowles *et al.*, 2003).

Grinde and Patil (2009) reviewed 50 studies for evidence to support the 'Biophilia' hypothesis – that a lack of visual contact with nature creates a discord that impacts negatively on health and well-being – and concluded that the concept had some merit. Bringslimark *et al.* (2009) critically examined a number of studies on the psychological value of indoor plants and found that they were very heterogeneous in nature and recommended more rigour in future experimental designs. EPA (2009) indicated that Americans spend 90% of their time indoors and Jenkins *et al.* (1992) that Californians over the age of 11, in 1987 and 1988 when their survey was carried out, spent on average 87% of their time in buildings and 7% in enclosed vehicles. Hitchings (2010) suggested there is now a preference for being indoors among many in the Western/developed world. Some 50% of the office workers he studied did not leave their building during working hours, the remainder doing so only briefly – and did not consider this unusual behaviour despite their indoor environment being 'unremarkably bland'. As vegetation is well known for having beneficial effects on the mental and physical

FIGURE 3.1 Indoor plants create atmosphere, improve aesthetics and deliver pollution-reduction services: left: in the home © John Dover; right: at work, this example is from Centrica's offices in Oxford, 600 people can see this wall, containing 9,000 plants, from their desks. © Biotecture Ltd

state of people (see below), as well as acting as air conditioners and cleaners, the presence of plants in the workplace and at home becomes even more important with the reluctance of people to leave buildings. But which plants are best? Wolverton (1997) tested 50 plants for a range of attributes and the top ten were given as: areca palm *Chrysalidocarpus lutescens*, lady palm *Rhapis excelsa*, bamboo palm *Chamaedorea seifrizii*, rubber plant, the dracaena *Dracaena deremensis* 'Janet Craig', English ivy *Hedera helix*, dwarf date palm *Phoenix roebelenii*, the Ficus *Ficus macleilandii* 'alii', Boston fern *Nephrolepis exaltata* and peace lily *Spathiphyllum* sp. The following sections examine the values that plants bring to the indoor environment.

## 3.2 Noise reduction

Costa and James (1995) suggest that plants are unlikely to operate in the same way as walls in terms of sound reduction. The sound insulation properties of a brick wall of 450 kg.m$^{-2}$ being given as 48 dB reduction and a plasterboard wall of 7.5 kg.m$^{-2}$ = 22 dB reduction; a (1906) estimate of sound insulation by plants was quoted at 11 dB. Plants were considered to reduce noise inside buildings through absorption, diffraction and reflection of sound – properties which would vary with species, size, plant architecture, potting media and containers, moisture content of soil, planting density, etc. Effects would be most noticeable and valuable in rooms with many hard surfaces and Costa and James (1995) considered that plants would perform best at reducing the reverberation time for noises of higher pitch with lower frequencies being diffracted or reflected and absorbed by other materials. However, Wong *et al.* (2010b) tested a green wall system, using the Boston fern *Nephrolepis exaltata*, in a reverberation chamber and showed that it had good sound-absorption qualities at a range of frequencies and that performance increased with the density of the planting. Praag (2011) cited in Rutgers (2012) compared the sound-reduction qualities of a living wall built on a wooden framework with one based on an open framework supporting plants in aluminium containers and with a glass façade. The wooden structure performed best, probably because the gaps between the aluminium containers allowed sound propagation. At 1 kHz the

FIGURE 3.2 Breakout space at Q-Free's headquarters in Trondheim, Norway. The wall was installed for both its aesthetic and noise-reduction properties. © Biotecture Ltd

wooden-framework living wall absorbed almost 80% of the sound, the best the glass façade could do was 10% at the 125 Hz band. Natural soil also appeared to be better than artificial media as a sound absorber (Rutgers, 2012). Green walls may be particularly useful in reducing noise in rooms, as Costa and James (1995) suggest that walls and corners of rooms are better places to site plants than in the centre of such spaces (Figure 3.2). Costa and James (1995) include a table that shows sound reduction by *Dracaena fragrans*, *Dracaena marginata*, *Ficus benjamina*, *Howea forsteriana*, *Philodendron* sp., *Schefflera arboricola* and *Spathiphyllum* sp. at a range of frequencies from 125 Hz to 4 kHz.

## 3.3 Plants and illness

Ulrich (1984) in a study based in a hospital in Pennsylvania, USA, showed how even visual access to vegetation could have positive outcomes on health. He showed that post-operative recovery periods were shorter in patients who had access to a window that looked out onto trees compared with those who had windows looking out onto a brick wall. These patients were also more likely to receive fewer doses of strong painkillers, being more likely to need asprin and acetaminiophen rather than potent narcotics. In Korea, Park and Mattson (2009) looked at post-operative recovery of 80 female thyroidectomy patients who either had flower and foliage plants in their rooms or no plants at all. All 'plant' rooms had an arrowhead vine *Syngonium podophyllum* 'Albolineatum', Cretan brake fern *Pteris cretica* 'Albolineata', variegated vinca *Vinca minor* 'Illumination' and yellow star jasmine *Trachelospermum asiaticum* 'Ougonnishiki' and two plants each of dendrobium *Dendrobium phalaenopsis*, peace lily

*Spathiphyllum aureum*, golden pothos *Epipremnum aureum* and kentia palm *H. forsteriana*. As with the study of Ulrich (1984), post-operative recovery periods were shorter and the use of analgesics lower for patients in rooms with plants; in addition, perceptions of pain, anxiety and fatigue were also lower and a more positive attitude evident in association with plants. Preventing fatigue and restoring the attention of patients undergoing serious illness through environmental stimulus was considered to be important in cancer patient recovery programmes (Cimprich, 1993). Breast cancer patients exposed to the natural environment for two hours per week during early therapy were better able to direct their attention than those from a control group not so exposed (Cimprich, 2003).

## 3.4 General health and comfort – and the workplace

Staff working in the Institute of Physics building in Berlin, which has a façade covered in wisteria, like seeing the plants from their windows and prefer it to having sunshades (Steffan *et al.*, 2010). The presence of plants in the work environment or at least views of nature is a recurring theme in the literature, some from *in situ* experiments and some lab-based. An example of the latter was that of Chang and Chen (2005) in Taiwan who reported responses to computer-manipulated images of various office environments in their laboratory and monitoring responses to the images using electromyographic, electroencephalographic, blood volume pulse and psychological state questionnaire techniques. The most positive office configuration (least stressful environment) was a window view of a nature scene coupled with indoor plants. Joye *et al.* (2010) suggested that incorporating greenery in retail outlets should have benefits to retailers, and Brengman *et al.* (2012) demonstrated, in an online photographic experiment in the Netherlands with 4,293 participants drawn from a 100,000-strong consumer panel, that 'in-store' greenery gave shoppers a degree of pleasure and stress reduction in complex store environments, but the effect was not significant in 'lean', or more sparse, settings. The boutique Anthropologie managed to get free full-colour coverage of the opening of its new store in Regent Street, London by a national newspaper partly on the basis of its stairwell being fitted with a 160 m², 200 m high interior green wall composed of 18,000 plants from 14 species (Cosgrave, 2009) (Figure 3.3).

In an early study Lohr *et al.* (1996) demonstrated the positive values that plants could have on people working in enclosed office environments without access to windows. Her team gave a computer-based task to 98 students and staff at Washington State University, USA, in a bland computer room with and without plants. Students working in an environment with plants demonstrated higher productivity as measured by faster reaction times, less stress (as recorded by lower systolic blood pressure) and, at the end of the task, higher attentiveness. Bringslimark *et al.* (2007) also found that plants had a positive effect on productivity. Dravigne *et al.* (2008) examined the responses of 449 self-selected full-time office workers in Texas and the Midwest, USA, to their working environment using an 80 question survey. Subjects were selected from a larger initial sample by matching for work schedule, ethnicity, gender and salary. Four unequal groupings were used (proportion of respondents in parentheses):

- no plants in the office or windows with views of green spaces (50.6%)
- plants in the office, but no views of green space (18.2%)
- no plants in the office, but views of green space (13%)
- plants in the office and views of green space (18.2%).

**FIGURE 3.3** Green Wall installed in the stairwell of Anthropologie's store in London. © Biotecture Ltd. Staff report that they go home with fewer headaches than when working in other retail spaces because of the 'pleasant atmosphere created by the living wall' (Richard Sabin, pers. comm.).

FIGURE 3.4 The author's old office before moving to a modern building with no external view. The author preferred his old office.

The results showed that office workers with plants and green views and just plants in their office had higher job satisfaction and quality of life scores than those without plants (Figure 3.4). Respondents with natural views in the absence of plants also had an overall more positive view of their quality of life than those with no plants or natural views. When mean job satisfaction scores were compared for gender, males responded positively to plants and plants+natural views, but females did not. Other than commenting that earlier studies had not found such a gender difference, no explanation was offered. Shibata and Suzuki (2002) also found a gender effect with males performing better when plants were present for 'creative' tasks.

Smith *et al.* (2011a) examined the impact of introducing plants into the open-plan offices of two floors of a large multinational financial services company based in Edinburgh: one floor had plants, the other did not. Initially (for 3.5 months) a low level of plants was introduced into the 'planted' office, followed by a further 2.5 months with a higher density. Staff sickness data were compared between experimental groups and with the same time period in the year prior to the study; perception data was gathered via a questionnaire survey carried out before and after plant installation. Results indicated that plants reduced feelings that the work environment contributed to work pressure and that health-related concerns were lower where there were plants. Workers in the control office also indicated that they would like plants. Morale was not improved in the office with plants but this was attributed to the nature of the work rather than to the presence of plants; this view might be corroborated by the very high sickness levels prior to plant installation. During January to June, on the 'planted' floor in 2007 (the year before plants were introduced) 1,351 days were lost to sickness compared with 340.5 days on the control floor. The figures for the equivalent period in 2008, when plants had been installed, was: 'planted' floor 701 days and control floor 425 days. The results suggest a very strong influence of plants in reducing sickness levels. Shibata and Suzuki (2002) and Hesselink *et al.* (undated) make a distinction between the impact of plants on 'creative' workers compared with 'production' workers engaged in routine tasks whereby plants enhance 'creative' activities but distract from 'production' activities – though noting that production workers would still benefit from the restorative presence of plants in areas away from their workstations. Hesselink *et al.* (undated) did not report on gender effects but did find that the subjects who reported themselves as being physically exhausted or with high work-related stress benefitted most from the presence of plants. Nieuwenhuis *et al.* (2014) in a series of three studies, which sequentially refined their experimental approach, concluded that plants in offices compared to offices without plants (or any other distractions: 'lean' offices) had beneficial effects on workers in terms of both perception of the work environment (air quality, concentration, workplace satisfaction) and their productivity. They considered that the improvement in the perception of the work environment persisted over time, rather than being just a brief response to a change in conditions. The productivity change was quantified, in their final study using tasks designed to test 'vigilance' and 'information processing', as a 15% increase in offices with plants compared to those without plants.

In Norway, Fjeld *et al.* (1998) investigated the effect of plants in the workplace on the health of 51 office workers. The work was carried out during three months in the springs of 1995 and 1996, with half the offices having plants and half without for one spring and then the situations being reversed the following spring. Staff filled in questionnaires on 12 symptoms (five neurophysiological: fatigue, feeling heavy-headed, headache, nausea/drowsiness, concentration problems; four related to mucous membranes: 'itching, burning, irritation of eyes', 'irritated, stuffy or running nose', 'hoarse, dry throat' and coughing; three skin-related: 'dry or flushed' facial skin, 'scaling/itching' of scalp or ears, hands with 'dry, itching, or red' skin). The plants used were Chinese, Philippine or golden evergreen *Aglaonema commutatum*, striped dragonpalm dracaena *D. deremensis*, golden pothos *Epipremnum aureum* and heart-leaf philodendron *Philodendron scandens* on the window sill and the 'Janet Craig' dracaena and *E. aureum* by the door. There was a significant, 23%, reduction in neuropsychological symptoms overall, principally due to a 30% reduction in perceived fatigue. In addition, mucous-related scores were significantly reduced by 24% overall, primarily due to a reduction in coughing by 37% and symptoms of dry/hoarse throat by 25%. Skin-related conditions were not significantly reduced when taken as a group, but 'dry or flushed facial skin' was significantly reduced by 23%. The causes of the improvements in health and reduction of discomfort were considered to probably be due to an improvement in air quality, and an increase in well-being mediated by a more congenial working environment. A questionnaire from this study was later reported by Fjeld (2000) with 82% of the participants in the study agreeing to the statement 'I feel more comfortable if I have plants in my office' and 82% to 'I would like to have plants in my office in the future'. These findings were confirmed by a later study in the diverse environments of a hospital X-ray department and a high school, the latter with students aged 14–16 (Fjeld, 2000). These studies differed from Fjeld *et al.* (1998) in that full-spectrum lights were installed as well as plants. In the X-ray department the benefits were assessed in relation to exposure time and were found to have increased with the amount of time spent during the day in the radiology room (decrease in symptoms by 34% working all day; 21% present half a day; 17% less than half a day); the benefits were confirmed to be still operating 11 months after the end of the study (with the plants still in place). The school classrooms were modified by having plants in three rooms (air was also passed through the growing medium/root zone), the control situation was a further three rooms without plants. Significant impacts included a reduction in reported headaches by 37%, dry, itching eyes by 30%, dry or hoarse throat by 36%. Perceptions of the working environment by the students were also improved by the presence of plants. Bringslimark *et al.* (2007) demonstrated, in a survey of 364 office workers, that having plants in view or in their own work area significantly reduced the amount of sick leave they took. A contrary finding was that plants in the area very near the workstation had a positive relationship with sick leave, though why such a result should be generated was not satisfactorily explained, although the distinction made between 'creative' and 'production' values of plants by Hesselink *et al.* (undated) might offer some insight.

## 3.5 Air conditioning

Cañero *et al.* (2012) cite the Spanish building regulations (MITTMH, 2009) which give acceptable ranges for both temperature (20–24°C) and humidity (30–70%) and note that in warm climates air conditioning, requiring significant energy consumption, is needed to maintain such regimes. Even in temperate countries such as the UK, peak summer temperatures can make office buildings uncomfortable, especially with the reliance on heat-emitting electronic equipment such as computers for everyday business activities. Lohr and Pearson-Mims (1996) found that interior plants improved the humidity levels of both computer labs and offices at Washington State University. Smith and Pitt (2011) found plants provided more comfortable levels of humidity when introduced into an open-plan office; installation of plants raised levels from below 30% to 40–60% relative humidity. Costa and James (1995) cite references detailing the evaporative cooling value of plants in buildings.

## 3.6 Microbial load reduction

One of the fears about plants in buildings is that they encourage mould organisms because of elevated humidity (Wolverton & Wolverton, 1996). However, Wolverton and Wolverton (1996) in a simple study found that rooms with plants could have 50% lower airborne microbes compared with plant-free rooms. Petri dishes with nutrient agar plates were exposed to the air in rooms with or without plants and cultured in an incubator and the resulting microbial colonies counted. Organisms identified included bacteria, actinomycetes (bacteria with fungal-like filamentous colonies) and moulds. Wolverton and Wolverton (1996) concluded that plants actively suppressed microorganisms and suggested that plant defence chemicals (allelochemicals) (Ehrlich & Raven, 1967; Vollmer & Gordon, 1975) such as terpenes and phenolic compounds on leaf surfaces and exudates from roots (Bakker *et al.*, 2012) were the primary source of control. Fjeld (2000) reported that including plants in a hospital radiology department did not result in an increase (or decrease) in the fungi/fungal spore content of the air. Burchett *et al.* (2010) in a study of offices in Sydney with and without plants showed no significant increase in fungal spores in offices with plants – and all offices had almost a 20-fold lower concentration of spores than outdoor air.

Curiously, although describing the chemical-degrading abilities of microorganisms associated with plants, Wolverton and Wolverton (1996) did not consider that such organisms could have a role in suppressing the microbial load in rooms. However, the surfaces of plant leaves and stems (the phylloplane) and the root zone (the rhizosphere) maintain communities of microorganisms potentially useful in removing airborne microbes. The rhizosphere, for example, acts as host to a wide range of organisms including bacteria, some of which help protect plants against disease not only by stimulating resistance, but also by the production of suppressive antibiotics and enzymes which kill other microbes, including fungi, by directly attacking cell walls of potential competitors (van Loon *et al.*, 1998; Whittaker & Feeny, 1971). Plants actively promote the growth of the root zone microbial community and tailor it to their needs via root exudates (Bais *et al.*, 2006; Bertin *et al.*, 2003). Likewise, the phylloplane hosts microorganisms that can, either by competition or direct antagonism, protect plants from disease (Blakeman & Fokkema, 1982) and direct inoculation with 'beneficial' organisms has been demonstrated in agricultural contexts (Li *et al.*, 2006) and also in laboratory studies on the degradation of volatile organic compounds (VOCs) (Sandhu *et al.*, 2007). The protective action of the microbial community of both phylloplane and rhizosphere may be disrupted by the use of fungicides (Blakeman & Fokkema, 1982), which probably should be avoided, if possible, during routine plant maintenance programmes.

## 3.7 Air pollution

### 3.7.1 Carbon monoxide, carbon dioxide and nitrogen dioxide

Wolverton *et al.* (1985) showed that the spider plant *Chlorophytum elatum* var. vittatum and golden pothos *E. aureus* could reduce levels of carbon monoxide; in the same study the spider plant was also shown to reduce nitrogen dioxide levels. Carbon monoxide and VOC levels were reduced in an open-plan office after introduction of plants compared with an 'unplanted' office (Smith & Pitt, 2011). Tarran *et al.* (2007) showed that carbon monoxide levels could be significantly reduced in offices by three *D. deremensis* 'Janet Craig' plants (in 300 mm pots). The scale of the reduction was impressive: 92% reduction (from 0.225 to 0.017 ppm) in an air-conditioned building, 86% (from 0.071 to 0.010 ppm) in naturally ventilated offices; carbon dioxide levels were also reduced by 10 and 25% respectively. Carbon dioxide removal by indoor plants is not a subject that has received extensive coverage, but Torpy *et al.* (2014) assessed eight indoor plant species for their ability to remove $CO_2$ and found, on a leaf area basis, that bamboo palm *Dypsis lutescens* was the most effective followed by dracaena 'compacta' *D. deremensis,* weeping/Benjamin's fig *Ficus benjamina* and kentia palm *H. forsteriana*. The first three are plants adapted to high-light conditions and the last adapted to low-light conditions. Torpy *et al.* also found that light levels affected uptake of $CO_2$, with low-light species such as dwarf mountain palm *Chamaedorea elegans*, Chinese evergreen *Aglaonema commutatum* and *H. forsteriana* having reduced uptake in high-intensity light conditions. The other two species tested were aspidistra *Aspidistra elatior* and the Moreton Bay chestnut *Castanospermum australe*.

### 3.7.2 Odours

Unpleasant odours in buildings, such as hydrogen sulphide, ammonia and methyl mercaptan, are clearly undesirable and some are characteristic of the specific environment; for example, ammonia is a typical odour present in hospitals, nursing homes, lavatories and refuse stores (Oyabu *et al.*, 2003). Oyabu *et al.* (2003) investigated the potential of plants for the removal of odours using golden pothos (*E. aureum*) in three different soils and also used a 'snake plant' and a 'rubber plant' as comparisons. Formaldehyde, acetone and ammonia were used as test odours. Whilst the authors noted the harmful nature of the chemicals, it was the subjective nature of the smell rather than the toxicity they were interested in. The plants were considered effective in odour removal, with time needed for removal being longer for higher concentrations of odours and odours of larger molecular weight. Soil type influenced odour removal; unfortunately no controls (using pots with soil but no plants) were used in the study.

### 3.7.3 Volatile organic chemicals

Volatile organic compounds (VOCs) are a group of about 18 million organic (carbon-based) chemicals (Golding & Christensen, 2000). The term 'volatile' relates to their ability to vaporise at typical room temperatures (for more precise definitions see Guieysse *et al.*, 2008). Many hundreds of VOCs have been identified from indoor air (Posudin, 2010) and are usually found at far higher concentrations indoors than out (Brown, 2002). Wolkoff (2003) points out that as building products constitute the largest indoor surface area, they are therefore also the source of most indoor VOCs. Wolkoff (2003) also documents the various EU and Nordic standards to minimise such emissions and gives examples of trade labelling schemes such as the Gemeinschaft Umwelt Teppisch Ordnung (Society for Testing Carpets in Europe). VOC sources include the resins used in chipboard wood products (Chiappini *et al.*, 2011) and are emitted from a wide range of building materials, electrical goods, furniture, paint, household products and print media (Posudin, 2010; Wang *et al.*, 2007; Wolverton & Wolverton, 1993). Many VOCs have known biological activity and can be found at concentrations potentially damaging to human health in indoor air (Wang *et al.*, 2007). VOCs have been implicated in 'sick building syndrome' (ALA, undated) and exposure can result in nausea, vomiting, abdominal pain and diarrhoea (Xu *et al.*, 2010). Whilst newly constructed buildings/products will have the highest concentrations of VOCs, the chemicals may be continually emitted 'outgased' at lower levels for years (Wolverton & Wolverton, 1993). The US National Aeronautics and Space Administration (NASA) initiated work on plants in the early 1980s to identify ways of purifying air as part of their research into independent structures such as space stations and also energy-efficient homes, which, by definition, have reduced ventilation and are sealed against unwanted draughts (Dimitroulopoulou, 2012; Wolverton *et al.*, 1984; 1989b).

### 3.7.4 Plants and mechanisms of VOC removal from air

A number of studies have indicated that plants are capable of removing VOCs from indoor air via a variety of mechanisms (Figure 3.5). Of the hundreds of VOCs contaminating indoor air, perhaps most work using plants has been carried out on formaldehyde and will be examined below – studies on other VOCs are briefly given in Table 3.1 and Table 3.2.

Plants are probably able to detoxify many compounds directly (Schaffner *et al.*, 2002), but Wolverton *et al.* (1989a), in a pioneering experiment with benzene and whole plants and plant roots of the dragon tree *D. marginata,* showed that the major site for removal of the compound by *D. marginata* was via the root zone. Wolverton and Wolverton (1993) investigated the differential proportions of formaldehyde and xylene removed by different plant parts and estimated that 33–49.5% was via the plant leaves and stems and the balance by the roots – or rather by some of the microorganisms growing in association with plant roots. Tests with the plants *F. benjamina* or *Spathiphyllum* growing in potting soil for five months showed the root zone removed 660 and 659 μg.h$^{-1}$ respectively more formaldehyde than potting soil left for five months with no plants; the equivalent figures for *S. trifasciata* and *Kalanchöe* were worse than potting soil on its own at -91 and -92 μg.h$^{-1}$ respectively. Not all plant species tested appeared to have VOC-degrading microorganisms which, in their tests, were shown to be gram-negative rod bacteria: *Kalanchöe* soil contained gram-positive bacteria whilst *Spathiphyllum* contained gram negative bacteria. Wood *et al.* (2002) demonstrated benzene and hexane removal by rhizosphere microorganisms with *Spathiphyllum wallisii* 'Petite' peace lily, *D. deremensis* 'Janet Craig' and Kentia palm *H. forsteriana* in both potted and hydroponic media. Chun *et al.* (2010) confirmed activity of root microorganisms by growing bacterial material removed from the rhizosphere and testing with the VOCs benzene and toluene. Inoculation of the cultured isolates into pots with living plants further confirmed the VOC-removing potential of the rhizosphere bacteria when compared with un-inoculated plants. Wolverton and Wolverton (1993) hypothesised that VOCs taken in by leaves and stems might be translocated by the plant to the root zone for removal by microorganisms, but recent work by Sandhu *et al.* (2007) has demonstrated degradation of phenol by microbial communities on leaf surfaces and Schaffner *et al.* (2002) demonstrated formaldehyde degradation directly by the plant. Tani and Hewitt (2009) isolated peace lily *Spathiphyllum clevelandii* and golden pothos *E. aureum* leaves from the rest of the plant and demonstrated uptake of 13 of 14 VOCs by the former and five of six VOCs by the latter (the exception in each case being acetone) (Table 3.1). The removal mechanisms postulated were assumed to be metabolism in the leaf or translocation to other parts of the plant following stomatal uptake. Wolverton and Wolverton (1993) also felt that the degradation rate may have improved over time during their experiments, an observation confirmed by Orwell *et al.* (2004) who coined the term 'induction' for the initial, slower, take-up period (over a 2–4 day period) and confirmed the activity of root-zone microorganisms. A lag in uptake, given the rapid ability of microorganisms to replicate, could have simply been through an increased population of microorganisms capable of breaking down the VOCs, but whilst Wolverton *et al.* (1989b) demonstrated increased bacterial populations after six weeks of periodic exposure to benzene (from 3.1 x 10$^4$ at the start rising to 5.1 x 10$^4$ cfu.g$^{-1}$ after six weeks) they felt that this correlation was not an adequate explanation of their observations. Wolverton and Wolverton (1993) suggested that the accelerated uptake after a lag-phase was via microbial adaptation, a suggestion repeated by Orwell *et al.* (2004) who considered that it was actuated by the induction of a specific biochemical pathway in the microbes or plant (or both). Kim *et al.* (2012) considered that the induction reflected a change in gene expression by the plants, microorganisms, or both, rather than simple population increase by the microorganisms. However, Wood *et al.* (2002) demonstrated that whilst the overall population size of the microbial community did not change following exposure to VOCs (in their study of benzene) the community *structure* did change with increases in population size

**TABLE 3.1** Studies on volatile organic chemicals other than formaldehyde in relation to plant uptake

| VOC studied | Study | Summary |
|---|---|---|
| **α-Pinene** | Yang et al. (2009) | See Table 3.2 for tests of plants for this VOC. |
| **Acetone** | Tani & Hewitt (2009) | Acetone exposed to peace lily Spathiphyllum clevelandii and golden pothos Epipremnum aureum leaf attached to the plant but contained within a teflon bag into which acetone was continuously introduced via a flow-through mechanism. The VOC uptake was estimated using proton transfer reaction mass spectrometry. The plants did not take acetone up consistently and only when concentrations were very low. |
| **Ammonia** | Wolverton and Wolverton (1993) | Eleven species of plant assayed for ammonia removal, rates varied from 7356 $\mu g.h^{-1}$ to 984 $\mu g.h^{-1}$ (see Table 3.3 for details). |
| | Cornejo et al.(1999) | Spider plant Chlorophytum comosum was shown to remove trichloroethylene. |
| **Benzaldehyde** | Tani & Hewitt (2009) | Benzaldehyde taken up by peace lily S. clevelandii leaf (see Acetone above for more detail). |
| **Benzene** | Wolverton et al. (1989a,b) | C. morifolium, D. deremensis 'Warneckei', D.massangeana, F. benjamina, G. jamesonii could all reduce benzene concentrations – albeit with differing levels of efficiency. See further Table 3.2 and also, for rhizosphere effects, Chun et al. (2010) under Toluene. |
| | Cornejo et al. (1999) | Eight plant species (flaming Katy Kalanchoe blossfeldiana, spiderwort Tradescantia fluminensis, dracaena D. deremensis, Chinese primrose Primula sinensis, spider plant Chlorophytum comosum, regal geranium Pelargonium x domesticum, strawberry saxifrage Saxifraga stolonifera, and Magnesia sp.) were tested for removal of benzene in a sealed chamber with 108 $\mu g$ benzene. On a whole plant basis the best were P. domesticum (95% removal), K. blossfeldiana (85% removal) and S. stolonifera (41% removal); the remaining five species had between 10 and 25% uptake. Expressed as $\mu g$ of benzene removed per gram of foliage over 24 h, the best species were P. domesticum (8.5 $\mu g$), K. blossfeldiana (3.4 $\mu g$) and S. stolonifera and Magnesia sp. (both 2 $\mu g$). Later tests with K. blossfeldiana and rubber plant Ficus elastica confirmed the activity of the former, but also the good removal efficiency of the latter. |
| | Orwell et al. (2004) | Seven plant species tested for benzene removal: kentia palm Howea forsterianai, peace lily Spathiphyllum floribundum var. Petite, S. floribundum var. Sensation, D. deremensis var. 'Janet Craig', D. marginata, devil's ivy E. aureum, and Queensland umbrella tree Schefflera actinophylla var. Amate. The results showed that most of the activity was via microbes in the root zone, but some species also removed benzene via the foliage. On a per plant basis, D. deremensis var. 'Janet Craig' performed best immediately after 'induction' and H. fosteriana worst but data suggests that removal rates may vary with time and plant species from induction. Data also presented on removal rate related to leaf area, dry weight of shoots, roots and pot mix. |

| VOC studied | Study | Summary |
|---|---|---|
| | Liu *et al.* (2007) | Seventy-three species of plant were screened for their ability to remove benzene from a 150 ppb benzene-in-air feedstock over a 2 h period: 20 species removed no benzene, 13 removed 0.1–9.99% benzene, 17 removed 10–20% and 17 removed 20–40%. Ten species were then chosen for further evaluation from the latter group. Plants were fumigated for 8 h/day for 2 days using the same feedstock concentration. On the basis of benzene absorption expressed as $\mu g.m^{-2}$ leaf area.day$^{-1}$ by far the most effective species was the jade plant *Crassula portulacea* (syn. *C. ovata*) at 724.9 $\mu g.m^{-2}.d^{-1}$ the next best being the hydrangea *Hydrangea macrophylla* at 293.7 $\mu g.m^{-2}.d^{-1}$. The worst two species of the ten were the Boston fern *N. exaltata* cv. 'Bostoniensis' and *D. deremensis* cv. variegata with 73.5 and 59 $\mu g.m^{-2}.d^{-1}$ respectively. The other plants were intermediate between these values (higher to lower: *Cymbidium* 'Golden Elf', *Ficus microcarpa* var. fuyuensis, *Dendranthema morifolium*, *Citrus medica* var. sarcodactylis, *Dieffenbachia amoena* cv. Tropic Snow, *Spathiphyllum* 'Supreme'). |
| **Crotonaldehyde** | Tani & Hewitt (2009) | Taken up by peace lily *S. clevelandii* leaf (see Acetone above for more detail). |
| **Diethyl ketone** | Tani & Hewitt (2009) | Diethyl ketone taken up by peace lily *S. clevelandii* and golden pothos *Epipremnum aureum* leaves (see Acetone above for more detail). |
| **Methacrolein** | Tani & Hewitt (2009) | Methacrolein taken up by peace lily *S. clevelandii* leaf (see Acetone above for more detail). |
| **Methyl ethyl ketone** | Tani & Hewitt (2009) | Methyl ethyl ketone taken up by peace lily *S. clevelandii* and golden pothos *E. aureum* leaves (see Acetone above for more detail). |
| **Methyl isobutyl ketone** | Tani & Hewitt (2009) | Methyl isobutyl ketone taken up by peace lily *S. clevelandii* and golden pothos *E. aureum* leaves (see Acetone above for more detail). |
| **Methyl iso-propyl ketone** | Tani & Hewitt (2009) | Methyl iso-propyl ketone taken up by peace lily *S. clevelandii* and golden pothos *E. aureum* leaves (see Acetone above for more detail). |
| **Methyl n-propyl ketone** | Tani & Hewitt (2009) | Methyl n-propyl ketone taken up by peace lily *S. clevelandii* and golden pothos *E. aureum* leaves (see Acetone above for more detail). |
| **Iso-butyraldehyde** | Tani & Hewitt (2009) | Iso-butyraldehyde taken up by peace lily *S. clevelandii* leaf (see Acetone above for more detail). |
| **Iso-valeraldehyde** | Tani & Hewitt (2009) | Iso-valeraldehyde taken up by peace lily *S. clevelandii* leaf (see Acetone above for more detail). |
| **N-butyraldehyde** | Tani & Hewitt (2009) | N-butyraldehyde taken up by peace lily *S. clevelandii* leaf (see Acetone above for more detail). |
| **N-valeraldehyde** | Tani & Hewitt (2009) | N-valeraldehyde taken up by peace lily *S. clevelandii* leaf (see Acetone above for more detail). |

| VOC studied | Study | Summary |
|---|---|---|
| **Octane** | Yang *et al.* (2009) | See Table 3.2 for plants tested for removal of this VOC. |
| **Pentane** | Cornejo *et al.*(1999) | Pentane showed to be removed by spider plant *C. comosum*, and rubber plant *Ficus elastica*. |
| **Proprionaldehyde** | Tani & Hewitt (2009) | Proprionaldehyde taken up by peace lily *S. clevelandii* leaf (see Acetone above for more detail). |
| **Toluene** | Kim *et al.* (2012) | Initial low levels of toluene shown to induce a stronger response in *Begonia maculata* (x3.85), *Ardisia crenata* (x3.18) and *A. japonica* (x2.52) after exposure. The enhanced response was short-lived requiring continual exposure to toluene or re-exposure. |
| | Chun *et al.* (2010) | Nine plant species tested for the ability of microorganisms in their rhizosphere for removal of VOCs; rhizosphere community of *Pachira aquatica* performed best and inoculated into media with other plant species. Demonstrated that rhizosphere microorganisms could remove removed toluene, benzene and xylenes from the air. See further Table 3.2 for plants tested for removal of this VOC. |
| | Orwell *et al.* (2006) | Toluene and m-xylene exposed to *D. deremensis* 'Janet Craig' and *Spathiphyllum* peace lilies 'Sweet Chico' at 0.2, 1.0, 10 and 100 ppm and in mixtures. VOC removal induced at 0.2 ppm and reduced below 20 ppb over 24 h. The presence of toluene also increased the rate of m-xylene removal. |
| **Trichloroethylene** | Wolverton *et al.* (1989a,b) | *Chrysanthemum morifolium, D. deremensis* 'Warneckei', *D. massangeana, F. benjamina, Gerbera jamesonii* could all reduce trichloroethylene concentrations – albeit with differing levels of efficiency. See also Table 3.2 for plants tested for removal of this VOC. |
| **Xylene** | Wolverton and Wolverton (1993) | Thirty plant species were tested for their ability to remove xylene, rates varied from 610 $\mu$g.h$^{-1}$ to 47 $\mu$g.h$^{-1}$ (see Table 3.3 for details). As with formaldehyde (see text) two Boston fern plants or three 'Janet Craig' dracaenas would be capable of removing the typical new office loading of 0.22 $\mu$g.l$^{-1}$ of xylene from a 22.32 m$^3$ office (493 $\mu$g xylene), although a single pygmy date palm or two dumb canes should achieve the same effect. See further Chun *et al.* (2010) under 'Toluene' for rhizosphere effects, and Orwell *et al.* (2006) for removal by *D. deremensis* and *Spathiphyllum* peace lilies and synergistic effects. |

Sources: as indicated.

TABLE 3.2 Performance of plants examined by Yang *et al.* (2009) for their ability to remove five VOCs from four classes of compound

| | Poor removal performance | Intermediate removal performance | Excellent removal performance |
|---|---|---|---|
| | *Peperomia clusiifolia* | *Ficus benjamina* | *Hemigraphis alternata* |
| | *Chlorophytum comosum* | *Polyscias fruticosa* | *Hedera helix* |
| | *Howea belmoreana* | *Fittonia argyroneura* | *Tradescantia pallida* |
| | *Spathiphyllum wallisii* | *Sansevieria trifasciata* | *Asparagus densiflorus* |
| | *Schefflera arboricola* | *Guzmania* sp. | *Hoya carnosa* |
| | *Codiaeum variegatum* | *Anthurium andreanum* | |
| | *Calathea roseopicta* | *Schefflera (Dizygotheca) elegantissima* | |
| | *Aspidistra elatior* | | |
| | *Maranta leuconeura* | | |
| | *Dracaena fragrans* | | |
| | *Ficus elastica* | | |
| | *Dieffenbachia seguine* syn. *amoena* | | |
| | *Philodendron scandens ssp. oxycardium* | | |
| | *Syngonium podophyllum* | | |
| | *Epipremnum aureum* | | |
| | *Scindapsus aureus* | | |
| | *Pelargonium graveolens* | | |

**Range of removal efficiency by VOC (mg.m$^{-3}$. m$^{-2}$.h$^{-1}$)**

| | | | |
|---|---|---|---|
| Benzene | 0.03 to 1.20 | 0.66 to 2.74 | 2.21 to 5.54 |
| Toluene | 1.54 to 3.18 | 3.60 to 5.09 | 5.81 to 9.63 |
| Octane | 0.00 to 2.03 | 0.65 to 3.98 | 2.76 to 5.58 |
| Trichloroethylene | 1.48 to 2.86 | 3.58 to 6.15 | 5.79 to 11.08 |
| α-Pinene | 2.33 to 4.61 | 4.30 to 8.68 | 8.48 to 13.28 |
| **Total VOCs** | **5.55 to 12.98** | **16.78 to 24.13** | **26.08 to 44.04** |

of some species, reductions in others and the appearance of some species in cultures only after benzene had been administered. Orwell *et al.* (2004) presented data on benzene removal by a range of plants (see Table 3.1) using a range of parameters: standard pot plant (12 months old, 0.3–0.4 m in height in 150 mm diameter pots with a defined potting medium), leaf area, of shoot dry weight, root dry weight, and dry weight of the potting mix and demonstrated that the ranking of best/worst in terms of removal efficiency varied depending on parameter and time since induction. Liu *et al.* (2007) screened 73 plant species for benzene removal (Table 3.1) but their initial screen was of very short duration (2 h) and they may have missed effective plants that needed an induction period. Kumar *et al.* (2011) suggest that simple compounds of low molecular weight and high solubility are those most easily degraded by microorganisms given that more complex compounds require more energy to break them down, a limitation that may not apply to degradation by the host plants themselves. Kim *et al.* (2008) using *Ficus benjamina* and *Fatsia japonica* plants convincingly demonstrated that the

**FIGURE 3.5** Routes and modes of uptake of volatile organic chemicals (VOCs) by pot plants. VOCs may be degraded by the microbial community living on the *Phylloplane* (leaf surface); VOCs may also adhere to the cuticle (the waxy outer surface of leaves) or penetrate directly through it. The *Stomata* are used for taking in carbon dioxide and giving off oxygen. VOCs can also enter by this route and be translocated throughout the plant to be metabolised. The *Rhizosphere* is the root zone and supports a microbial community which contains species which can breakdown VOCs. © John Dover

TABLE 3.3 Plants surveyed by Wolverton and Wolverton (1993) for their ability to remove the volatile organic chemicals (VOC) formaldehyde, xylene, and ammonia from the air★

| | | | Removal rate of VOC $\mu g.h^{-1}$ and rank (x) | | |
|---|---|---|---|---|---|
| Species type | Scientific name | Common name | Formaldehyde | Xylene | Ammonia |
| **Bromeliad** | | | | | |
| | Aechmea fasciata | Urn plant | 234 (31) | | |
| | Guzmania cherry | Cherry guzmania | | 146 (26) | |
| | Neoregelia cv. | | | 47 (30) | |
| **Bulb** | | | | | |
| | Tulip 'Yellow Present' | Tulip 'Yellow Present' | 717 (16) | 229 (14) | 2815 (7) |
| **Climber** | | | | | |
| | Cissus rhombifolia | Venezuela treebine | 376 (24) | | |
| | Hedera helix | English ivy | 1120 (6) | 131 (27) | |
| | Syngonium podophyllum | Arrowhead plant | 341 (25) | 220 (16) | |
| **Ferns** | | | | | |
| | Nephrolepis exaltata 'Bostoniensis' | Boston fern | **1863 (1)** | 208 (18) | |
| | Nephrolepis obliterata | Kimberley queen fern | **1328 (5)** | 323 (6) | |
| **Herbaceous foliage plants** | | | | | |
| | Aglaonema sp. 'Silver Queen' | Chinese evergreen 'Silver Queen' | 564 (21) | | |
| | Anthurium andraeanum | Flamingo flower/Tail flower | 336 (26) | 276 (8) | **4119 (4)** |
| | Calathea ornata | Prayer plant | 334 (27) | | |
| | Calathea vittata (elliptica) | | | | 3100 (6) |
| | Chlorophytum comosum 'Vittatum' | Spider ivy 'Vittatum' | 560 (22) | 247 (12) | |
| | Cyclamen persicum | Cyclamen | 295 (29) | 173 (21) | |
| | Dieffenbachia camille | Dumb cane | 469 (23) | **341 (2)** | |
| | Dieffenbachia maculata | Dumb cane | | **325 (4)** | |
| | Dieffenbachia sp. 'Exotica compacta' | Dumb cane | 754 (15) | | |
| | Dracaena deremensis 'Janet Craig' | Janet Craig dracaena | **1361 (4)** | 154 (25) | |
| | Dracaena deremensis 'Warneckei' | Striped dracaena | 760 (12) | 295 (7) | |
| | Dracaena marginata | Dragon tree | 772 (11) | **338 (3)** | |
| | Dracaena fragrans (deremensis) | Cornstalk dracaena | 938 (9) | 274 (9) | |
| | Euphorbia pulcherrima | Poinsettia | 309 (28) | 116 (28) | |
| | Ficus benjamina | Benjamin tree | 940 (7) | 271 (10) | 1480 (9) |
| | Ficus sabre | Saber Ficus | 692 (17) | | |
| | Homalomena sp. | Homalomena | 668 (18) | **325 (5)** | 5208 (2) |
| | Kalanchöe | Kalanchoe | | 170 (22) | |
| | Liriope spicata | Creeping lilyturf | 758 (13) | 230 (13) | **4308 (3)** |
| | Rhododendron indicum | Azalea | 617 (20) | 168 (23) | 984 (11) |
| | Sansevieria trifasciata | Snake plant or mother-in-law's tongue | 189 (32) | 157 (24) | |

| | | | Removal rate of VOC μg.h$^{-1}$ and rank (x) | | |
|---|---|---|---|---|---|
| Species type | Scientific name | Common name | Formaldehyde | Xylene | Ammonia |
| | Senecio cruentu (Pericallis cruenta) | Cineraria | | 115 (29) | |
| | Spathiphyllum sp. 'Clevelandii' | Peace lily | 939 (8) | 268 (11) | 1269 (10) |
| **Herbs** | | | | | |
| | Chrysanthemum morifolium | Hardy garden mum | **1450 (2)** | 201 (19) | **3641 (5)** |
| **Orchids** | | | | | |
| | Dendrobium sp. | Orchids | 756 (14) | 200 (20) | |
| | Phalaenposis sp. | Moth orchid | 240 (30) | | |
| **Succulent** | | | | | |
| | Aloe barbadensi (vera) | Barbados aloe | 188 (33) | | |
| **Woody foliage plants** | | | | | |
| | Chamaedorea elegans | Parlour palm | 660 (19) | 223 (15) | 2453 (8) |
| | Phoenix roebelenii | Pygmy date palm | **1385 (3)** | **610 (1)** | |
| | Rhapis excelsa | Lady palm | 876 (10) | 217 (17) | **7356 (1)** |

Notes

The rank of the plants for efficiency is given in parentheses and the top five in removal efficiency are given in bold for each pollutant.

The temperatures when measurements were made varied between 21.9 and 26.9°C and pot sizes varied from as little as 8.0 cm and 12.7 cm with all the rest in the range 15.2 to 35.6 cm, so the ranking is only indicative – for exact details see the original paper.

*Common names have been taken from the general literature and may not be accurate; likely synonyms are given in brackets after the scientific name. Plants have been grouped in a similar way to that of Kim et al. (2010).

leaves and root zone took in similar amounts of formaldehyde during the day when the plant was actively taking in air via the stomata, but at night when the stomata were closed the majority (ratio about 1:11) of the formaldehyde was removed by the root zone. Kim et al. (2008) suggested that formaldehyde removed by the leaves at night was via the cuticle – they did not consider phylloplane effects. Yoo et al. (2006) also showed differential uptake between day and night for the above-ground parts of English ivy Hedera helix, peace lily Spathiphyllum wallisii, S. podophyllum and grape ivy Cissus rhombifolia when exposed to benzene or toluene; Cissus rhombifolia was the least effective of the four. The uptake was greater during the day for all species with the exception of H. helix with benzene which appeared to remove comparable amounts of benzene whether during the day or night. Differences in day:night removal were less pronounced than implied by the findings of Kim et al. (2008). Yoo et al. (2006) commented that the ability of H. helix to remove similar amounts of VOCs during the day or night indicated that the major removal route was mainly by adhesion to the cuticle or cuticular absorption, but did not consider microbial degradation at the phylloplane.

Wolverton and Wolverton (1993) demonstrated that a particular plant species may not be uniformly good at removing all pollutants; their best plant at removing formaldehyde, the Boston fern, was 18th at removing xylene from substantially the same group of plants tested. Likewise, the best plant at removing ammonia in their assays, the lady palm *Rhapis excelsa*, was 10th out of 33 for formaldehyde removal and 17th out of 30 for xylene removal (Table 3.3). Cornejo *et al.* (1999) found selectivity in uptake of VOCs when presenting a mixture of benzene and toluene to flaming Katy *Kalanchöe blossfeldiana*; whilst benzene was removed by the plant, toluene was not. In other tests with the spider plant *C. comosum* and mixtures of trichloroethane, benzene and pentane the uptake of benzene and pentane appeared to be depressed by the presence of trichloroethane. Yoo *et al.* (2006) also found that mixtures of VOCs (benzene and toluene) affected uptake by *H. helix*, *S. walisii*, *S. podophyllum* and *C. rhombifolia* – interestingly *H. helix* outperformed the other three species in the removal of both benzene and toluene from a mixture of the two VOCs. Differential uptake was also found by Yang *et al.* (2009) who screened 28 plant species for their ability to remove five VOCs from four classes of chemical: benzene and toluene to represent the 'aromatic' hydrocarbons (chemicals based on a benzene ring), octane to represent the straight-chain organic hydrocarbons, trichloroethylene to represent organic compounds that have a halogen group (in this case chlorine) and $\alpha$-pinene, a terpenoid compound found in many essential oils emitted by plants in scents and also used as defence chemicals. Plants were compared on an uptake/leaf area basis. Of the 28 plants, 16 were considered poor at VOC uptake, seven 'intermediate' and five 'superior' (Table 3.2). Four plants removed all VOC classes effectively: red flame ivy *Hemigraphis alternata*, English ivy *H. helix*, the wax plant *Hoya carnosa*, and plume asparagus/asparagus fern *Asparagus densiflorus* removed all five compounds well and purple spiderwort *Tradescantia pallida* was effective against all but octane. Other plants had more selective responses: the nerve plant *Fittonia argyroneura* removed the two aromatic hydrocarbons and the halogenated hydrocarbon; the Benjamin tree *F. benjamina* removed octane and the terpene; the Ming tree *Polyscias fruticosa* only removed octane.

Plants and their associated microbial communities are clearly capable of making a valuable contribution to indoor air quality, but the removal methods, detoxification mechanisms, time dependency of response and involvement of unquantified microbial communities of potentially plant species-specific nature make for a complex system that has not yet been fully elucidated, but holds considerable promise for improvement in indoor air quality and significant opportunities for development of novel systems such as biofiltration (Darlington *et al.*, 2000; Guieysse *et al.*, 2008; Kumar *et al.*, 2011) and indoor living walls (see below). As yet there appears to be no information on the impact of controlled-release fertilisers, slow-release fertilisers or liquid applications of fertiliser on the microbial communities of the VOC-degrading rhizosphere of indoor potted plants or green walls (Šrámek & Dubský, 2007; Tan *et al.*, 2012), although benzene uptake by the arrowhead vine *S. podophyllum* grown in hydroculture has been shown to be almost as good as that with conventionally grown plants (Irga *et al.*, 2013). Whilst it is known that plants expend considerable energy by diverting up to 45% of photosynthetic output to foster rhizosphere microorganisms, hypothesised signalling by the plant to elicit a tailored VOC removal has not yet been confirmed or rejected (see further, and references, in Wood *et al.*, 2002). Recent work has indicated that 'plant growth-promoting rhizobacteria' emit VOCs themselves (including ethylene) which appear to induce systematic resistance in plants to a range of negative factors including insect pests, disease, heavy metal contamination and drought (Farag *et al.*, 2013).

## Formaldehyde

Formaldehyde ($CH_2O$) is a common indoor pollutant causing a range of symptoms including eye conditions, skin and respiratory distress, nausea, vomiting and diarrhoea and is a potential carcinogen (Dingle *et al.*, 2000), although recent work (Golden, 2011) suggests that the causal link with cancer is not proven and that indoor limits should be set at 0.1 ppm. Pollution sources include chipboard and plywood glues, foam house insulation, soft furnishings, paper, cosmetics, smoking and gas fires (see references in Dingle *et al.*, 2000). Wolverton *et al.* (1984) tested the ability of golden pothos (*Scindapsus aureus*), the arrowhead plant (nephthytis) *S. podophyllum* and spider plants *Chlorophytum elatum* var. *vittatum* to remove formaldehyde from sealed plexiglass chambers; experimental controls using pots with soil, but without plants, were used to examine the impact of soil microorganisms on formaldehyde removal. In these experiments the controls and experimental plants removed formaldehyde. The soil-only control reduced formaldehyde from a starting concentration of 15 to 10 ppm over a 24 h period whilst golden pothos and nephthytis plants reduced

TABLE 3.4 Indoor plants graded as 'Excellent' (removing 1.2 $\mu g.m^{-3}.cm^{-2}$ of leaf area over a 5 h period) by Kim *et al.* (2010) in removing formaldehyde

| Type | Scientific name | Common name | $\mu g.m^{-3}.cm^{-2}$ |
|------|-----------------|-------------|------------------------|
| **Woody foliage plants** | | | |
| | *Psidium guajava* | Guava | 2.39 |
| | *Rhapis excelsa* | Lady palm | 1.67 |
| | *Zamia pumila* | Coontie | 1.32 |
| **Herbaceous foliage plants** | | | |
| | *Chlorophytum bichetii* | Airplane plant/Siam lily | 1.25 |
| | *Dieffenbachia amoena* | Dumbcane | 1.24 |
| | *Tilandsia cyanea* | Pink quill | 1.23 |
| | *Anthurium andraeanum* | Flamingo flower/Tail flower | 1.22 |
| **Korean native plants** | | | |
| | *Nandina domestica* | Heavenly/sacred bamboo | 1.58 |
| | *Dendropanax morbifera* | Korean dendropanax | 1.50 |
| | *Ardisia crenata* | Coralberry | 1.46 |
| | *Laurus nobilis* | Sweet bay | 1.40 |
| **Ferns** | | | |
| | *Osmunda japonica* | Japanese royal fern | 6.64 |
| | *Selaginella tamariscina* | Resurrection fern | 4.84 |
| | *Davallia mariesii* | Squirrel's foot fern | 4.15 |
| | *Polypodium formosanum* | Caterpillar/grub fern | 3.62 |
| | *Pteris dispar* | Amakusa fern | 1.95 |
| | *Pteris multifida* | Spider brake | 1.92 |
| | *Microlepia strigosa* | Lace fern | 1.49 |
| | *Botrychium ternatum* | Winter flowering warabi | 1.42 |
| **Herbs** | | | |
| | *Lavendula* spp. | Lavender | 2.12 |
| | *Pelargonium* spp. | Geranium | 1.87 |

it from 18 to 6 ppm over the same period. In initial tests with a starting concentration of 14 ppm, the spider plant was so effective that a second set of experiments was conducted with a higher starting concentration of 37 ppm; after 24 h the concentration was below 2 ppm. Whilst these values are impressive, Wolverton *et al.* (1984) estimated that a typical 418 m$^3$ energy-efficient home would need some 70 spider plants to purify the air of the 7,000–8,000 μg of formaldehyde a day generated by the building (a figure that includes an estimate of the formaldehyde released by cooking). Wolverton *et al.* (1989a) subsequently showed that corn/mass cane *Dracaena massangeana*, ficus *F. benjamina*, gerbera daisy *Gerbera jamesonii*, pot mums *Chrysanthemum morifolium* and the striped dracaena *Dracaena deremensis* 'Warneckei' could all reduce formaldehyde concentrations in the air of sealed chambers. Wolverton and Wolverton (1993) subsequently published data on 33 plant species for their ability to remove formaldehyde (Table 3.3); the best, the fern *N. exaltata* 'Bostoniensis', was shown to remove the pollutant at the rate of 1,863 μg.h$^{-1}$ and the worst, the succulent *Aloe barbadensi* (*Aloe vera*), ten times less at 188 μg.h$^{-1}$.

Kim *et al.* (2010) examined 86 species of plant for their ability to remove formaldehyde and found a range of plants which were 'excellent' (Table 3.4) – defined as removing at least 1.2 μg.m$^{-3}$ per cm$^2$ of leaf area over a 5 h period. As a group, ferns were best at formaldehyde removal, but herbaceous foliage plants were also effective. As different researchers tend to use different methodologies, it is difficult to make comparisons between them, there tend to be few species in common, and very few assay large numbers of species. It is almost certainly coincidental that the 10th 'best' species identified by Kim *et al.* (2010), *R. excelsia*, is also the 10th 'best' species of 33 identified by Wolverton and Wolverton (1993) and *Anthurium andraeanum* 21st in the list of Kim *et al.* (2010) and 26th in that of Wolverton and Wolverton (1993). However, it is interesting that, as with Kim *et al.* (2010), Wolverton and Wolverton (1993) found their most effective formaldehyde remover was the Boston fern *N. exaltata*. Using data from the US Environmental Protection Agency (EPA) for formaldehyde emissions in new office buildings (0.173 μg.l$^{-1}$) Wolverton and Wolverton (1993) estimated that a 22.32 m$^3$ office with a floor area of 9.3 m$^2$ would contain about 3,916 μg of formaldehyde which could be removed by two Boston fern plants or three 'Janet Craig' dracaenas.

Portable buildings are particularly susceptible to formaldehyde pollution and Dingle *et al.* (2000) examined the potential of plants to remove the pollutant in five such structures in Perth, Australia. Five plants (one from each of the following species: spider plant *C. comosum*, fig tree *Ficus* sp., cast-iron plant *Aspidistra elatior*, dumb cane *Dieffenbachia amoena* and arum ivy *Epipremnum aureum*) were placed as a group in each building. An additional group of five plants was added every second day until 20 plants had been added. Formaldehyde levels were monitored throughout the experiment and showed that plants did not reduce levels significantly from the 856 ppb (parts per billion) starting concentration without plants until the maximum of 20 plants had been added, when an 11% reduction was recorded. At this time the average concentration of formaldehyde was 761 ppb, which was still 661 ppb over the occupational guideline exposure level for non-industrial buildings. The number of plants to achieve this reduction was equivalent to one per 1 m$^3$ and reduction below this level would clearly require an impractical number of plants on horizontal surfaces, although indoor green walls may be of considerable value in such contexts. Dingle *et al.* (2000) concluded that plants were not an effective way of reducing formaldehyde in the air of portable office buildings. The experiment and interpretation suffered from a number of limitations: 1) the short duration of the experiment whereby it was assumed that a two-day period would be sufficient for a group of plants to exert a noticeable impact on a large volume of air with continual outgasing; 2) the assumption that all plants were equal in their formaldehyde extraction efficiency; and 3) that in a non-portable office building or home situation a reduction of 95 ppb might be considered a useful impact, especially as their study of 18 conventional offices showed background formaldehyde concentrations in the range 10 to 78 ppb (mean = 22 ppb).

Xu *et al.* (2010) examined the effect of a biofilter on formaldehyde removal from air. The filter was composed of compost, vermiculite powder and ceramic particles and formaldehyde-laden air entered from below the cylindrical filter and exited at the top. The filter was operated with and without a spider plant grown in the filter medium. The presence of the spider plant changed the profile of accumulation of formaldehyde in the biofilter and Xu *et al.* (2010) suggested that this was because the spider plant was actively removing the chemical via its root system and also that the microbial community surrounding the roots may be involved in formaldehyde removal. The activity was later confirmed by Xu *et al.* (2011). Aydogan and Montoya (2011) examined four species of plant for their ability to remove formaldehyde and showed that the root zone was more effective than the leaves. Kim *et al.* (2008) showed that the ratio of formaldehyde removal in percentage terms by the aerial parts compared with the root zone for *F. japonica* was 61:39 during the day but 2:98 at night, and for *F. benjamina* the ratios were 43:57 during the day and 6:94 at night. Very little (about 10%) of the formaldehyde removed by the root zone was considered to have been removed by the potting medium – the majority was via microorganisms. Aydogan and Montoya (2011) also demonstrated that some plants were better than others in their VOC removal efficiency. Schaffner *et al.* (2002) showed that plants can detoxify formaldehyde directly via enzymatic action (in this case via a 'glutathione-dependent formaldehyde dehydrogenase').

### 3.7.5 The real world and VOC removal

The majority of work on VOCs has been in laboratory situations; studies such as Dingle *et al.* (2000) using buildings are relatively rare and even their study was atypical as most people do not work or live in portable buildings. Wood *et al.* (2006) examined total VOC (TVOC) removal by two species of plant – *D. deremensis* 'Janet Craig' and the peace lily *Spathiphyllum* 'Sweet Chico' – in 30–50 m³ staff offices in three buildings at the University of Technology in Sydney, Australia. The VOCs present in buildings used in the study were identified as: acetone, dodecane, ethanol, ethylbenzene, limonene, methylbenzenes, methylbutane, methylecyclopentane, 2-methylpantane, n-decane, n-hexane, n-pentane, toluene and xylenes. In their first experiment, air-conditioned and non-air-conditioned offices were provided with either three or six plants (plants were individually potted in 300 mm diameter pots) of *D. deremensis* and compared with offices without plants; weekly measurements were taken of TVOC concentrations for nine weeks. The treatments (0, 3, 6 plants/office) were then randomly reallocated and measurements taken for a further nine weeks. The results showed that indoor TVOC levels were far higher than outdoors. When all offices were analysed together there was an obvious trend in TVOC reduction with increasing number of plants, although the difference was not statistically significant. When only offices with initially very high concentrations (>100 ppb) were compared, it was clear

**TABLE 3.5** Impact of *Dracaena deremensis* plants on total VOC levels in Australian university staff offices

| Treatment | All offices TVOC (ppb) | High VOC offices TVOC (ppb) |
|---|---|---|
| Outside | 50±15 | 50±15 |
| Inside 0 plants | 110 ±15 | 190±40 |
| Inside 3 plants | 80±7 | 105±15 |
| Inside 6 plants | 80±7 | 100±10 |

Source: Wood et al., 2006.

that substantial reductions in VOCs had occurred in offices with plants (Table 3.5) and that there appears to be a 'trigger' concentration. The scale of TVOC removal depended on whether the offices were in an air-conditioned building or in a naturally ventilated building. TVOC reductions were evident in the air-conditioned building when plants were present, but the reductions were not statistically significant. In the naturally ventilated building, the impact of plants on TVOC levels was substantial, and statistically significant, with reductions of the order of 75% whether three or six plants were used (decline from 280 ± 120 ppb to 65 ± 10 ppb).

In their second experiment, Wood *et al.* (2006) used smaller *Dracaena* plants and a mixture of the latter with peace lilies. The design was simpler with either six plants (in individual 200 mm diameter pots: five peace lilies + one *Dracaena*) per office or no plants. Subsequent air-quality monitoring showed some interesting deviations from the first experiment. The air-conditioned building had higher indoor VOCs than outside, but the naturally ventilated one did not have levels significantly different from outside. When all offices were compared in the air-conditioned building, TVOC levels did not differ with or without plants – probably because the peace lilies were producing their own VOCs as they were flowering. However, when TVOC concentrations exceeded the threshold of 100 ppb in control offices, offices with plants had 70% lower TVOC levels. In the naturally ventilated building TVOC levels never exceeded the threshold for TVOC reduction. Both experiments showed that plants could remove VOCs under realistic office conditions and that the VOC removal was triggered by exposure to a threshold level of VOCs which then reduced to below the trigger threshold; the latter was later confirmed in laboratory studies by Orwell *et al.* (2006).

### 3.7.6 Dust and particulates

The negative impact of very fine dust particles from diesel vehicles and other combustion sources is well known (see Chapter 2) but exposure to particulate matter indoors results from external sources as well as internal ones such as cooking and fires (Hänninen *et al.*, 2011). The infiltration of particulates from outside to inside varies seasonally, with summer having the highest loads with close to twice the concentration of winter, which experiences the lowest levels; spring/autumn have particulate infiltration in between the two extremes – probably as a result of the temperature-related incidence of window opening in naturally ventilated buildings (Hänninen *et al.*, 2011). Lohr and Pearson-Mims (1996) found that dust loads on horizontal surfaces were reduced when plants were used in indoor environments. House dust has also been shown to contain high molecular weight phthalates used as plasticisers; diisononyl phthalate and diisodecyl phthalate have been tentatively linked with asthma in a study of Norwegian children (Bertelsen *et al.*, 2013). House dust has been shown by Rudel *et al.* (2003) to carry at least 66 endocrine disrupting chemicals (EDCs) including penta- and tetrabrominated diphenyl ethers used as fire retardants and a banned carcinogen fire-retardant precursor 2,3-dibromo-1-propanol, 27 pesticidal compounds and synergists. Fulong and Espino (2013) showed that decabromodiphenyl ether – a fire retardant used in a range of electrical and electronic devices, building materials and fabrics – was present in dust taken from various indoor sites at the Diliman Campus of the University of the Philippines. The compound has been found in indoor dust in studies in Belgium, Canada, China, Hong Kong, Japan, the Philippines, Portugal, the UK, USA and Thailand (see references in Fulong and Espino, 2013). EDCs are compounds that affect hormonal functions, including sexual behaviour, and the development of the male and female reproductive systems; they are also implicated in the initiation of various cancers (Crisp *et al.*, 1998; Diamanti-Kandarakis *et al.*, 2009). Particulate sampling by Lohr and Pearson-Mims (1996) was fairly crude, using 60 mm diameter aluminium weighing dishes to capture air-deposited dust. Each collecting dish was placed inside an open container 70 mm in diameter by 45 mm deep to reduce dust remobilisation and accidental spillage and left for one week before dust loads were estimated. Plants used were primarily smooth-leaved and included Chinese evergreen *Aglaonema* sp., bamboo palm *Chamaedorea seifrizii*, dragon tree *D. marginata*, golden pothos *E. aureum* and peace lily *Spathiphyllum* sp. In a computer lab, with 2% of the room filled with plants, dust load was significantly reduced by 15% by the presence of plants. In a second experiment, an office environment, plants were located around the edge of a room and took up 5% of the space; dust was collected as before (except the surrounding cans were omitted) and was significantly reduced by 21%. Whilst Lohr and Pearson-Mims (1996) did not characterise the particulate size ranges captured in their study, it has been shown by Hänninen *et al.* (2011) that indoor air can include the dangerous $PM_{10}$ and $PM_{2.5}$ particulate fractions and indoor plants are likely to assist in their removal.

## 3.8 Indoor green walls

Green walls inside buildings can be considered as part of a building's air-conditioning system, helping to cool the air and reduce the energy requirement for conventional air-conditioning systems (Franco *et al.*, 2012), remove air pollutants, and reduce noise and potentially dust and microbial loads (see above). Indoor green walls can be constructed using any of the approaches used for external green walls (Chapter 5), but most are of the 'living wall' type using a felt mat or modular approaches that hold plants and rooting media; water and nutrients are typically delivered using an irrigation system (see Chapter 5). The latter requires electrical and water connections. Variants have been developed that are free-standing with water troughs to capture and recycle irrigation water to permit more flexible placing, but still require an electrical connection (Figure 3.6). However, new designs are constantly emerging such as self-contained free-standing room dividers and wall modules that incorporate water tanks and use capillary action, avoiding the need for irrigation pumps or permanent piped water (Figure 3.6). Small systems such as LivePicture need re-watering only once every four to six weeks. Such passive systems have obvious aesthetic value as well as providing some air conditioning and pollutant removal services (section 3.7). More active systems have also been developed which use the walls as biofilters to remove pollutants and to cool and condition the air; such systems promote air flow through the green walls (Franco *et al.*, 2012).

Passive living wall systems are typically attached to structural and internal walls, and damp penetration from the green wall to the building walls is prevented through the use of water-proof membranes. In the case of active green walls, this waterproofing is offset to create a route for the recirculation of the processed air. Typically such internal green walls have the warmer air from the building forced through the rear of the growing media (through the root zone first, then the leaf zone), so conditioned air emerges from the wall into the building envelope (Figure 3.7) (Franco *et al.*, 2012). External green walls can also be used as active air-conditioning systems. Franco *et al.* (2012) showed that the choice of growing medium in living walls had an impact on the amount of water retained, with polyurethane retaining most in their comparison test with polyamide-polypropylene and polyester; the latter held least water and had a more hydrophobic character. As might be expected, resistance to airflow through the substrate was increased by the presence of plants, by the level of saturation with water and by higher windspeeds. The polyurethane substrate had the highest airflow – attributed to its more porous nature; with vegetation the overall airflow rates were (highest–lowest): polyurethane>polyester>polyamide-polypropylene. The crucial feature of the system is, obviously, how saturated the air is when it leaves the green wall: fast-moving air picks up less water, so the relationship between saturation of the growing medium and its porosity combined with the airflow-reducing impact of vegetation are all important contributing factors to the eventual air saturation. In their tests, Franco *et al.* (2012) considered air speeds of 0.25–0.5 m.s$^{-1}$ to be optimum, and in that range found that polyamide-polypropylene was the best at promoting air saturation (whilst also being the most frugal in water consumption) followed by polyurethane and finally polyester. Whilst the work of Franco *et al.* (2012) is helpful, the results may not reflect reality after the walls have been established for some time. Their test sections of substrate were planted up with only one species of plant (*Scindapus aureus*), covering only 50% of the test area, and no details were given of the length of time the plants had been left to grow *in situ*. Green walls are typically grown with a mixture of plant species to add a range of visual textures (growth form, leaf morphology) and colours; this variability, combined with additional leaf and root-zone growth, may strongly impact on water retention and pressure drop.

In a later study, Cañero *et al.* (2012) examined the performance of an approximately 8 m$^2$ living wall installed in the hall of a building at the University of Seville (195.36 m$^3$). Four rooting substrates (two organic: coconut fibre and 'Xaxim' composed of fern root; two synthetic: 'Epiweb' (polyetylentereftalat) and a geotextile (acrylic textile fibre mix on a polypropylene base)) were used with the plants inserted into growing 'pockets'. The wall was planted-up with 24 indoor-tolerant species in 2008 and evaluated in 2009 when the plants had established. Plants were maintained using built-in irrigation and additional lighting was provided. Over a 15-week study period, the wall moderated internal temperatures, reducing them, on average, by 4°C and increasing humidity by 15%. The thermal performance of the substrates was similar, with the geotextile giving the best result. The mean room temperature at the far end of the hall, some 11 m away, was the control (27.1°C); the average air temperatures 30 cm from substrates was 23.8°C for the geotextile, 24.1°C for both Xaxim and coconut fibre and 24.2°C for the Epiweb. The estimated energy saving on air conditioning

**FIGURE 3.6** Indoor green walls. Top left and right: before and after installation of Mobilane LivePicture wall modules at Staffordshire University's Science Centre in Stoke-on-Trent, UK. Middle row left: installation requires only three bolts; middle right: installing a pre-planted cassette into LivePicture; bottom left: indoor living wall showing water reservoir at base (Bin Fen system); bottom right: free-standing room divider (Mobilane); above: LivePicture planted up with arums for colour. © John Dover

was 20%, although it was not clear how far the cooling effect extended from the living wall. The organic media appeared better for plant growth than the synthetics and Xaxim, being composed of fern root, released spores which germinated on its surface. However, the durability of most of the media appeared similar with the exception of coconut fibre, which, despite being reinforced, was clearly degrading. The synthetic media retained less water and dried faster than the organics, leading to higher water consumption and irrigation requirements. The authors note that although Xaxim performed well, it is not recommended due to the endangered nature of the species used in its construction.

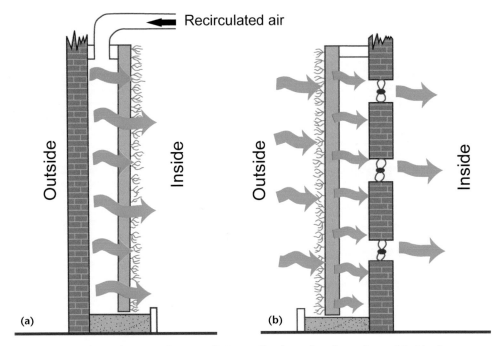

**FIGURE 3.7** Airflow schematics for active living walls where a) walls are located inside the building and indoor air is recirculated through the living wall, and b) walls are located outside the building and air is drawn from the outside of the building, through the living wall and into the building. Redrawn from *Franco et al.* (2012).

## 3.9 Summary

- Vegetation in buildings has a positive benefit on psychological well-being and has a positive effect on work activities.
- Vegetation in buildings has the potential to remove harmful air pollutants – in particular VOCs that contribute to 'sick-building syndrome'.
- There are indications that dust and microbial loads in indoor air may be reduced by plants – but more work is needed.
- Plants, typically in the form of living walls, can be incorporated into the air-conditioning systems of buildings, actively modifying temperature, humidity and pollutant levels.
- Novel wall systems have been developed that allow substantial numbers of plants to be grown in offices and workplaces which would otherwise take up large areas of desk or floorspace.

# 4

# PERMEABLE PAVEMENTS

This chapter will:

- introduce the concept of permeable pavements
- describe the impacts that permeability has on stormwater runoff
- examine the effects of the subsurface microbiology of permeable pavements on pollutants
- examine the way permeable pavements are being used to modify various common urban elements including alleys, parking areas and roads.

## 4.1 Introduction to permeable pavements

If a ground surface is made of impermeable material such as concrete or asphalt, the amount of rainwater deposited on it that is diverted to drains can be around 95% (Schuler, 1994) with the remainder presumably evaporating from the wet surface. In contrast, with a vegetated surface, runoff can be very low, around 10%, with a high proportion being evapotranspired by plants, and the remaining water taken up by the soil (Arnold & Gibbons, 1996). Schuler (1994) gave a specific example of a meadow considered to be 1% impervious with a runoff of 6%. Different proportions of vegetation cover and impermeable surface result in different levels of infiltration (Arnold & Gibbons, 1996) (Table 4.1).

TABLE 4.1 Effect of different proportions of sealed surface on runoff, evapotranspiration and infiltration

| Source | Land cover | Runoff (%) | Evapotranspiration (%) | Infiltration (%) |
|--------|-----------|-----------|------------------------|------------------|
| 1 | Natural vegetation | 10 | 40 | 50 |
| 2 | 10–20% sealed surface | 20 | 38 | 42 |
| 3 | 35–50% sealed surface | 30 | 35 | 35 |
| 4 | 75–100% sealed surface | 55 | 30 | 15 |
| 5 | 100% | 95 | not given | not given |

Sources: (1–4) data from a figure in EPA (1993) cited in Arnold and Gibbons (1996), and (5) Schuler (1994).

TABLE 4.2 Proportion of surface covered by impervious material depending on land use and area (data from USA)

| Land use | Area of plot (ha) | (acre) | Sealed surface (%) |
|---|---|---|---|
| Residential | 0.809 | 2.00 | 12 |
| | 0.405 | 1.00 | 20 |
| | 0.202 | 0.50 | 25 |
| | 0.135 | 0.33 | 30 |
| | 0.101 | 0.25 | 38 |
| | 0.051 | 0.13 | 65 |
| Industrial | | | 72 |
| Commercial and business | | | 85 |
| ★Shopping centres | | | 95 |

Sources: Cronshey (1986) and ★Arnold & Gibbons (1996)

The quality of a running water body is related to the proportion of its catchment that is composed of impermeable surfacing. Arnold and Gibbons (1996) indicate that it takes as little as 10% sealed surface for stream health to become 'impacted', with 'degraded' status after only 30% of a watershed has become sealed. So the use of a permeable material to replace impermeable surfaces should not only improve infiltration, and reduce runoff to sewers and drains, but also improve urban water quality including reducing impacts on aquatic macroin-vertebrates and fish (Schuler, 1994). Contaminants, such as heavy metals (e.g. copper and zinc) and polyaromatic hydrocarbons (PAHs), have been shown to be substantially reduced in water percolating through permeable pavements compared to runoff from sealed surfaces (Boving et al., 2008; Brattebo & Booth, 2003). Of course, the degree to which an area is sealed depends on its function (Table 4.2), with the more intensive non-residential uses (industrial, commercial, business and shopping centres) having the highest levels of sealed surface and residential areas becoming increasingly sealed with decreasing curtilage (plot) area.

Whilst the basic concept of increasing surface permeability by substituting sealed surfaces with permeable ones (to promote infiltration of rain *in situ*) is a simple one, there is a range of approaches that can be employed (Scholtz & Grabowiecki, 2007). These include:

- *loose gravel*: cheap and simple, but with a tendency to scatter and develop ridges – especially if the base underneath the gravel surface is not adequate to bear loads;
- *permeable asphalt*: which looks very similar to ordinary tarmac, except when water is thrown onto it – with one you get puddles, with the other you don't as the water drains through pores in the surface;
- *porous concrete*: no fine aggregates are used so gaps between stones bound by cement provide drainage pathways;
- *permeable blocks*: these look like normal blocks used in hard landscaping but rain can drain through pores in the material and/or around the block edges;
- *wheel tracks*: where only the area driven on is a hard surface and the rest is planted-up;
- *open concrete blocks or plastic blocks*: here the load-bearing surface is covered with open

**FIGURE 4.1** Top: open concrete blocks seeded with grass in the car park of the cable car station at Fuente De, Picos de Europa, Spain; bottom left: interlocking plastic grid system; bottom right different grades and styles of turf reinforcement mesh. © John Dover

blocking into which gravel or a plant-growing medium is inserted – such areas are often planted up with grass (Figure 4.1);

- *reinforcing mesh*: a variation on the open concrete/plastic blocks approach used on grassed surfaces or to retain gravel. A wide range of semi-rigid plastic mesh is available. Such materials when used with grass are often coloured green so that if exposed by hard wear they do not look unsightly (Figure 4.1).

The range of colours, types and shapes of surfacing materials is very wide, so there is no real reason why the majority of hard landscaped surfaces should not now be constructed with a permeable action. Of course the surface type, colour, etc. will be chosen to complement the context, but the longevity and success of the installation will depend on the design of the sub-base, which will in turn depend partly on the surface type, the soil type (clays cause problems as they are impermeable), the load to be supported and the frequency of use. The intensity of rainfall may also determine which system, if any, is most appropriate; Brattebo and Booth (2003) demonstrated extremely good performance of permeable pavement systems in the Pacific Northwest, where their peak rainfall intensity was 7.4 mm.h$^{-1}$, but noted that their experience might not be directly applicable to areas with higher rainfall intensities.

The basic concept of the permeable pavement is capable of further refinement; for example, Tota-Maharaj *et al.* (2010) have experimented with combining permeable pavements with geothermal heat pumps – an approach which seeks to gain additional benefits by recycling the water for non-potable use (e.g. toilet flushing, car washing, garden watering).

Scholtz and Grabowiecki (2007) reviewed the literature on permeable pavement systems and noted that they have the following attributes, which are superior to using roadside drains/gully systems:

- better collection and infiltration of rainfall and stormwater runoff
- evaporation of water stored in pores
- recharge of groundwater
- prevention of pollution (removal of heavy metals and hydrocarbons)
- water recycling
- reduction of Biochemical Oxygen Demand (BOD), Chemical Oxygen Demand (COD), and ammonia levels, in runoff
- reduction in suspended solids
- low maintenance.

As permeable pavements are essentially a modification of surfaces that would normally be impermeable, they can be incorporated in areas where space concerns would mean that additional structures such as swales, used to transfer water to storage basins, could not be used (Starke *et al.*, 2010).

The sub-base of a permeable pavement is typically made up of graded stones, and the air space between them allows water to freely drain (and also acts as a water store during rainfall (Fassman & Blackbourn, 2010)). Sub-bases are usually at least 200 mm deep and are placed on top of a geotextile membrane (such as Terram® www.terram.com/) that is laid on the soil surface to help spread the load (Figure 4.2).

This basic design (DCLG, 2008) will need to be modified depending on the properties of the installation site; for example, if placed on clay subsurface, drainage pipes may need to be

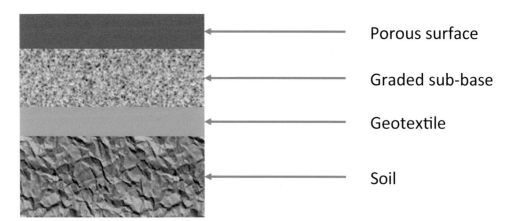

FIGURE 4.2 Schematic cross-section of a permeable driveway, pavement or parking area; there may also be a geotextile above the sub-base (DCLG, 2008; Fassman & Blackbourn, 2010).

incorporated to take the water away to a soakaway, rain garden or pond. In more industrial situations, where there is the potential for pollution, the water may need to be isolated from the soil by an impermeable membrane and drainage pipes laid to direct the water to a water treatment area, which may be, for example, a reed bed, before final infiltration or ponding. Technical specifications have been drawn up for performance of permeable systems; for example, German regulations specify the hydraulic conductivity of such pavements (Starke et al., 2010).

There are some reports of maintenance and construction issues with respect to permeable pavement systems (see below). However, a study of four different systems, and a conventional sealed asphalt surface, installed at the car park at the King County Public Works in Renton, Washington (USA) demonstrated continued permeable function after six years of constant use (Brattebo & Booth, 2003). The four systems used were as follows (primary source: Brattebo and Booth, 2003):

- Two plastic grid systems – one with sand in the grid cells on which grass was grown (Grasspave2®), the other composed of the same cell type but filled with gravel (Gravelpave2®). The plastic grid provides good load-spreading action with very low impermeability (see ISI, 2014).
- A concrete 'honeycomb' system with topsoil and grass in the void (Turfstone®, see Tobermore, 2014), with about 60% impermeability.
- A block paving system (UNI Eco-stone®), 88% impermeable, with the between-block gaps filled with gravel (UNI-Group, 2014).

Visual comparison of the surface of the permeable pavement systems and the asphalt control after six years showed minimal wear and tear, with only a couple of incidences where the Grass-/Gravelpave2® systems had shifted. Water runoff from the asphalt control was rapid and followed the rainfall patterns; in the permeable systems almost all water was infiltrated from the same rainfall events. The only water runoff from the permeable systems that could not be attributed to leaks in water conduits probably came from localised saturation of the Grasspave2® system where heavy rain ran off the parked cars and an incidence with the same system where 121 mm of rain fell in a 72 h period resulting in 4 mm of runoff. This storm was the most extreme event of the six-year study and runoff, as a percentage of the total rainfall event, was a mere 3%. The permeable pavement systems also delayed the peak flow by about 1 h (Brattebo & Booth, 2003).

Permeable pavements which are composed of porous blocks or porous asphalts have been reported as having surface clogging (including at block joints) by sediments and as a result of mechanical action breaking up the pavement surface (this latter where there is frequent heavy traffic). Clogging occurs typically around three years after installation (Scholtz & Grabowiecki, 2007), which can reduce efficiency and result in additional servicing costs to open the surface up again. Grass-based systems may get clogging of the geotextile membrane. Brattebo and Booth (2003) noted that Grasspave2® and Gravelpave2® systems might require some maintenance if exposed to more frequent vehicle movements than typical in their six-year study of a works car park. Boving et al. (2008) found that the geotextile under the porous pavement they were studying appeared to impede infiltration. The need for maintenance of permeable pavements may affect the decision to install such a system in the first place; factors promoting clogging appear to be the proximity to very fine soil, catchment characteristics (including leaf fall), and fine particles from adjacent bitumen asphalt (Fassman & Blackbourn, 2010). Boving *et al.* (2008) found that material brought into a permeably paved car park from the road system on vehicles contributed to clogging, especially sand.

## 4.2 Permeable pavements and street trees

Permeable pavements might be expected to have benefits for street and other trees over impermeable surfaces because the latter are known to provide poor growth conditions. However, recent work by Viswanathan *et al.* (2011) comparing permeable and impermeable concrete systems on root growth by the American sweetgum (*Liquidambar styraciflua*) found no benefit. Both systems resulted in less root growth than a soil-surfaced control, probably due to high $CO_2$ soil concentrations. However, as the experiment was performed on a low permeability, clay-based soil, the response may not be generaliseable to all soils. Given that the soil surface around street trees is frequently capped with a resin-bonded aggregate, information on the impact of such systems is essential (for more information see section 7.11).

## 4.3 Microbiology of permeable pavements

### 4.3.1 Hydrocarbon management (especially in car parks)

One of the major issues in dealing with permeable pavements in areas such as car parks is the potential for pollution by hydrocarbons from lubricating oil reaching the soil beneath the pavement and potentially contaminating aquatic systems. Fortuitously, it appears that a hydrocarbon-utilising microbial community can develop on the geotextile mat below the sub-base (Pratt *et al.*, 1999). The hydrocarbons are metabolised into sugars and act as a carbon source for the community which, in composition, can be extremely diverse consisting of bacteria, fungi and bacteria-eating protozoa (Coupe *et al.*, 2003). Oil conversion efficiency can be extremely high; Coupe *et al.* (2003) suggest up to 98.7%. The degrading community does require aerobic conditions however (Newman *et al.*, 2006; Pratt *et al.*, 1999), so maintaining good air spaces in the sub-base is probably critical and may be facilitated by the protozoa in the oil degrading community (see refs in Coupe *et al.*, 2003). Whilst new permeable pavements can be inoculated with a commercial starter culture, over time the microbial community changes from that in the inoculum and it appears that after about 39 weeks an un-inoculated and inoculated pavement are equally good at degrading hydrocarbons (at least under laboratory conditions) (Newman *et al.*, 2006, 2002). After five weeks protozoan communities, whilst initially richer for the first four weeks in inoculated systems, became richer in un-inoculated systems (Coupe *et al.*, 2003). The oil-degrading community may need inputs of nitrogen and phosphorus to function (Newman *et al.*, 2006), and fertilisers are often added, but Rushton (2001) in her study of a car park at the Florida Aquarium in Tampa showed that rainfall was a good source of nitrogen (but not phosphorus). Tota-Maharaj and Scholz (2010, 2013) pointed out that dog faecal material would be a likely component of 'real-world' water entering permeable pavements and that this would carry with it a range of nutrients. Their experimental permeable pavement systems substantially reduced Biochemical Oxygen Demand by 99%, ammonium-nitrogen by 97% and orthophosphate-phosphate by

95% (Tota-Maharaj *et al.*, 2010). Infiltrated water from permeable systems compared with runoff from a sealed asphalt surface showed significantly lower levels of lubricating oil in a six-year car park study (Brattebo & Booth, 2003). Newman *et al.* (2004) noted that different geotextiles perform differently in their efficiency of oil retention, and also that membrane type becomes less important over time.

### 4.3.2 Microbiology of permeable pavements: pathogens

Whilst the ability to host a hydrocarbon-degrading microbial community is an extremely positive aspect of permeable pavements, Scholtz and Grabowiecki (2009) articulated fears that the real-world contaminants in water which infiltrated into permeable pavements, such as dog faeces, might lead to contamination and multiplication of bacteria associated with public health risks – especially if the water was recycled for general non-potable use. The risk of microbial contamination in permeable pavements when combined with a geothermal heat pump system might be considered more risky than conventional permeable pavements as the former would have warmer subsurface conditions than the latter, promoting microbial growth. Their experiments (using plastic 'wheelie bins') thus used material sourced from gully pots (the 'trap' part of a drain which holds water from, for example, road runoff), which would contain a range of contaminants, mixed with de-chlorinated tap water and fresh dog faeces. Whilst potentially pathogenic organisms were recovered from the permeable pavements, the conclusions were not particularly helpful in determining real risk as organisms were not identified to species. The experiment was continued and refined, however, using more advanced techniques of microbial identification using DNA analysis (Tota-Maharaj *et al.*, 2010). Standard bacterial analyses from culturing inflow ('runoff') and outflow (infiltrate) on specialist agars (for details, see Tota-Maharaj *et al.*, 2010) showed very high reductions in bacterial contamination, especially for *Escherichia coli*. The more exact DNA analyses carried out on some bacterial colonies derived from infiltrate did not contain the classic human health-related species that might have been expected with dog faecal contamination of water (*Salmonella*, *E. coli*, faecal *Streptococci*) and *Legionella* – this latter was assayed due to the warm nature of the system Tota-Maharaj *et al.* were investigating which combined a permeable pavement with a geothermal heat pump (Tota-Maharaj *et al.*, 2010). Tota-Maharaj and Scholz (2010) showed that, on average, the permeable pavement system removed 98.6% of total coliforms, *E. coli* and faecal *Streptococci*. Despite these impressive results, Tota-Maharaj and Scholz (2013) noted that potential adopters of permeable pavement systems used to process and recycle water for non-potable domestic use might be sufficiently risk averse to consider the reductions they had previously demonstrated insufficient. They therefore experimented with a disinfection phase whereby remaining microbial contamination could be removed and demonstrated that a combination of suspended titanium oxide powder ($TiO_2$) and UV light could completely remove total coliforms, *E. coli* and faecal *Streptococci* after 80–100 minutes of exposure.

## 4.4 Permeable pavements and the heat island effect

As Starke *et al.* (2010) note, the evapotranspiration component of vegetation in urban areas not only contributes to the reduction in runoff, but also has a cooling effect. The opportunities for evaporation of water from permeable pavements (and transpiration from any plants

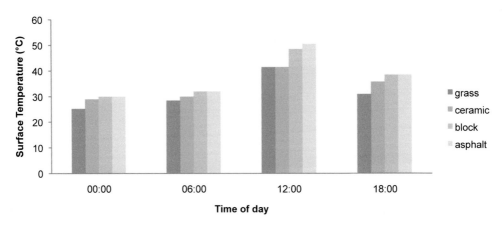

FIGURE 4.3 Surface temperatures in °C of different ground cover materials: grass, porous ceramic, porous block, impermeable asphalt on a warm summer's day in Tokyo with air temperature of 35°C at 12:00. Re-drawn from data in Figure 5 of Asaeda and Ca (2000).

incorporated in some designs) has the potential to reduce their surface temperature compared to similar sealed surfaces and hence reduce the amount of heat stored during the day and subsequently released at night. As a result, permeable pavements have the potential to reduce the heat island effect.

Asaeda and Ca (2000) explored the heat-reducing potential of permeable pavements in Tokyo in 1994 and 1995 using ten 2 x 2 m plots composed of a number of materials including a porous block pavement, normal (impermeable) asphalt, a permeable ceramic and grass. The porous blocking had large pores that water drained through rapidly, whilst the ceramic had a range of pore sizes, some of which were very fine, and retained water within its structure for long periods (days). They recorded surface and below-ground temperatures (and other parameters) but here we examine just the surface temperature data. The data presented were from a 30 h period on 9–10 August 1994 (Figure 4.3). The peak air temperature was 35°C at 12:00. At the same time-point the porous block was over 48°C, similar to that of the normal asphalt which was just in excess of 50°C; the ceramic and grass were nearly indistinguishable with a surface temperature of about 43°C (Figure 4.3).

The take-home message from this study was that grass was always the coolest surface, but that a porous structure that retains some water for evaporative cooling comes a pretty close second and is sometimes as cool as grass. Impermeable surfaces and porous structures that do not retain water have similar thermal properties and are warmer. Asaeda and Ca (2000) suggested that the good performance of the ceramic material might also be, in part, because some of the pores were fine enough to act as capillaries taking up water from the substrate below it and enhancing its evaporative potential.

Starke *et al.* (2010) compared evaporation from two concrete block systems – one permeable, the other impermeable – and found that overall evaporation was 16% greater using the permeable system, and that the evaporation took place over several days compared with the relatively fast evaporation of the small amount of water on the impermeable surface. In the former, the local air quality was moderated over a longer period of time than the latter.

At the other end of the temperature scale, a study in Lulea, northern Sweden, demonstrated that permeable pavements are less prone to freezing, and thaw more quickly, than impervious pavements (Bäckström, 2000). However, the capillary action that seems to work so well in surface temperature reduction work by Asaeda and Ca (2000) may not be as helpful as it may seem when used under cold conditions, as capillarity is deliberately avoided in road surfacing to prevent damage through freeze–thaw action (Starke et al., 2010).

## 4.5 Front gardens

Perry and Nawaz (2008) used aerial photographs to track the change in surface sealing of domestic gardens in a 1.16 km$^2$ area of Leeds in the UK. They found a 13% increase over a 33-year study period (1971–2004), three-quarters of which was attributable to the paving-over of front gardens, presumably for off-road car parking. Smith et al. (2011b) showed that front gardens in Greater London made up 25% (9,400 ha) of the area of garden surrounding houses compared with 63% for back gardens from a total estimated area of 37,900 ha (24% of Greater London). Of the front garden area, 63% (5,900 ha) was estimated to be 'hard surfaced' in 2006–2008 and is probably primarily impermeable (sealed) surfacing; this contrasts with an estimated 4,200 ha (45%) in 1998–1999.

On 1 October 2008 a new amendment to the Town and Country Planning (General Permitted Development) Order 1995 came into force in the UK (DCLG, 2008), as a result of the trend of converting the front gardens of houses into parking areas for cars and the effect such sealed surfaces had in increasing the pressure on sewer systems – with increased potential for flooding and the transport of car-related pollutants.

Before this change property owners could convert gardens to parking areas without needing planning permission under 'permitted development rights'; following the change this was still possible, but required planning permission unless the parking area was permeable to water, or was less than 5 m$^2$ in area, or that water from the sealed surface was directed to some form of soak-away such as part of the garden (DCLG, 2008). This regulation change was a response to the substantial flooding that many areas of the UK had experienced in 2007, and a recognition that further climate change would exacerbate the situation. Interestingly, the regulation also affects replacement driveways (DCLG, 2008). At the watershed level, increasing levels of sealed surface also impact on the ecology of freshwater systems with negative effects apparent with as little as 10% sealed surface and severe effects being apparent at 30% (Stone, 2004).

Before the change in regulations in the UK, the concept of sustainable urban drainage would probably have had a very limited 'circulation' in the country, restricted to those charged with designing, implementing and building urban infrastructure; after the change, knowledge of the concept (if not the terminology) potentially expanded to anyone with a front garden who wanted to park their car on it! The Royal Horticultural Society in the UK has even produced a guidance note explaining the issues and giving example layouts whereby car parking, drainage and greenery can be integrated in front gardens (RHS, undated). Stone (2004) noted that in Madison, Wisconsin, planning regulations for residential areas at the time had the unfortunate effect of resulting in larger areas of sealed surface than necessary compared with alternatives. He also noted that typical driveways in Madison made up 20% of the sealed surface of a residential plot and suggested the approach that was implemented in the UK four years later. Interestingly, Stone (2004) suggested a tax on sealed surfaces, which would have impacted on new driveways and also promoted retrofitting of permeable

TABLE 4.3 Attributes of common approaches to creating permeable driveways to avoid need for planning permission in the UK (DCLG, 2008)

| | Planning permission needed? | Installation | | Use on slopes | Environmental features | | | Maintenance/ Management issues | Wheelchair accessibility |
|---|---|---|---|---|---|---|---|---|---|
| | | Cost | Construction | | Visual | Pollution control | Local microclimate | | |
| Loose gravel | None | Inexpensive | Simple | No | Integrates well with garden design | | | Gravel scatters and ruts form, will need sweeping | Unsuitable |
| Wheel tracks | None, if less than 5 m² | Inexpensive | Simple | | Integrates well with garden design | | | Simple but regular | |
| Reinforced grass | None | Moderate | Depends on system | | Integrates well with garden design | Absorbs pollutants, dust, reduces noise | Reduces local temperature | Regular mowing. Grass will die if vehicles left in same place for long periods. | |
| Reinforced gravel | None | Moderate | Depends on system | OK, but see maintenance | Integrates well with garden design | | | Will need sweeping to avoid gravel spreading especially on slopes | |
| Impermeable block paving with gaps between, permeable blocks, permeable asphalt | None | Expensive | Experienced/ specialist contractor needed | | Wide variety of block types and colours | | | Low maintenance, long life | |

surfaces. In some ways the UK approach is more egalitarian, as a tax has the potential to impact low-waged families more than the well off. However, a tax might be expected to have an immediate stimulatory impact on reducing the levels of existing residential sealed surface whereas the UK approach will take much longer as existing sealed surfaces wear out and are replaced with permeable ones (the regulation requires replacement driveways to be permeable).

The nature of the sub-base is quite critical to the success of a permeable pavement and in the context of front gardens in the UK the Environment Agency has made specific recommendations (Table 4.3). It is important that the sub-base does not contain fine material that will be compacted and fill the gaps between the spaces of the larger stones. In the UK the 'old' type of hardcore used in driveways was called 'MOT Type 1' and this should not be used – two suitable alternatives being '4/20' and 'Type 3' (DCLG, 2008).

## 4.6 Alleys

Chicago has a large network of narrow accessible alleyways (3,058 km or 1,900 miles) and has an active programme to convert them from sealed surfaces to permeable. *The Chicago Green Alley Handbook* (CDT, undated) has been created to explain and promote the concept. For Chicago this is particularly important as many of the alleys were not designed with connections to the city's sewers and frequently flood. Making the alleys permeable to water is seen as a less expensive flood control solution than upgrading with standard drainage solutions whilst also providing significant environmental benefits including:

- recharging groundwater;
- reducing local temperatures (new surfaces are designed to have a higher albedo);
- using recycled material in the sub-base reducing pressure on landfills;
- reduction of light pollution.

Because lighting systems are upgraded at the same time with modern energy-efficient, low-glare, 'dark sky compliant' systems, light pollution is reduced with the tantalising suggestion that it will be possible 'to see the stars' again. Citizens are also encouraged, as part of the programme, to:

- create rain gardens to prevent runoff from gardens into the alleys;
- carry out recycling including composting organic material;
- plant garden trees to shade the alleys;
- use native plants and trees in their gardens that are adapted to local (northern Illinois) conditions to cut down on the need for garden watering;
- collect roof water in barrels to water the garden;
- use permeable paving materials for paths and parking areas;
- install green roofs on buildings and especially garages;
- create water retention ponds for stormwater control;
- create vegetated swales (which they also call 'bioswales') (CDT, undated).

## 4.7 Roads

Fassman and Blackbourn (2010) monitored a 395 m$^2$ experimental road section (including 200 m$^2$ of permeable pavement and 195 m$^2$ of path, driveway and grass) and an adjacent conventional asphalt area of the road (850 m$^2$ including associated features) along Birkdale Road in Auckland, New Zealand between 2006 and 2008. This time-period included 81 storms amounting to 1,128.1 mm of rain. The permeable pavement was constructed of impermeable blocks with 10 mm joints between them filled with aggregate chippings. The design specification was for a ten-year life and capability of coping with a two-year 24 h 'annual recurrence interval' peak flow. Downslope of the permeable section, a heavy-duty perforated drain pipe was positioned to remove underdrain water (i.e. water that passed through the permeable pavement and sub-base). The results showed that surface runoff, over 24 h periods, from the conventional road section mirrored the peak rainfall events, but in the case of the permeable pavement the feature measured (the underdrain discharge) was considerably reduced in intensity with a substantial lag in the case of small (25 mm) and two-year storm events (63 mm) progressively becoming shorter with more intense storms (five-year events (98 mm) and ten-year events (152.3 mm)). Some rainfall was probably held in the sub-base air pockets, some evaporated off, and some infiltrated into the relatively impermeable (clay) soils, all contributing to a reduction in the volume of water in the underdrain discharge. Essentially the permeable pavement was able to replicate pre-development conditions – an aim of the local Department of Environmental Resources Programs and Planning Divisions to reduce impacts on stream integrity and habitats (Fassman & Blackbourn, 2010).

## 4.8 Swales and strands

Swales are typically shallow vegetated channels; strands are larger channels that collect water from a number of swales. As with permeable pavements, they are often included as components of sustainable urban drainage systems frequently leading to retention ponds. They not only convey excess water away from sealed surfaces, but, as they are vegetated, they also act as infiltration structures and can remove pollutants (Rushton, 2001). Rushton (2001) demonstrated that a permeable-surfaced area of a car park, coupled with a swale, could reduce runoff by 50% compared with an asphalt-surfaced area with no swale; a swale reduced runoff by 32% even when coupled with a sealed parking area. However, runoff reduction was best during 'small storms'; for large rain events there was less of a benefit. A number of small structures have been used or suggested to help reduce runoff; an early example was the use of small sunken vegetated islands in car parks, whereby water would run off the car park surface into the basins for infiltration and treatment (Bitter and Bowers, 1994 cited in Arnold and Gibbons, 1996).

## 4.9 Summary

- Permeable pavements can reduce the overall volume of water reaching drains, and increase the time between rainfall and peak loads reaching those drains.
- Permeable pavements can be made out of a range of materials which incorporate vegetation or inorganic materials which allow infiltration.

- Because of the variety of materials that are available, the majority of ground surfaces could be made permeable, though there may be maintenance issues.
- The biological component of permeable pavements is not restricted to vegetation, but includes a subsurface community of microorganisms that can degrade pollutants.

# 5

# GREEN WALLS

This chapter will:

- define the different kinds of green wall
- identify the structural issues that relate to different green walls
- describe green wall components/media
- cover planting and maintenance
- identify biodiversity and other ecosystem services
- consider the issue of sustainability.

## 5.1 Introduction

The green wall concept is essentially very simple: instead of growing plants on roofs or on the ground where they take up space, grow them up unused space on walls. Just in terms of aesthetics, green walls are far superior to green roofs as they are visible to a much greater proportion of the population. As for wider environmental benefits such as reducing the heat island effect and surface temperatures (Wilmers, 1988), there is a far greater opportunity for delivering some environmental services with green walls than with green roofs. Köhler (2008) estimated that walls represented double the ground footprint of a building, in urban centres, although this seems likely to be on the conservative side, with Peck *et al.* (1999) estimating it at between four and twenty times the roof area depending on the height of the building, and much of the space on flat roofs is taken up with other infrastructure, reducing growing opportunities (Cheng *et al.*, 2010). On the negative side, environmental conditions become increasingly harsh for wall vegetation at higher elevations (Peck *et al.*, 1999) and at lower elevations salt spray during winter may cause damage (Whittinghill & Rowe, 2011). Green walls have been known and grown for thousands of years; Köhler (2008) identifies vines grown in palace gardens in the Mediterranean some 2,000 years ago as the earliest deliberately planted examples. Whilst many green walls establish naturally, there is increasingly global interest in incorporating them in building and city design because of their multifunctional benefits (e.g. Köhler, 2008; Loh, 2008;

**FIGURE 5.1** In this student accommodation development (The Minories, near Tower Bridge, London) the 191 m² Biotecture Living Wall has been used as the external rain cladding of the building allowing the costs of the green wall to be offset against using conventional cladding. © Biotecture Ltd

Sheweka & Magdy, 2011) and because retrofitting green walls is relatively straightforward compared with green roofs.

Most early published work on green walls was in German with ecological, rather than purely botanical, publications only dating from the end of the 1970s (Köhler, 2008). Köhler (2008) tracked green wall publication numbers from the 1850s to 2005 and showed an explosion in publications in the 1980s and 1990s; since that paper a flurry of studies devoted to green walls has been published in English (Chiquet *et al.*, 2013; Francis & Lorimer, 2011; Franco *et al.*, 2012; Lindberg & Grimmond, 2011; Melzer *et al.*, 2011; Perini *et al.*, 2011b; Sternberg *et al.*, 2011b; Viles *et al.*, 2011; Wang *et al.*, 2011; Whittinghill & Rowe, 2011; Xia *et al.*, 2011). It is easy to get swept away with enthusiasm for green walls, but it is perhaps worth remembering that they are not cost free. Establishment costs will vary with the different types of wall (see below for different types) and can be completely free if natural colonisation of existing structural walls is used, or can be in the many hundreds of € per square metre for the most sophisticated living wall systems. Perini *et al.* (2011b) give estimated costs for a range of systems (Table 5.1). Much of the technology for the more complex systems is relatively new and low volume; it is likely that costs will decrease over time. However, substantial cost savings are possible

**TABLE 5.1** Example costs for different types of green wall system (mostly after Perini *et al.*, 2011b)

| System type | Approach | Materials | $€.m^{-2}$ |
|---|---|---|---|
| Direct | Plants growing on wall surface, planted in ground | Climbing plants | 30–45 |
| Indirect (façade) | Plants growing up support materials, planted in ground | Climbing plants + support e.g. stainless steel wire | 40–75 |
| Indirect (façade) | Plants growing up support materials, planted in planter boxes filled with soil. Planters can be at different heights up a building to increase coverage | Climbing plants + • galvanized steel support • coated steel • HDPE | 600–800 400–500 100–150 |
| Living wall | Plants growing in a module subdivided into smaller planters. Plants rooted in a soil or similar substrate (Greenwave system) | Plants + HDPE planter module + support structure | 400–600 |
| Living wall | Plants growing into panels containing hydroponic stonewool substrate (Biotecture system) | Plants + panels + support structure | 500–1000 |
| Living wall | Plants growing rooted directly in a panel which provides support with nutrients and water supplied wholly hydroponically (Fytowall system) | Plants + modules using a foam-based substrate in a steel meshwork + support structure | 750–1200 |
| Living wall | Plants growing in pockets between two-layers of a felt mat medium; hydroponic | Plants + felt layers + support structure | 350–750 |

NB: living wall systems have more flexibility in the range of plants used than direct and façade systems, but will require complex automatic irrigation and in-line fertiliser control. Planter box systems will also need irrigation and fertiliser such as used for the more complex living wall systems if mounted above ground level, but could be manually serviced if at ground level only.

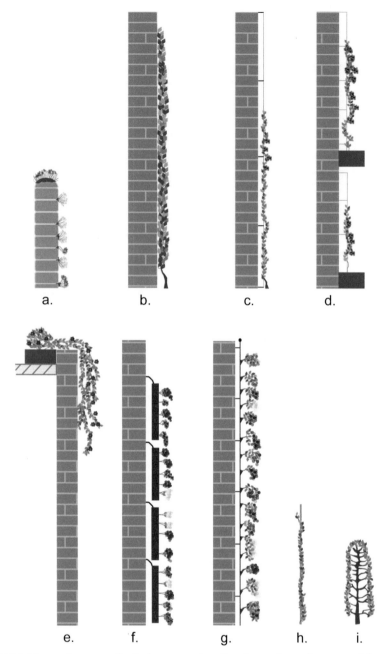

FIGURE 5.2 Types of green wall: a) plants growing directly on the wall or rooted into cracks; b) climbing plants adhering directly to the wall surface; c) climbing plants growing up a support structure not directly on the wall; d) climbers growing in planters typically on support structures but could also use plants that adhere directly to the wall (irrigated); e) plants cascading over the top of a wall from a planter (irrigated); f) modular 'living wall' where modules contain a growing medium (irrigated); g) hydroponic 'vertical garden': plants grown in felt pockets either pre-formed or pockets cut into felt layers *in situ* (irrigated); h) free-standing green screen – plants grown up a metal mesh support; i) hedge (after Ottelé *et al.*, 2011).

if the green wall is designed as part of the external cladding of the building rather than as an 'add-on' (Figure 5.1).

Maintenance costs also need to be considered: depending on the wall type it could be a quick annual survey to check that undesirable plants have not established on the wall, or that climbing plants have not obscured windows, got into the eaves or blocked gutters (Köhler, 2008), perhaps nothing more than a simple matter of a bit of cutting back. As wall sophistication increases, so will management costs; for example, for irrigation and fertilisation of living wall systems, as well as checking and managing plant growth, integrity of substrate, etc. Set against this will be the many positive values of the walls, but it is best to know exactly why you are creating green walls, what attributes you value in them (or why you are installing them), and their potential establishment and maintenance costs.

## 5.2 Types of green wall

### 5.2.1 The basics

There are several different types of green wall (or 'vertical greenery system' (Wong *et al.*, 2010a)); the simplest is a building or curtilage wall that plants have colonised naturally; the most complex is one which is composed of a pre-planted modular framework of subdivided plastic partitions, fed water and nutrients through a network of pipes, and problems with irrigation detected by sensors and transmitted to the installation/maintenance company so engineers can be dispatched to repair the fault (Figure 5.2). Naturally the continuum between these extremes also reflects a continuum in establishment and maintenance costs and also the complexity and range of plants that can be incorporated in designs. Perez *et al.* (2011b), like others (e.g. Loh, 2008), have attempted to bring some order to the classification of green walls to enable like-for-like comparisons of, for example, energy conservation values. Unfortunately, they have omitted some categories and their approach is not entirely satisfactory as they class vegetation growing up a range of support systems, which they call 'double-skin' green walls, as 'extensive', grouping them with climbers which adhere directly to built walls.

Ottelé (2011) developed a classification for 'vertical green' which made a distinction between vegetation growing directly on the wall (direct greening) and that separated from the wall surface (indirect greening) – a useful distinction (see below). However, Ottelé (2011) identifies indirect greening by having an air gap between the wall and the vegetation – given that this air gap can be filled with insulation, this distinction is unduly restrictive and I will use the convention that the vegetation simply does not contact the wall surface directly. 'Direct' and 'indirect' can be useful terms, but unfortunately terminology overlaps so 'green façades' can include both direct and indirect approaches: compare Figure 5.2b, a direct greening approach, with Figures 5.2c and d, indirect approaches, and with 5.2e, somewhat intermediate as the vegetation may touch the wall but will not necessarily adhere to it.

## 5.2.2 Direct greening

Structural walls made of stone, brick or other material where flora are:

- either rooted directly in or on the wall (Figure 5.2a, Figure 5.3)
- or surface climbers: rooted in the ground, but adhere to the vertical surfaces directly through adhesive pads (Figure 5.2b, Figure 5.3) – a form of green façade.

These two forms can develop naturally, or can be deliberately planted. Whilst flora will naturally colonise the surfaces of buildings and free-standing walls such as boundary walls, given suitable conditions and enough time, there is little indication that walls intended to be structural components of buildings which have been deliberately created with pockets for plants to be grown in them has been taken up as a concept. An exception to this is the 'Growcrete' concept which facilitates growth of plants on exterior surfaces of buildings – see below. Free-standing walls, such as boundary walls and retaining walls, may well be designed with plant 'pockets'.

FIGURE 5.3 Direct greening of walls. Left: Red valerian *Centranthus ruber* plants directly rooted in a wall in Slapton, Devon, UK; right: Boston ivy *Parthenocissus tricuspidata* adhering to a building in Durham, UK. © John Dover

## 5.2.3 Indirect greening

These are typically structural walls, composed of the same material as for directly greened walls; here the flora are grown on support infrastructure of various kinds which are fixed to the wall but where the vegetation does not touch the wall surface:

- Green façades: rooted in the ground or in planters. Plants twine around support materials fixed to the vertical surfaces but are offset from them (Figure 5.2c, d; Figure 5.4), these may also have roof planters to 'cascade' vegetation down the walls from above (Figure 5.2e).
- Living walls – modular type: rooted into soil or other substantial substrate and water delivered through an irrigation system, along with additional nutrients (Figure 5.2f; Figure 5.4).
- Living walls – mat type: rooted into a thin, synthetic, 'felt-like' blanket and maintained wholly hydroponically (Figure 5.2g; Figure 5.4).

In all cases the structures are separated from the supporting wall (by an air space or insulation) and, in the case of the latter two, by a waterproof membrane.

## 5.2.4 Free-standing 'curtain' walls and hedges

Another class of green wall can be identified: where the vegetation is supported by a framework, temporary or permanent, which is designed to be free-standing. An example of this approach is the green wall in Shepherd's Bush, London, which is used to screen the Westfield Shopping Centre (Figure 5.5) and uses the ANS modular system. Such walls can use directly adhering plants, support structures for twiners, or living wall approaches as the underlying support structure can be custom designed. Free-standing green walls have a wide range of applications and some designs have the advantage of taking up very little ground space (Figure 5.2h and Figure 5.6).

There is, of course, another type of green wall: a hedge, which provides its own support mechanism (Figure 5.2i; Figure 5.7). A hedge can also be grown against a wall as well as being completely free-standing. There is surprisingly little known about the value of hedges in the urban environment (Chiquet *et al.*, 2012), which contrasts strongly with the wealth of knowledge about their country cousins (e.g. Barr and Petit, 2001; Dover, 2012).

FIGURE 5.4 Indirect greening approaches. Top left: Green façade but with support infrastructure for plants so they are held away from the wall © John Dover; top right: plastic modular unit holding organic substrate © John Dover; bottom: Patrick Blanc felt-based system (on a road bridge in Aix-en-Provence, France) © Caroline Chiquet.

FIGURE 5.5 Two faces of the same green wall © John Dover. The free-standing 'curtain' green wall at the Westfield Centre, Shepherd's Bush, London, has different flora either side reflecting the requirements of the different aspects, but also the need for high visual quality on the 'shopping side' of the structure (top). The plants for this particular wall are grown in the ANS modular system. The ridged stonework at the base of the wall on the shopper's side is actually a water feature designed to stop passers-by picking vegetation, but when this photograph was taken the water had been turned off and the structure was being used as welcome seating and picnicking space by shoppers.

FIGURE 5.6 The installation of a Mobilane 'Green Screen' at the Staffordshire University campus in Stoke-on-Trent, UK. The structure is free-standing, requires little space, being extremely narrow, and the root system is contained in a 'U'-shaped bend in the supporting meshwork. © John Dover

FIGURE 5.7 Hedges can be completely free-standing to mark boundaries or divide areas into intimate plots or can be grown against a wall – evergreen species can be very effective in hiding ugly structures. © John Dover

## 5.3  Directly greened walls: rooted directly on the wall

### 5.3.1 Introduction

These walls are effectively direct analogues of natural vertical features such as cliff faces and ravines; they are also subject to the same limitations as these natural features (Larson *et al.*, 2000, 2004). A number of publications examine the severe environmental constraints imposed on plants that colonise and root into walls (e.g. Woodell, 1979; Gilbert, 1996; Francis, 2010) and the 'urban cliff hypothesis' compares the conditions in extreme natural conditions with that pertaining in urban areas (Lundholm, 2006). Essentially these constraints break down into structural, environmental and biotic considerations.

### 5.3.2 Structural issues

Unbroken, sheer cliff faces are inhospitable to plants because gravity is a great motivator: without any cracks or ledges plants cannot find a purchase. In the same way, a new wall made of natural stone, brick, concrete or plastic material with a fine surface and well mortared presents little opportunity for colonisation. The acidity–alkalinity of the building material will also influence plant colonisation, especially by lichens and mosses, although Gilbert (1996) considers this to be of minor importance for most other species. In the case of both cliff and wall, time and erosive processes (e.g. alternating heat/cold cycles, frost, rain (with dissolved acids), wind-blasting with sand or other particles, and mechanical damage) eventually cause cracks to appear, small ledges to develop, and aggregations of dust or eroded material to accumulate (Woodell, 1979), providing plants with both colonisation niches and substrate. Lisci *et al.* (2003) summarise the process of plant succession once erosive processes have provided colonisation niches (Figure 5.8).

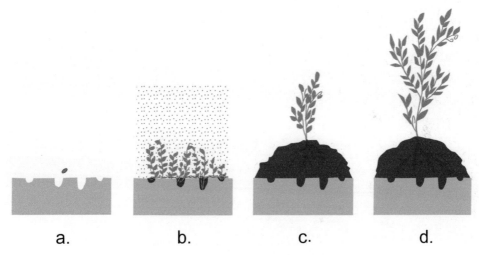

a.                      b.                      c.                      d.

FIGURE 5.8  The stages of wall colonisation following initial disruption of the wall surface: (a) a moss spore falls on porous or damaged stone; (b) it develops into a plant which accumulates atmospheric dust; (c) the moss-captured dust acts as growing medium for a plant seed which germinates and grows; and (d) the moss does not damage the substrate, but the roots of the growing plant penetrate it. Redrawn from Lisci *et al.* (2003).

FIGURE 5.9 Walls in the village of Slapton, Devon, UK. Clockwise from top left: boundary (curtain) wall showing plants colonised in aggregated material at the base of the wall, specialised wall flora such as navelwort on the main surface of the wall and ivy and fern on the top of the wall; capstone colonised by lichen, ferns and mosses; top of wall colonised by brome grass; soil-retaining wall completely covered in a diverse flora and especially red valerian. © John Dover

With walls, the mortared joints are typically the weak points where erosive processes act first, unless the construction materials are naturally friable or of poor quality. It is general practice to make the mortars of a similar strength as the building materials (Busby, 1960) or slightly weaker. Of course, not all walls have mortared joints, some being fitted 'dry'. From about 1870 lime-based mortar, which is susceptible to weathering, started to decline in use (Gilbert, 1996) and since 1930 most new buildings in the UK have been made with much stronger non-lime mortars (BDA, 2001). This change in bonding material means that plant colonisation is less likely on new walls, or at least will take considerably longer (Woodell, 1979): of the order of 40–80 years (Gilbert, 1996). There are three main classes of mortar: lime as the bonding agent plus an aggregate such as sand; cement mortars typically using ordinary Portland cement and sand; and mortars containing both cement and lime as bonding agents with sand as aggregate (Busby, 1960). In London, it appears that some species are now restricted to very old walls because of the change to non-lime mortars (Gilbert, 1996). Repairs to soft, old or easily eroded structures, if done with inappropriate mortars (e.g. cement rather than lime), can result in the mortar eventually protruding from the eroding face, creating colonisation ledges. Walls are often constructed with unintentional built-in colonisation ledges, typically at the tops of boundary or retaining walls where there are capstones (Figure 5.9), but often near the ground where there are sloping 'knees'. Vertical walls are more challenging for plants to colonise than sloping walls, and the flora of vertical

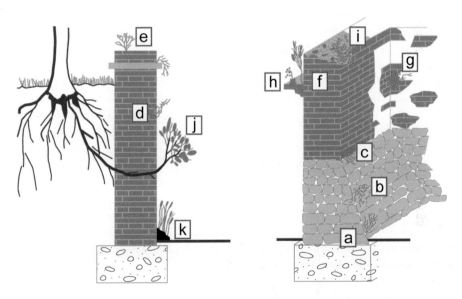

FIGURE 5.10 The main microsites for plant growth. Gaps in stone/brickwork at: (a) ground level, (b) on a slope, (c) where two different materials meet, (d) on a vertical surface, (e) on a horizontal surface, (f) at the junction of vertical and horizontal faces, (g) at the junction of two vertical faces. Additionally, growth may take place on: (h) a porous horizontal surface (see further Figure 5.8) and (i) on broken or damaged surfaces. Also, rhizomes may penetrate retaining walls (j), and (k) soil accumulations at the base of a wall may provide growing media. Adapted from Lisci *et al.* (2003).

walls is much more distinctive (Gilbert, 1996). Aggregations of soil particles can also build up at the base of walls where a wider variety of plants can colonise (Figure 5.9). Lisci *et al.* (2003) produced a nice diagram of potential wall colonisation sites, redrawn and modified here as Figure 5.10.

### 5.3.3 Environmental issues

If structural issues were the only factors involved in plant colonisation of walls, then almost all older walls would be covered in plants! The best place I know for plant colonisation of stone walls in England is the pretty coastal village of Slapton in south Devon (Figure 5.9) and even Slapton hardly approaches that ideal! The local abiotic environment of the wall is crucial, and in particular that which relates to moisture (Lisci & Pacini, 1993). There are several ways that plants on walls can obtain moisture: directly from rain, from highly humid environments (e.g. Francis & Hoggart, 2008, 2009), from seepages or water spilling over wall surfaces, from capillary action from the wall foundations, and from lateral transmission through the wall if it acts as a soil-retaining wall (e.g. as part of a garden) (Lisci & Pacini, 1993; Woodell, 1979). Different parts of the wall will experience different moisture regimes in any given situation. If water comes from capillary action, it will not penetrate very far up the wall, and not at all if a damp-course is installed. Walls are unlikely to be uniformly inundated by spills or seepages, sheltered parts of walls may not receive as much moisture as more exposed areas, etc. A soil-retaining wall (Figures 5.9 and 5.10) is perhaps that which has most consistent access to moisture (Woodell, 1979), and even then the upper part of the wall will have less as it is likely to be at least at soil surface level if not somewhat higher. Whatever the source of moisture, it will be modified by the thickness of the wall, the aspect (with north-facing walls the dampest and south-facing the driest in the northern hemisphere) and exposure to the prevailing winds. Some plants are resistant to desiccation and may appear dead and shrivelled during the dry season, especially on more exposed parts of walls, but return to good condition as the seasons change (Woodell, 1979) (Figures 5.9 and 5.11). Geographical location also plays a part. Within the UK the eastern half of England is less suitable for wall flora than the western half, Wales and Scotland (Gilbert, 1996). In continental Europe, the suitability of walls as plant habitat declines with increasing distance eastward from the coast (Gilbert, 1996). Nutrients for plants on walls can come from material in solution, depending on the water source, or from dust, soil, guano, etc. Essentially, there is a tension between moisture and elements that reduce moisture; in principle this means that sheltered and shady places should produce a good flora. However, plants need light to grow and so another axis enters the equation: with dark, moist walls having a limited range of shade-tolerant plants capable of utilising the moisture (Woodell, 1979) just as bright, exposed walls have few plants capable of colonising them (unless there is a constant water source).

### 5.3.4 Biotic issues

The specific make-up of the flora on any particular wall or part of a wall will be influenced by a) the structural and environmental issues already discussed and b) the adaptations of individual plants to those environmental and structural issues (Lisci *et al.*, 2003); because of this, wall ecologists recognise discrete zones which have distinctive floral communities (Figure 5.11).

**Top:** sometimes with capstones and ledges. Plenty of gaps for establishment

**Upper and middle:** little water, highly stressed environment, distinctive vegetation

**Lower:** damper (capillary action), splashes from passing vehicles

**Base:** soil build-up from dust, fertile

**FIGURE 5.11** The main areas identified by wall ecologists and the different environmental conditions and flora they support: a) the wall top typified by a range of species including sedums and woody plants (e.g. red valerian), lichens, ferns, grasses, and flowers (e.g. Mexican fleabane *Erigeron karvinskianus*); b) the upper and middle zones which have the most specialist wall flora, e.g. plants such as ivy-leaved toadflax, navelwort, yellow corydalis; c) the lower zone which is typically damper and can thus support plants which are less tolerant of stressful conditions; and d) the base of the wall where a wide range of plants can grow, though they will become subject to drought stress in warm weather. Adapted from a diagram in Gilbert (1996). Images © John Dover

TABLE 5.2 Examples of papers with data on naturally colonised wall flora

| Town/City | Country | Authors |
|---|---|---|
| Pavia, Rome, Siena | Italy | Lisci & Pacini (1993); Lisci *et al.* (2003) |
| Srinagar, Kashmir | India | Ahmed & Durani (1970) |
| London | England | Francis & Hoggart (2009); Woodell (1979) |
| Berwick-on-Tweed, Chester, Chichester, Norwich, Southampton, York | England | Gilbert (1996) |
| Caernarvon, Chepstow, Conway, Denbigh, Tenby | Wales | Gilbert (1996) |
| Concarno, Brittany | France | Gilbert (1996) |

There is a third dimension which is also important: a source of propagules. As is well understood from restoration work in grasslands (Bekker *et al.*, 2000), if there is no nearby source of colonists, then the flora will remain very simple – even so, because of the stressed nature of the wall environment, only a subset of plants would be able to colonise. So dispersal mode of plants and dispersal distances becomes an important consideration, especially in very urban areas. Woodell (1979) notes that more than 1,200 species of plant have been recorded from walls and also that the majority of plants probably colonise because they have unspecialised dispersal mechanisms, with wind being the most likely dispersal mode (e.g. Ahmed & Durrani, 1970; Lisci & Pacini, 1993). Ants are also known to be important in the dispersal of some wall flora (Ahmed & Durrani, 1970), which is facilitated by the presence of a food 'reward' for the ants on the seeds: the elaiosome (Gilbert, 1996). Similarly, bird guano may be an important 'carrier' of the seeds of fleshy fruits (Ahmed & Durrani, 1970; Lisci *et al.*, 2003). It should be obvious from Chapter 2 that the city environment is different from that of the countryside, and it can be inimical to plants, with some (e.g. lichens) extremely sensitive to pollution whilst others (e.g. hartstongue fern *Phyllitis scolopendrium*) are remarkably resilient (Woodell, 1979). Whilst the flora of most walls is likely to be composed largely of relatively common and widespread species, albeit ones capable of tolerating tough conditions, walls are known to harbour some rarities (Woodell, 1979); for example, the plants yellow whitlow-grass *Draba aizoides*, wall bedstraw *Galium parisiense*, perfoliate penny-cress *Thlaspi perfoliatum*, London rocket *Sisymbrium irio,* the moss *Trichostomopis umbrosa* and the lichens *Calicium corynellum* and *Roccella phycopsis* (see further Gilbert, 1996). In the UK, about 20 species characteristic of cliffs and rocky outcrops (half native, half introduced) are mainly found on walls (Gilbert, 1996). Table 5.2 gives a number of references to wall flora studies. There appears to be little information on the value to fauna of naturally colonised green walls, but see the work of Attrill *et al.* (1997) on river walls along the Thames, UK (Table 5.8).

## 5.3.5 New walls

Naturally colonised walls can be beautiful or nondescript, be vertical carpets or simply a varied patchwork of bare wall and many or few plants. What is undoubted is that natural colonisation by flora takes a considerable time. In terms of green infrastructure, walls with existing flora should be cherished and maintained as appropriate (see Gilbert, 1996), but, generally speaking, developing new walls and waiting for natural colonisation will not address the needs of the present. We know from references in Woodell (1979) that deliberate inclusion of plants in walls has been going on for some considerable time; my favourite is the charming instance of a man planting tulip, iris, fritillary and daffodil bulbs in a mud wall in Harwan, Srinigar, in Kashmir (Ahmed & Durrani, 1970). There is no reason why new boundary, garden and retaining walls should not be designed to incorporate vegetation that is deliberately planted and Kontoleon and Eumorfopoulou (2010) use the term 'Landscape Walls' to describe such structures. For suitable plants and design ideas, Jekyll (1982) (initially published in 1901 and revised eight times) is as good a place as any to start. Whilst her book is concerned with stone walls in ornamental garden settings, many of the principles will be relevant to green infrastructure plantings whether of native or more horticultural flora. Clearly the issues that impact on existing naturally colonised walls in terms of structure, environment and plant suitability need to be addressed, but designing a wall with planting specifically in mind should allow a wider range of species to be used. It may even be possible to design walls specifically for the conservation of rare flora. Soil or other rooting media can be incorporated in the core of new walls and cavities included for direct planting – low boundary walls should pose no particular safety issues, but the integrity of larger walls would probably need the input of a structural engineer at the design stage (see also Woodell, 1979; Francis, 2010). We know that soil-retaining walls are likely to provide good opportunities for lateral water movement into the wall, so free-standing walls are likely to pose the greatest challenge and may need to incorporate piped irrigation for use during very dry periods – incorporation of sensors for automatic watering would be an obvious refinement. For river walls, irrigation is unlikely to be a significant problem; but the smooth materials used in modern construction are not plant-friendly and Francis and Hoggart (2008) describe approaches that can encourage colonisation.

Rock gabions, metal mesh cubes filled with rocks, are often used to reinforce steep banks or to prevent erosion, and plants can be grown on soil used to cover the horizontal surfaces. A different approach is to use strips of high density polyethylene (HDPE) which are linked together to form a honeycomb and which are filled with soil on site. The 'Webwall' system produced by ABG Ltd (ABG, undated) can be stacked in an offset to form a stepped gradient to a bank. Substantial-sized banks can be stabilised in this way using a geotextile behind the walls to provide extra stability. Webwall panels come in standard heights of 250 or 500 mm, in 4 m lengths, and widths between 1 and 2 m. The advantage of this type of system is that on-site material can be used to fill the honeycomb, reducing labour, fuel and transport costs, and the open upper surface of the steps can be planted up with vegetation to hide the HDPE. The deeper soil profile means a wider range of species (grass to trees) can be planted compared with rock gabions, and with better soil moisture retention.

Another new approach to growing wall flora is the use of the Growcrete system investigated by Ottelé et al. (2010a) at Delft University in the Netherlands. These are essentially panels of concrete with an upper (or outer, when mounted) surface with a very open granular structure created by using 32 mm diameter lava stones. This open structure (depth: 80 mm, on an 80 mm structural base) is used to hold a growing medium. The pH of the soil medium is influenced by the highly alkaline nature of the concrete/lava substrate and in tests increased from a starting level of pH 7.2 to pH 9.2 over a three-month period. As with non-retaining walls, watering is via precipitation and, because of the extreme environment, they are likely to support the classic wall vegetation of standing or building walls. Ottelé et al. (2010a) found, as might be expected, that of their seven test plantings (wild or creeping thyme *Thymus praecox,* sedum *Sedum* spp., grey cranesbill *Geranium cinereum subcaulescens,* maidenhair spleenwort *Asplenium trichomanes,* hart's tongue fern *A. scolopendrium,* ivy-leaved toadflax *Cymbalaria muralis* and aubretia *Aubretia* sp.), ivy-leaved toadflax and the sedums did best and aubretia, grey cranesbill and the spleenwort were not suitable. Because this wall is designed with a rugose surface and with potentially high plant coverage, it should be quite effective in particulate removal compared with naturally colonised walls that tend to have sparse vegetation, but data are not yet available.

## 5.4 Direct greening: surface climbers

This section deals with plants that are rooted in the ground but climb up the wall surface.

### 5.4.1 Plants that adhere to the vertical surfaces directly through adhesive pads

Whilst many plants can be grown against vertical surfaces, only those, like ivy, that are capable of adhering to building surfaces through aerial rootlets which exude an adhesive substance (Endress & Thomson, 1976; Steinbrecher *et al.,* 2010; Xia *et al.,* 2011) can do so without additional support. As well as the visual and environmental improvement of urban walls, climbing vegetation is also useful in the stabilisation of slopes (see Wang *et al.,* 2009). There are about 16 species of ivy (Ackerfield & Wen, 2003), of which *Hedera helix* (with over 400 varieties) is the most dominant the UK (Sternberg *et al.,* 2011a). A single ivy plant can reach 30 m in height and cover 600 m² of wall (Dunnett & Kingsbury, 2004). Other self-adhering species commonly found on buildings and walls include: Boston ivy *Parthenocissus tricuspidata,* trumpet creeper *Campsis radicans,* Chinese Virginia creeper *Parthenocissus henryana* and climbing hydrangea vine *Schizophragma hydrangeoides* (Dunnett & Kingsbury, 2004; Sternberg, 2010; Wang *et al.,* 2009). Species that can attach themselves to walls but which are better for having some support include: Virginia creeper *Parthenocissus quinquefolia,* winter creeper *Euonymus fortunei* and climbing hydrangea *Hydrangea petiolaris* (Dunnett & Kingsbury, 2004). Ahmed and Durrani (1970) indicate that species of *Lonicera* are self-supporting on the walls of Srinigar. Dunnett and Kingsbury (2004) note that most climbers have a strong tendency to grow towards light (positive phototropism) but that this tendency is quite weak in *P. tricuspidata* – useful for ensuring the whole wall surface is covered rather than just the middle and upper parts.

Witter (1986) described a project initiated by the Kassel City Council in Germany to encourage the greening of walls in the city. The project 'went live' in 1984 with much publicity including an A2 poster, planting tips, access to further council advice services and even a procedure for requesting planting trenches to be dug by the council along footpaths. Despite the fanfare, and interest, nothing spectacular in the way of uptake resulted. Whilst the public were being wooed, plantings were made at the base of particularly ugly civic buildings and it appears that in 1985, when the project was re-launched, the combination of publicity and examples worked. When Witter (1986) reported on the project, there were almost 40 public buildings with climbing plants. At pretty much the same time, Berlin had developed an incentive scheme for green walls which resulted in 245,584 m² being grown over the 15 years of the programme (1983–1997) (Köhler, 2008).

## 5.4.2 Do climbing plants damage walls?

The answer to this is 'yes and no'. Woodell (1979) points out that plants cannot establish until some, typically abiotic, structural change has been made in a wall to allow colonisation; this appears to include lichens, once regarded as primary colonisers (Lisci *et al.*, 2003). Warscheid & Braams (2000), however, note that microbial biofilms establish quickly when stone surfaces are exposed to the environment with equally speedy surface-deteriorating effects including providing moisture for freeze-thaw cycles and facilitating the acquisition of atmospheric pollutants. Ironically, Warscheid and Braams (2000) also note the deliberate use of some microbial agents in creating protective surfaces on stone and even in the removal of some surface pollutants. Once in, plants can accelerate degradation through acidic exudates (Lisci *et al.*, 2003) and mechanical effects via root growth forcing gaps to expand (Lisci *et al.*, 2003; Viles *et al.*, 2011). Colonisation niches may also allow ingress of pests into buildings. Fortunately, after this catalogue of horrors, Woodell (1979) pulls things back into perspective by noting that any structural damage by plants is likely to be mostly through tree and shrub growth rather than by herbaceous species, and is likely to happen extremely slowly, allowing remedial action to be taken before real damage occurs. Viles *et al.* (2011) convincingly demonstrated that ivy (*Hedera helix*) rooted directly in cracks in walls, as opposed to growing as a climber up walls, can have serious structural impacts – and that failed attempts to control ivy may actually promote damage (see also Lisci *et al.*, 2003). Some ancient monuments may be susceptible to degradation by physical and chemical (acid) attack from lichens. In such situations, elimination of some species of lichen may be advisable, but for others it may not, as removal can accelerate deterioration (Lisci *et al.*, 2003). For non-heritage walls, lichen damage is likely to be so slow as to be irrelevant.

## 5.5 Indirect greening of walls

### 5.5.1 Types of indirect greening systems

'Indirect' green walls are composed of three classes of wall: green façades and two types of living wall (mat and modular) (section 5.2.3).

The advantages of these 'designed' systems are obvious:

- they allow a greater degree of control over which part of a building or wall the plants grow over (over and above simple pruning);
- they allow the use of a much wider range of plant species;
- they have the great advantage over green roofs in that they are very easy to retrofit and can, potentially, cover a greater area than roofs.

The living wall systems allow greater artistic expression in the planting designs, with Patrick Blanc's creations being quite startling and dramatic in their exposition (Blanc, 2008).

### 5.5.2 Plants that twine around support materials fixed to vertical surfaces

Plants that adhere directly to wall surfaces (see section 5.3.1) have an advantage in establishment as they require no additional infrastructure: simply prepare the ground appropriately, plant them and off they go; in many cases they do not even seem to need planting but occur without human intervention! They do have downsides: many people fear that they will damage the wall surface and there is a relatively limited range to choose from (compared with plants that could potentially grow in living wall systems). Dunnett and Kingsbury (2004) give details of 55, but note that there are over 200 species of Clematis, and the number of English ivy varieties is in the hundreds. Climbing plants that cannot adhere directly to wall surfaces but need something to climb up obviously require a greater investment in support infra-structure, although this can be minimal (Figure 5.12). Adhering climbers can also be trained up support structures. There are advantages to using support infrastructure for green walls: the support can be offset from the wall such that there is an additional air gap which poten-tially improves the insulating properties of the planting and growth is more easily directed. Non-adhering climbing species have different attachment mechanisms, by a) twining the stems around supports, b) sending twining tendrils from the stems, or c) using leaf stems to twine around supports (see Figure 5.13, Table 5.3) and these require different support systems depending on the plant and the architectural effect intended (Kontoleon & Eumorfopoulou, 2010). Dunnett and Kingsbury (2004) provide outline information on a wide range of support approaches, and the website www.fassadengruen.de/eng/indexeng.htm describes a variety of different climbing plants with a very useful graphic giving recommendations

FIGURE 5.12 Wisteria grown up a building along a busy street in Köln (Cologne), Germany. Support infrastructure is required for this stem-twining plant, but is minimal compared with that required for living wall systems. © John Dover

**FIGURE 5.13** Different forms of attachment to trellis/wire supports used by climbing plants. Left: leaf tendrils; middle: stem tendrils; right: twining stems. © John Dover

for the most suitable attachment designs related to the growth form of the plants. Wooden trellises may be appropriate for typical residential building use, or more intimate contexts such as shading outdoor seating and dining areas (Figure 5.14). However, wood suffers from inevitable decay problems and, where strength and durability is required, metal can be substituted (Figure 5.14). In most commercial applications, the material of choice will probably be one of stainless steel rope, rods or wire mesh (e.g. Figure 5.15). Rigidly framed modular mesh panels of plastic-coated, painted or galvanised metal are also available. Dunnett and Kingsbury (2004) discuss materials in depth, but, of course, specialist advice is recommended for any but the simplest of projects, and there are increasing numbers of companies offering such services.

**FIGURE 5.14** Left: wooden trellis used to support wisteria up a domestic wall; right: metal frame used to support climbers to help shade a window and garden seat. © John Dover

FIGURE 5.15 A Jakob © stainless steel vertical rope and horizontal rod system. Left: installation at Staffordshire University, UK, May 2011; right: planted up and growing July 2011. © John Dover

Dunnett and Kingsbury (2004) note that whilst some climbing plants can potentially grow to 24 m, in practice 10 m is more likely – so for buildings over eight floors, a more creative approach would need to be employed to cover the whole surface. Installation of planters at intervals up the building which are supplied with irrigation (and nutrients) is the obvious solution, but access for maintenance, unless installed on balconies or an accessible framework, may require the use of 'cherrypicker' machines (Figure 5.16) or for very tall buildings rope or cradle access from the roof. Planters can also be positioned on flat roofs so that plants cascade down the walls (Van Bohemen *et al.*, 2008) – this gives another aesthetic dimension to work with, and also allows the greening of moderate-sized buildings which would otherwise need intermediate wall-mounted planters (Figure 5.2c, e). Cascades can also be used in combination with other wall greening approaches including naturally adhering plants, green façades and living walls. For example, the Crédit Suisse building 'Uetlihof' in Zurich had some of the faces of its internal courtyards set with planter boxes at different heights so as to allow a continuous green curtain to fall down from roof level to the ground. Plants included in this project were Virginia creeper *Parthenocissus quinquefolia*, perfoliate honeysuckle *Lonicera caprifolium*, goldflame honeysuckle *L. hekrottii*, chocolate vine *Akebia quinata,* sweet autumn clematis *Clematis paniculata* (Badeja, 1986). The Institute of Physics at Humboldt University in Berlin-Adlershof (Lise-Meitner-Haus) has incorporated 150 planters at different levels on the exterior of nine façades to investigate the environmental value of green walls. The planters are fed rainwater stored in cisterns with excess filling a courtyard pool (BSDUD, 2009, undated). The plant that appears to do best in the planters (of ten tested) is wisteria but the plants do best in planter boxes that have had insulation installed to protect against

**FIGURE 5.16** Planter boxes with support structures for plants used to cover the side wall of an underground car park, Monaco. © Mobilane UK Ltd

temperature extremes and particularly against low winter temperatures (Schmidt, 2009). A harsher environment for planters is on the Avenue Princesse Grace in Monaco where Mobilane installed a 20 x 150 m planter system in 2006 up the south-facing wall of the car park under the Boulevard du Lavotto. The average temperature is 40°C in August but there are also strong winds and, being coastal, the potential for salt corrosion of the infrastructure. As a result the steelwork has been galvanised and powder coated, the irrigation control system can detect leakage and monitors soil moisture, and the system is also claimed to be earthquake resistant (Mobilane, undated-a) (Figure 5.16).

### 5.5.3 Living wall – vertical garden/mat

In this system, plants are rooted into a thin, synthetic, horticultural felt and maintained wholly hydroponically. This type of wall was pioneered by Patrick Blanc (Blanc, 2008), who calls it a 'Mur Vegetal' or 'Vertical Garden'. This style of green wall is essentially a layer of felt (Blanc uses polyamide, but other materials can be used) stapled or otherwise fixed onto a waterproof layer and held in a metal framework. Water and nutrients are fed through an irrigation system; plants are inserted into cuts in the felt (Blanc, undated). Walls of this type are typically offset from the structural wall providing an air pocket. This basic concept has been immensely successful with high-profile installations all over France; for example, the 198 m frontage of the well-known Musée du quai Branly (37, quai Branley) in Paris. A more recent, and even more dramatic, installation can be found on a bridge in Aix-en-Provence which is 15 m high, 650 m$^2$ and has between 30 and 35 plants.m$^{-2}$ (or about 20,000 plants) (Figure 5.4). This approach has even been used to decorate pubs such as the Driver in London (Figure 5.17). Plants can be pulled out of their 'pockets' for replacement or, if required, sections of mat can be cut-out and replaced.

TABLE 5.3 Characteristics of a range of common climbing plants

| Common name | Scientific name | Growth habit | Deciduous/ Evergreen | Max. height | Position | Flower | Flower colour | Berries/Seeds | Notes | Source |
|---|---|---|---|---|---|---|---|---|---|---|
| English Ivy | Hedera helix | A | Evergreen | | Full sun/semi-shade | September | | Black berries | Many varieties available including variegated forms | FG |
| Dutchman's Pipe | Aristolochia durior syn. macrophylla and A. tomentosa | T | Deciduous | | Full sun/shaded | June/July | | Cylindrical fruit | | FG |
| Clematis | Many Clematis species | LT | Deciduous | 5m | Moderate sun/semi-shade | | Varies | Varies; fine seed heads e.g. Old Man's Beard C. vitalba | Many species and varieties | FG |
| Grape | Vitis vinifera | ST | Deciduous | | Full sun | | Inconspicuous | Edible grapes | Many varieties | FG |
| Grimson glory grape | Vitis coignetiae | ST | Deciduous | 15m | Full sun/semi-shade | | Inconspicuous | Small black berries | Red, orange, gold foliage in autumn | Köhler (2008); BBC; RHS |
| Boston ivy/Japanese creeper/Japanese ivy | Parthenocissus tricuspidata | A | Deciduous | | Full sun/semi-shade | | Inconspicuous | Blue-black berries | Leaves turn red in autumn before shedding | FG |
| Virginia creeper/American ivy/Five-leaved ivy | Parthenocissus quinquefolia | A + ST | Deciduous | | Full sun/semi-shade | | Inconspicuous | Blue-black berries | Leaves turn red in autumn before shedding | FG |
| Thicket creeper/Virginia creeper/Grape woodbine | Parthenocissus inserta | ST | Deciduous | | Full sun/semi-shade | | Inconspicuous | Blue-black berries | Leaves turn red in autumn before shedding. Adhesive pads may occasionally form in deep shade | FG |
| Chinese Virginia creeper | Parthenocissus henryana | A | | 10m | | | | | | Dunnett & Kingsbury (2004) |
| Trumpet vine/creeper | Campsis species e.g. C. tagliabuana | A | Deciduous | | Full/part sun | | Red/orange 'trumpets' | Bean pods | | FG; Köhler (2008) |
| Climbing hydrangea | Hydrangea petiolaris | A | Deciduous | | Semi-shade/shaded | | White | Small, inconspicuous | Adhering, but extra support recommended | FG; Köhler (2008) |

| | | | | | | | | | | Dunnett & Kingsbury (2004) |
|---|---|---|---|---|---|---|---|---|---|---|
| Climbing hydrangea vine | *Schizophragna hydrangeoides* | A | Deciduous | | Full sun/semi-shade | | Red | Cylindrical fruit | | FG |
| Akebia/ Chocolate vine | *Akebia quinata/ A. trifoliata* | T | Deciduous | | Full sun/semi-shade | | Green | | | FG |
| Hop | *Humulus lupulus* | T | Deciduous | | Full sun/semi-shade | | | Loose, open, green cones | Male and female plants | FG |
| Wisteria | *Wisteria species* | T | Deciduous | 20m | Full sun | | Blue, white, pink cascades | Pods | | FG |
| Honeysuckle | *Lonicera species* | T | Deciduous | | Full sun/semi-shade | | White-red | Black, red, orange | Some perfumed | FG |
| Evergreen honeysuckle | *Lonicera henryi* | T | Evergreen | | Semi-shade but can tolerate full sun | June/July | Orange | Blue-black berries | | FG |
| Silver lace vine | *Polygonum aubertii* | T | Deciduous | | Full sun/semi-shade | July–September | White | Usually none | | FG |
| Kiwi vines | *Actinidia arguta, A. deliciosa, A. kolomikta* | T | Deciduous | | Full sun | | Yellow-white | Kiwi fruit | | FG |
| Bittersweet | *Celastrus orbiculatus and C. scandens* | T | Deciduous | | Full sun/semi-shade | | Green | Red/yellow berries | Leaves yellow in autumn; male and female plants | FG |
| Wintercreeper | *Euonymus fortunei* | A | Evergreen | | Full sun/semi-shade | | Inconspicuous | Yellow berries | | FG |
| Wild grapes | *Vitis ripara, V. berlandieri* | ST | Deciduous | | Full sun/semi-shade | | Green-yellow | Usually none | | FG |
| Kolomikta | *Actinida komomikta* | T | Deciduous | 8m | Full sun | | Inconspicuous, white | | Dioecious; leaves green and pink/white; drought sensitive; flowers scented | Köhler (2008); RHS |
| Devil's Ivy/Golden pothos | *Scindapsus aureu/ Epipremnum aureum* | AR | Evergreen | 8m | Full sun/semi-shade | | Inconspicuous | Inconspicuous | Indoor plant | Franco *et al.* (2012); RHS |

**Growth habit:** A (Adhere): using adhesive pads; T (Twine): stems twine around supports; ST (Stem Tendril): slender curling outgrowths from stems that twine around supports; LT (Leaf Tendril): as for stem tendrils but the leaf stems twine around supports rather than specialist structures; AR (Aerial Roots): aboveground roots used as attachment mechanism. Excluded here are rambling species like roses or shrubs frequently grown up walls but that need human intervention to fix them to supports. Source key: BBC: www.bbc.co.uk/gardening/plants/plant_finder; FG: www.fassadengruen.de; RHS: www.rhs.org.uk

**FIGURE 5.17** Even pubs can have living walls. This is the Driver Public House in London. Left: maintenance work using a cherrypicker exposing the waterproof backing; right: the upper floors covered in lush well-textured vegetation. © John Dover

### 5.5.4 Living wall – modular

There is considerable variation in design in this category, but essentially plants are rooted into a soil, or soil-like growing medium or other substantial substrate such as Grodan© rockwool. Water and additional nutrients are delivered through an irrigation system. Such systems typically require more substantial support infrastructure because of their weight. Advantages include:

- a deeper rooting medium – some module designs use horizontal or sloping top edges to help retain material and give extra planting depth;
- easy replacement of individual dead or diseased plants (as roots are constrained within individual planting units (cells) as opposed to infiltrating a supportive mat);
- designs with interlocking panels of modules can be easily replaced/removed, so planting designs can be modified 'on the hoof' and pre-grown before replacement.

Living walls of this type typically consist of modules that are divided into a number of individual planting cells and are designed for easy removal and replacement on a structural framework. An example of the type is manufactured by the ANS group (www.ansgroupeurope.com): modules are made of High Density Polyethylene (HDPE) and divided

**FIGURE 5.18** Rockwool-based green walls (Mobilane LivePanel System): a) immediately after planting May 2012 at Staffordshire University, Stoke-on-Trent, showing the green-faced rockwool panels; b) a developing 49 m² wall at a Waitrose supermarket in Bracknell, UK (photo: mid-May 2012; planted November 2011 (Cronin, 2012)). © John Dover

FIGURE 5.19 Top left and middle: plug plants being grown on in crumbled Grodan® rockwool; top right: plants newly inserted into green wall modules in the nursery; bottom: once planted up modules are allowed to gain full coverage before installation on-site (Biotecture system). © John Dover

FIGURE 5.20 Modular system (ANS) planted up for indoor use showing the flexibility of such systems in creating colourful designs with structural and textural complexity. © John Dover

**FIGURE 5.21**  A simple concept, but effective. The VertiGarden system is essentially a set of modified seed trays linked together with an irrigation system. Top left: high-impact design, composed entirely of petunias; top right: detail of the support mechanism for the growing trays; bottom: back and front detail of irrigation system. © John Dover

into individual cells filled with a soil-based substrate; walls are pre-planted and grown in greenhouses for two months prior to erection on site. Each module is irrigated and individual planting units within the module contain capillary matting.

Some companies plant on site (e.g. Mobilane; Figure 5.18) whilst others pre-plant in glasshouse conditions (e.g. ANS, Biotecture; Figure 5.19). Pre-planting allows complete control over establishment and guarantees an instant 'full-coverage' of the wall with well-developed plants on installation day. Planting on-site allows the development of the vegetation to be experienced by the client and public, which can give considerable pleasure, and may reduce nursery costs, although with less control over establishment conditions.

Walls of this type are suitable for indoor and outdoor use and have been used in many high-profile locations (Figure 5.5) including the Olympic stadium in London. A set of units planted up for the Ecobuild trade show in London in 2014 shows how startling colour combinations can be created (Figure 5.20). Different levels of sophistication have been achieved with different designs; for example, integrating irrigation pipework within the planter module as in the VertiGarden system (Vertigarden, 2011) (Figure 5.21). Companies have also started to offer bespoke irrigation solutions for green wall systems (Watermatic, 2011) and some work has been done to examine the effect of uneven irrigation (Cheng *et al.*, 2010). The internet has played its part in improving green wall irrigation systems: Biotecture Ltd, for example, monitors and controls living walls in the UK, USA, UAE and Norway from the corner of an office in West Sussex, UK (Richard Sabin, pers. comm.).

At my institution we have installed an experiment using the LivePanel system designed by Mobilane BV in Leersum, Netherlands, which comes in standard (600 mm tall x 1000 mm wide x 130 mm deep) or bespoke panels held in place by an aluminium frame with integrated automated irrigation and drainage (Figure 5.18). The growing panels are faced, front and back, with a hydrophobic mineral wool with a core of growing medium (substrate), the mineral wool faces being reinforced internally with mineral wool bracers (Mobilane, undated-b); each panel takes 40 plants. In some respects the Mobilane system, with its sandwiched substrate, is a hybrid between the HDPE-based modular walls and the Blanc Vertical Garden 'mat' style. Modular walls, having more substrate, are likely to be more resilient to temporary interruption in irrigation than the latter which employ thinner substrate mats. Mat-style walls have the advantage over modular walls that they take up less space and use less in the way of materials, but are probably a bit more vulnerable if the irrigation system fails as they have less growing media to hold water. Blanc (undated) estimates that his Vertical Garden system weighs less than 30 kg.m$^{-2}$ which compares with up to 60 kg.m$^{-2}$ for the Mobilane system (Mobilane, undated-b); the ANS modular system compares favourably with the Mobilane system at 64 kg.m$^{-2}$ (Scotscape, undated).

## 5.5.5 Plants that do not climb but are often used on walls

Some shrubs, despite having no natural method of wall attachment, are commonly grown up walls and could thus be seen as a form of green façade. Fruit trees such as peach (*Prunus persica*) are grown against walls, and in the past in some country estates special double-skinned heated walls were constructed with a fireplace at one end and a chimney at the other to help them grow in the UK's cool climate. Fruit trees, whilst yielding useful fruit and summer shade, do not give extensive coverage, whilst other species such as firethorns (*Pyracantha*) can give dense foliage, attractive fruit for wildlife, and because of their very sharp thorns a certain amount of additional security (Figure 5.22). Training against walls can be as simple as tying back to a trellis or even lead-headed nails. The first house I bought was cursed with an ugly facing of stone put up by a previous owner – rather than remove it, and potentially damage the brick wall behind it, I grew a dense pyracanthus up the ground floor wall and used lead-headed nails and wire to train it: it looked lovely. Of course, when I sold the house the next owner preferred the stone facing and took the pyracanthus down!

## 5.5.6 Free-standing/curtain walls

Curtain walls are free-standing green walls that are not attached to a structural wall; they incorporate their own structural infrastructure and can be permanent or temporary structures and single or double sided. The simplest of these would be a line of wire ropes or rope mesh up which climbing plants could grow. A variant on this basic concept is pre-grown ivy screens: ivy is grown up a galvanised steel meshwork frame. In the Mobilane system each 1.2 m long panel has about 65 English ivy plants grown in a coconut-fibre lined 'U'-shaped channel at the base of the meshwork that supports the roots. The units are installed on site supported by, and bolted to, tubular steel uprights concreted into the ground (Figure 5.6). The screens provide a more aesthetically pleasing result than simple metal mesh fencing, whilst providing security. Panels such as this are ideal for housing developments where a visual and security barrier is required at boundaries; they last longer and are more aesthetically pleasing than

FIGURE 5.22 Top left: espalier fruit trees trained against a garden wall at Erddig, Wales; top right: espalier trees around a modern domestic building in Exeter, Devon, UK; bottom left: pyracanthus in flower around a door in Exeter, Devon, UK; and bottom right: pyracanthus in berry around a door in the village of Keymer, Sussex, UK. © John Dover

**FIGURE 5.23** Van Gogh's *A Wheatfield, with Cypresses* reproduced as a free-standing living wall outside the Tate Gallery in Trafalgar Square, London. © ANS Group (Europe)

wooden panels such as larch lap. The maximum height commercially available pre-planted is 3 m, but extension meshwork is available for higher screens (HederaScreens, 2009). More sophisticated curtain walls can be created using the modular living wall systems, an example being the free-standing living wall at the Westfield Shopping Centre in Shepherd's Bush, London (Figure 5.5). This wall is 170 m long, 4 m high and contains 200,000 plants in 5,000 ANS modules. The design also incorporates a water feature to add interest and seating and to reduce leaf-picking by the public. Irrigation is sensor-driven and is needed on about 150 occasions a year using approximately 3 litres of water per $m^2$ each time; fault detection in the irrigation system is automated with error messages texted to the maintenance contractor (Grant, 2010). Free-standing green walls have also been used as high-impact marketing/publicity tools such as the large-scale replica of Van Gogh's *A Wheatfield, with Cypresses* which was erected outside the National Gallery in London during the spring and summer of 2011. This wall used 8,000 plants in an ANS system (Anonymous, 2011) (Figure 5.23). On a smaller scale, a modular wall using the VertiGarden system planted up with *Petunia x hybrida* Easy Wave red, white and blue illustrating part of a Union Jack flag was erected at the two-day Cheshire County Show at Tabley, UK, in June 2012 by the Cheshire West and Chester Borough Council (Figure 5.21). The wall used 48 VertiGarden modules, 16 plants to a module or 768 plants in total. Free-standing green walls have also been installed as visual barriers to hide major construction works. When Birmingham's new library was under development, screens 5 m high were erected over a three-year period in Centenary Square which used a mixture of Mobilane's Hedera green screens and LivePanel systems creating a visually stunning alternative to the usual construction works.

A pergola is a form of hollow 'curtain' green wall where trees (fruiting or non-fruiting), shrubs, climbers, vines, roses and many other plants are, typically, trained over a wooden or

FIGURE 5.24 Pergolas do not have to be as grand at this one in the King's Garden (Jardins du Roy) in Blois, France. Pergolas create shade and contemplative spaces as well as giving aesthetic pleasure and structural variety. © John Dover

metal framework to make a green tunnel. Pergolas are known to have been used in ancient Egyptian gardens and were introduced into Europe during the Renaissance in Italy (Hansen, 2010) becoming particularly popular in the 18th and 19th centuries (Johnston & Newton, 2004). The latter authors note that whilst pergolas can be 'stand-alone' structures, they can also be attached to a wall or span the gap between buildings. From the outside a pergola looks like a curtain green wall, from the inside it is a shaded delight of cool dappled green light sometimes scented by the plants used to make it (Figure 5.24).

## 5.6 Specific values of green walls

### 5.6.1 Aesthetics

From plants growing adventitiously on walls such as ivy-leaved toadflax and yellow corydalis *Corydalis lutea*, to wall planters, window boxes and flowering climbers to the extravagant foliage and flowers possible with living wall systems, green walls provide visual amenity – and all its associated benefits (see Chapter 2). Moreover green walls are usually, although not always, at street level and are far more visually accessible than green roof vegetation. Köhler (2008) reporting earlier work by Köhler *et al.* (1993) made the point that only 5% of all façades in Berlin were greened at the time of their study – if we assume this is a typical figure for urban centres, then the potential for improving the street-scene with green walls is immense.

### 5.6.2 Insulation

The insulation potential of green walls has been known for some time; Minke and Witter (1982) (cited in Peck *et al.*, 1999), for example, noted that a 4 cm air gap between an insulated wall and a 16 cm layer of vegetation could increase the thermal insulation qualities of the wall by up to 30%. Bartfelder and Köhler (1987) (cited in Stülpnagel *et al.*, 1990) claimed a reduction of up to 15°C in the daily temperature range of a wall covered in ivy compared with a bare wall, with the effect still evident, though reduced to 4°C, at 1 m distance from the wall. Bass *et al.* (2003) working in Toronto, Canada, tested the thermal performance of a 'vertical garden' compared with bare walls in a small experiment and concluded that: 1) they could reduce the energy needed for summer cooling, and 2) they were more effective than a green roof studied at the same time. The plant chosen in their system was the fiveleaf aralia *Acanthropanax sieboldianus*, an upright shrub which can grow to about 3 m high and 3 m wide (MSU, 2008). In this study, four plants were used, they were 1.3 m high and installed in pots. It appears from the description in Bass *et al.* (2003) that the shrubs were effectively being used in the mode of a hedge to shade a metal wall. The average unshaded wall temperature was 43°C compared with the shaded wall at 26.8°C and the leaf surfaces 26.1°C. The reduction in energy use for cooling was simulated for a one-storey office building (3,000 m²) in Toronto and estimated at 23%, with a further reduction for the energy use of fans by 20%. Other electrical consumption was also estimated (lights and equipment) leading to an assessment of overall annual energy saving for the building due to vertical gardening of 8%. Modelling for other Canadian cities suggested that in Vancouver or St Johns, energy savings on cooling might reach 30%.

In the UK, Johnston and Newton (2004) suggested it is probably best to have evergreen species (such as ivy) on walls exposed to the coldest conditions (north-facing and most west-facing situations) and use deciduous species on the sunniest (warmest) walls (typically south-facing) – so that leaves act as shade in summer, but being shed in the autumn, expose the building to the full force of sunlight in the winter; for east-facing walls the choice of evergreen or deciduous would depend on how much sun a particular wall gets.

Recent research by Viles and Wood (2007) and Lee *et al.* (undated), commissioned by English Heritage (the statutory advisory body to the UK government on the historic environment), has shown that covering the top surface of walls of ruined historic monuments with a layer of growing vegetation, rather than having negative effects, is potentially an effective conservation measure, acting as a thermal blanket, and may be superior to hard-capping in controlling moisture ingress. The insulation value of vegetation is also helpful to householders and businesses, with green wall vegetation reducing the need for air conditioning in the summer and heating in the winter (Köhler, 2008). Franco *et al.* (2012) point out that in hot countries such as Spain a high proportion of the country's current energy demand is for cooling. Increases in mean temperature are predicted by climate models and, even for temperate countries such as the UK, this implies increased future energy use for summer cooling (Franco *et al.*, 2012). It has been known for some time that ivy-covered walls have lower external temperatures than similar bare walls. Köhler *et al.* (1993) (cited in Köhler, 2008) claimed that a green wall could make a poorly insulated building effectively better insulated than a new build built to higher insulation standards! The shading effect of ivy was reported as resulting in 3°C lower temperatures in summer and the insulation effect 3°C warmer in winter. Thermal imaging of buildings has also been used to demonstrate, and confirm, the insulating properties of vegetation (Köhler, 2008; Ottelé, 2011). Sternberg *et al.* (2011a), working on historic walls at five sites in the UK, demonstrated that ivy growing on historic walls protected them from extremes of humidity and especially extremes of temperature. Exposed walls had, on average, 36% higher daily maximum and 15% lower minimum temperatures than ivy-covered walls.

Wong *et al.* (2009) carried out simulations of a ten-storey building of footprint 30 x 30 m with 4 m between each storey. The simulations included vegetated and non-vegetated façades with conventional windows and further simulations of the same building but with full glass façades and no, 50% and 100% vegetation. A constant working temperature inside the building of 24°C throughout the year was assumed to be maintained Monday–Friday from 9am to 6pm using air conditioning. The simulations showed that greening the façade could lower the mean radiant temperature and the energy cooling load of a building by blocking incoming solar radiation; the thicker the leaf cover the greater the impact. The 'envelope thermal transfer value' of a glass-fronted building with 50% vegetation cover and a shading coefficient of 0.041 was shown to be reduced by 40.7%. In addition, greening façades were shown to reduce the minimum air temperature of the local area.

Cheng *et al.* (2010) examined the insulation value of a modular turf-based green wall in Hong Kong. Modules were constructed from 100 x 50 x 7.3 cm aluminium frames enclosing a hydroponic rockwool substrate (Grodan©) with an integral irrigation line. A grass, *Zoysia japonica*, was grown on the modules in a nursery before wall installation; the grass was mown four weeks before experiments started and the modules installed in three, four and five-panel configurations on the concrete exterior of a flat (apartment). The green walls had lower temperatures than bare walls, and lower temperatures were achieved by larger numbers of modules. Average module temperatures, whilst lower than ambient air, showed modest differences (about 1°C) but peak differences could reach 14°C in the afternoon, though the panels tended to be up to 2°C warmer than ambient at night. The impact of panels on heat fluxes from the exterior to the interior of the building were more dramatic with the daily total heat flux with vegetation being $182.2 \pm 83.0$ Wh/m$^{-2}$ compared with $595 \pm 201$ Wh.m$^{-2}$ without. This reduced heat flow resulted in less energy use for cooling in the 9.35 m$^3$ room behind the vegetated panels by $1.4 \pm 1.85$ kWh (the air conditioner was set to maintain room temperature at 25.5°C). Thermal flux, and hence power consumption for cooling, was affected by climate conditions, as might be expected.

Perini *et al.* (2011a), working in the Netherlands, compared the insulation value of a bare wall, a wall colonised by English ivy, a wall shaded by an offset façade with English ivy, grape (*Vitis* sp.), clematis, jasmine (*Jasminum* sp.) and pyracanthus (*Pyracantha* sp.), and a living wall composed of Greenwave fibreglass/HDPE planter boxes (www.greenwavesystems.nl). This latter system is modular and is made up of planter panels which contain four planting units within a panel 200 mm wide by 600 mm long and 515 mm deep. The planter is filled with a growing medium and water and nutrients are supplied by irrigation, although the design does facilitate some watering via rain. The panels are hung on support racks which results in a 40 mm air gap between the planter and the support wall. The study was hampered by no direct measurement of an equivalent bare surface, and a bare surface temperature was estimated from the other studies. They found only small differences between the surface temperatures of a building wall with and without English ivy growing directly on it (1.2°C lower behind the ivy). The temperature difference was greater (2.7°C) on the building with the ivy growing up a metal trellis and a 200 mm gap between it and the building wall. With the living wall system a figure of 5°C difference was estimated. The impact of the green walls was more complex when studying wind speed, with it dropping close to zero within the foliage of the ivy growing directly on the wall and in the air gap of the living wall system. For the trellis-growing ivy, although wind speed dropped within the foliage, it increased again within the air gap between trellis and wall – although still about half that of the speed in front of the building. Essentially the results show that green walls do improve the energy efficiency of buildings; the authors suggested 200 mm is probably too large an air gap for the trellis system and suggested that a 40–60 mm gap would produce better results.

Eumorfopoulo and Kontoleon (2009) examined the impact of Boston ivy on temperatures of a six-storey rendered concrete building in July and August 2006 in Thessaloniki, Greece. The first three storeys were covered in ivy, whilst the upper three had bare walls; in the study, storeys three and four were compared. The results showed: 1) more stable internal temperature conditions where vegetation was used, 2) vegetation reduced peak temperatures, and 3) differences in temperatures between bare and vegetated walls were highest on the warmest days. Subsequently, Kontoleon and Eumorfopoulou (2010) modelled the effect of a 250 mm layer of Boston ivy around a building in the urban zone of Thessaloniki during the summer and demonstrated that orientation had a strong effect on the external temperature reduction effect of vegetation. The position of insulation on a building wall also has an impact on temperatures, with higher external temperatures experienced when insulation is placed on the outside of the building compared to cavity-filled or internal insulation (external>cavity>internal); the situation is reversed for internal temperatures, with external insulation producing the coolest interior temperatures (internal<cavity<external). The study compared bare surface conditions with 100% vegetation cover (for a range of wall constructions) and showed average peak temperature reductions of West 16.85°C, East 10.53°C, South 6.46°C, North 1.73°C. Peak interior wall temperature reductions followed

a similar trend W 3.27°C, E 2.04°C, S 1.06°C, N 0.65°C. The figures for the individual wall construction types are given in Table 5.4, and it is clear that whilst the position of the insulation within a wall does have an impact on both internal and external temperatures, such effects are swamped by the impact of orientation. The study also demonstrated that the cooling effect of vegetation increased linearly with percentage cover of the wall, and that interior temperature reductions resulted in energy savings from reduced air-conditioning requirements.

Wong *et al.* (2010a) tested eight different planting systems (Table 5.5, Figure 5.25) in Singapore's HortPark against a bare concrete wall control for efficiency in temperature reduction under tropical conditions in 2008. The walls (4 m wide by 8 m high) were not replicated and were equivalent to boundary walls rather than building walls, with no 'interior'. Different planting systems were shown to have big differences in insulation efficiency: plants climbing up a trellis did not perform well, probably due to poor coverage and also the lack of a substrate acting as a wall insulator. The two best-performing systems, in terms of average wall surface temperature reduction, were the living wall 'Parabienta' system produced by Japan's Shimizu Corporation and another modular system, with plant panels embedded in a stainless steel mesh. These systems gave maximum reductions of over 10°C. The Parabienta system and a modular panel system which combined 'versicells' with a slotted planter had the narrowest average range in daily temperatures of the eight wall types tested. Some of the green walls were also shown to reduce local air temperatures.

Living wall systems, in themselves, act as insulation because of the plants and growing medium, and also because they usually incorporate an air gap between the growing medium and the supporting wall (or adjacent wall if self-supporting). This air gap has the potential to have insulation inserted to further improve the thermal efficiency of the living wall system Ottelé (2011).

Ottelé *et al.* (2011) estimated the temperature reductions and energy savings (heating and air conditioning) of four of the green wall systems described in Table 5.1 compared with

TABLE 5.4 Peak summer temperature reductions on the external and internal surfaces modelled for a building in urban Thessaloniki, Greece, with bare walls compared with walls completely covered in Boston ivy. Three alternative wall constructions were considered with the position of building insulation varying in location from external, to cavity, to internal

| Insulation position | Temperature °C | | | |
| --- | --- | --- | --- | --- |
| | **N** | **E** | **W** | **S** |
| *Exterior temperature* | | | | |
| External | 1.87 (34.85) | 12.65 (47.10) | 19.01 (53.45) | 7.54 (41.40) |
| Cavity | 1.7 (34) | 9.32 (43.09) | 15.99 (49.76) | 5.93 (39.11) |
| Internal | 1.62 (33.69) | 9.61 (43.15) | 15.56 (49.11) | 5.89 (38.85) |
| *Interior temperature* | | | | |
| External | 0.58 (27.67) | 1.84 (27.87) | 2.86 (27.96) | 0.93 (27.77) |
| Cavity | 0.67 (28.12) | 2.06 (28.31) | 3.5 (28.48) | 1.09 (28.23) |
| Internal | 0.69 (28.59) | 2.23 (28.80) | 3.46 (28.92) | 1.15 (28.71) |

Source: Kontoleon and Eumorfopoulou (2010)
Note: temperatures in brackets = peak temperatures on bare walls.

TABLE 3.3 Eight systems used to test thermal and noise properties of green walls in Singapore's HortPark. Source: Wong et al. (2010a, 2010b)

| Wall number | Type of wall | Details | Substrate | Mean substrate width (m)* | Mean plant width (m)* | Total plant + substrate width (m)* | Plants | Orientation of plantings (in relation to ground) | Long axis = plant stem, short axis = planting medium |
|---|---|---|---|---|---|---|---|---|---|
| 0 | Concrete base wall | Control wall | Brick core surrounded by reinforced concrete frame | | | | | | |
| 1 | Living wall | Modular 'Versicell' + slotted planters | Green roof media + soil wrapped in a geotextile | 0.250 | 0.100 | 0.350 | Small to medium planted directly into the wall substrate | Roots parallel to ground, 90° to wall | |
| 2 | Façade | Modular trellis | Planter boxes at base of wall | 0.080 | 0.010 | 0.090 | Climbers | Vertical | |
| 3 | Living wall | Modular growing panels with stainless steel mesh grid | 'Mixed substrate' | 0.230 | 0.120 | 0.350 | Small plants rooted directly into wall substrate | Parallel to ground, 90° to wall | |
| 4 | Living wall | 'Parabentia' modular panel system – growing medium sandwiched within a stainless steel cage | Proprietary growing medium: a 'composite peat moss' | 0.080 | 0.120 | 0.200 | Small plants rooted directly into wall substrate | Parallel to ground, 90° to wall | |
| 5 | Living wall | Moulded UV-resistant plastic planter panel | 'Green roof substrate' | 0.070 | 0.110 | 0.180 | Small plants rooted directly into wall substrate | Angled from horizontal | |
| 6 | Living wall | Individual planters on a stainless steel frame | Soil | 0.065 | 0.055 | 0.120 | Small plants rooted directly into wall substrate | Vertical | |
| 7 | Living wall | Initially ceramic based tiles with pre-grown moss, replaced part-way through with flexible mat panels | Soil sandwiched between two-layers of moisture-retaining 'mat' | 0.060 | 0.120 | 0.180 | Small to medium plants rooted directly into wall substrate | Vertical | |
| 8 | Living wall | 'Plant cassette' – planters secured to wall via hinges | Lightweight growing medium | 0.280 | 0.200 | 0.480 | Small to medium-large plants rooted directly into wall substrate | Vertical | |

* Width here refers to the depth of the substrate and/or plant material in front of the concrete wall surface.

FIGURE 5.25 Green wall systems in HortPark in Singapore (see further Table 5.5). © Alex Tan

a bare wall of similar construction. The green wall types tested were: direct greening with English ivy, an indirect façade system with English ivy growing on a steel mesh, a living wall HDPE modular system subdivided into planter boxes (Greenwave system) in which the fern *Pteropsida* was planted, and a system based on layers of felt into which *Pteropsida* was planted. The energy savings for the different wall types were estimated for both temperate and Mediterranean climates (the latter approximating the climate conditions in the northern Italian city of Genoa), but savings on air conditioning were not calculated for the temperate climate. The green walls were estimated to reduce temperatures by 2.6°C in temperate zones and 4.5°C in the Mediterranean. There was no difference in the percentage energy saving in heating between the temperate and Mediterranean climates, but the systems delivered different savings: direct 1.2%, indirect (façade) 1.2%, Greenwave (planter) 6.3% and felt 4.5%. The energy savings due to reduced air conditioning in the Mediterranean were, however, uniform for all systems at 43%.

## 5.6.3 Shade

Different plants have different shading characteristics governed by such aspects as size and morphology of leaves, and density of growth – and these will be further affected by specific site conditions such as planting aspect, nutrient and water availability, and climate. As Pérez et al. (2011a) note, it is important to know which plants grow best in the locality. They tested four species – two evergreen (English ivy and honeysuckle *Lonicera japonicum*) and two deciduous (Virginia creeper and clematis) – in an experiment carried out in Puigverd de Lleida in Catalonia, Spain (an area with a dry continental Mediterranean climate). Plants were grown for a year, up an experimental mesh trellis, and then their shading ability measured with a light meter in July. They found the deciduous clematis did not grow well in the local conditions, and though Virginia creeper did cast strong shade, it did not grow well on the trellis. The two plants that grew best were the ivy and honeysuckle. Average daily light transmission (smallest figures = most shade) were: Virginia creeper (0.15) > honey-suckle (0.18) > ivy (0.2) > clematis (0.41). Vegetation performed as well at casting shade as building features incorporated by architects such as slats and awnings. Virginia creeper was also the plant studied by Ip *et al.* (2010) as window shade. The two study windows were single-glazed, measured 1.2 m high x 2.1 m wide and were fitted to a southeast-facing wall; the room volume was not given but the floor dimensions were 3 m x 3 m. In their study their 'bioshader' (plants grown up a trellis system from window planter boxes on the exterior of the building) reduced peak internal summer room temperatures by 4–6°C. However, energy savings on air conditioning may be negated by additional lighting requirements if shading is too effective, although the authors suggest design modifications may minimise this effect. Ip *et al.* (2010) also noted that as the season progresses Virginia creeper changes leaf colour before they are shed, allowing solar radiation gain in the winter months. The colour change may also have an interesting effect on mood by room occupants, though this was not mentioned or commented on.

   Wisteria growing on a metal meshwork trellis around three sides (SW, SE and NW) of a reconditioned building in Golmés in Lleida, Spain was monitored over a 12-month period, and was found to perform well in casting shade during the summer months. When there were no leaves, light penetration was reduced by 10,000–30,000 lux, but when leaves were present and fully expanded, light penetration was reduced by a massive 80,000 lux (Perez *et al.*, 2011a). Temperature records of the building surface were only made for the months May to October, but show that unshaded areas were on average 5.55°C warmer than shaded; in late summer (August and September) shaded areas were up to 15.18°C cooler than unshaded. Air temperatures inside the gap between the plant trellis and the building's walls did not show dramatic differences, but the trend was for air temperatures to be slightly warmer when leaves were not present (best: 1.6°C, SW facing) and cooler when leaves were present (max 3.8°C). Differences in relative humidity (RH) were also low in the gap between trellis and building walls, with RH being lower without leaves and higher with leaves.

The Institute of Physics building in Berlin-Adlershof, which uses wisteria grown in planters as façade shading, is interesting because the plant chosen is deciduous so that shade is cast in summer, but the maximum heat gain from the sun is obtained in winter. In this experimental design, the heat reduction through evapotranspiration is estimated by the water supply to the planter boxes (which, having a small surface area compared to the leaf area of the wisteria grown in them, contribute little to the evaporative cooling). Measurements for the south face of the building between July and Sept 2005 showed that evapotranspiration varied with the height above ground (floor of the building) but was equivalent to a mean cooling requirement of 157 kWh a day (Schmidt, 2009).

### 5.6.4 Local air temperatures/urban heat island

Generally speaking, vegetation should contribute to a lowering of local air temperatures and ultimately to a reduction in the urban heat island effect, although the impact of any one green wall would be rather small. Little information is available on this aspect, but Alexandri and Jones (2008) modelled the impact of green roofs and green walls on temperatures during the hottest month in nine cities with different climates (Athens, Beijing, Brasilia, Hong Kong, London, Montreal, Moscow, Mumbai and Riyadh) and showed that green walls, on their own, could make a modest contribution to reducing air temperatures. Hotter countries should benefit most from greening and this is borne out by the results. The predicted daytime surface temperature reduction of a south-facing wall in Riyadh was predicted to be 18.7°C at maximum with a daily average temperature reduction of 14.3°C; air temperature reductions in the canyon were predicted to be 5.1°C maximum and average 3.4°C. For cooler Moscow, the equivalent reductions were 9.8°C and 5.6°C for surface temperatures and 2.6°C and 1.7°C for air temperatures. Green walls can also contribute to reducing the surface temperature of asphalted ground and, whilst modest (Riyadh: maximum 2.0°C and average 1.3°C), helps reduce air temperatures further.

### 5.6.5 Damp

Vegetation stops walls getting wet which contributes to reduced heating costs; well-developed vegetation completely stops even the heaviest rain reaching the wall surface (Rath and Kiessl, 1989, cited in Ottelé, 2011). But also, if there is a problem with damp on interior walls, vegetation can help eliminate this by stopping lateral transmission from the outside to inside the building (Johnston & Newton, 2004), and may help by reducing cold spots where condensation facilitates mould formation.

### 5.6.6 Protection from degradation

Doernach (1979) reports that plants growing directly on walls (for up to 60–70 years) 'in many cases' protects rather than damages rendering, claiming similar houses needed walls re-rendering three to four times over the same period. However, he notes that in some cases, on north and west walls, damage has been reported and recommended that evergreens should not be grown directly on west-facing walls. Viles *et al.* (2011) showed that the aerial rootlets of ivy do not harm the surface of a range of stones, including sandstone and limestone, but protect them from erosive and degradative forces (but see also section 5.3.2 for the impact of true roots on walls). Minke and Witter (1982, cited in Ottelé, 2011) record that a bare wall can heat to +60°C and rapidly cool to -10°C, whereas a wall covered with vegetation has an equivalent range of +30°C to +5°C – reducing damage caused by thermal expansion and contraction. Peck *et al.* (1999) also noted that in addition to moderating the freeze–thaw cycle, green walls also potentially reduce damage caused by acid rain, ice and pollution.

### 5.6.7 Air pollution

Air pollution has strongly negative effects on buildings and monuments (Grossi & Brimblecombe, 2007; Warscheid & Braams, 2000). Warscheid & Braams (2000) illustrate this dramatically with two contrasting photographs of the same stone angel mounted on the 'Peters'-Portal of the cathedral at Köln (Cologne): first in its crisp pristine state in 1880 and then 113 years later in 1993 as a scarcely recognisable mess after weathering and staining. The nature of air pollution has changed over time with, for example, high levels of sulphate deposition and soot from coal burning (domestic and commercial) in the 19th to 20th centuries giving way to particulates and nitrogen oxides emitted from diesel vehicles in the mid-20th to early 21st centuries (Grossi & Brimblecombe, 2007).

There is little information available on the removal of gaseous air pollutants by green walls, but Morikawa *et al.* (2003) suggested that a wall of petunias could remove up to $1.57 \times 10^4$ tonnes of nitrogen dioxide a year due to its $NO_2$-phillic nature. Early work by Köhler *et al.* (1993, cited in Köhler, 2008) suggested that 4% of the particulate pollution in an inner-city area could be captured by green walls if all possible surfaces were used. Currie and Bass (2008) attempted to model the impact of juniper (*Juniperus* sp.) hedges placed 3 m around residential buildings in Midtown, Toronto, and showed modest levels of pollutant removal (Figure 5.26). The use of junipers in the model was hardly realistic, but was determined by other aspects of the study aims.

Köhler (2008) summarised research by Thoennessen (2002) which showed Boston ivy could act as a trap for a range of metals (Al, Cd, Co, Cr, Cu, Fe, Ni, Pb) deposited as dust (see also Bruse *et al.*, 1999, and Thönnessen and Werner, 1996). The measurements were taken from a house on a busy street at heights ranging from 2 to 13.5 m in Düsseldorf. Most dust was trapped on leaves 2 m above ground, and the leaves continued to act as a trap over the growing season, peaking in the autumn. Bruse *et al.* (1999) showed that streets with the highest traffic density (50,000–55,000 vehicles/day) also had the highest particulate emission rates (10 μg.s$^{-1}$); streets with relatively low traffic volumes (7,700–8,300 vehicles/day) but a high proportion of 'trucks' (41.2–47.2%), also had high particulate levels (5.1 μg.s$^{-1}$). Plant leaves sampled in a narrow street had the highest concentration of particulates despite low levels of traffic (12,000 vehicles/day) with the narrowness leading to much recirculation of the particulates compared to more ventilated streets. Sternberg *et al.* (2010) showed that English ivy, covering historic walls in Oxford, acted as an effective particulate trap, removing up to 2.9 x $10^{10}$ particles per m$^2$ of leaf including those most implicated in human health impacts (i.e. $PM_{10}$ and, especially, $PM_{2.5}$ and $PM_1$). More particulates were captured by ivy where traffic volume was highest; at a rural site the same study showed much lower particle densities (1.2 x $10^4$ m$^{-2}$). An elemental analysis of the particulates indicated that particles at the rural site were composed of organic materials derived from sources such as pollen, whilst the particulates at the urban sites originated from diesel and coal combustion. Ivy was considered to have value in protecting historic walls from air pollution as well as value in removing substances inimical to human health. Ottelé *et al.* (2010b) carried out a similar study to Sternberg *et al.* (2010) near Bergen op Zoom in the Netherlands: particulate deposition was compared between ivy growing on a sound barrier next to a busy road and ivy growing up a tree in a woodland. The overall results were broadly similar, with particulate loads being higher at the roadside (1.47 x $10^{10}$ particles per m$^2$) compared with the wood (8.72 x $10^9$ m$^{-2}$), although these figures include particulates trapped on the underside of leaves as well as the upperside. The upper surface trapped about twice as many particulates as the underside, and there were a high proportion of fine and ultrafine particulates trapped on leaves. English ivy was found to be better than painted metal, aluminium, glass and paper in absorbing particulates (Ottelé, 2011).

One insight by Sternberg *et al.* (2010) gives particular pause for thought: they note that although trees are well known as particulate traps, climbers, such as ivy, remove pollutants from people's immediate environment, and where much of the pollution is generated. In this respect, hedges may be of particular interest, though there is little information available. Varshney and Mitra (1993) showed that hedges of *Bougainvillea spectabilis*, *Duranta plumieri* and *Nerium indicum* planted along roads in New Delhi, India, could improve air quality by removing in excess of 30% of particulates. Particulate removal varied between species, with *D. plumieri* being most efficient, probably due to differences in hedge structure and

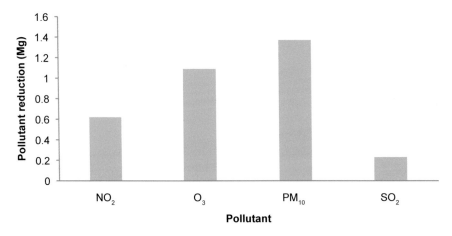

FIGURE 5.26 Annual predicted pollutant reduction in Midtown, Toronto, Canada, by planting juniper hedges around residential buildings. Source: Currie and Bass (2008).

leaf characteristics. Tiwary *et al.* (2008) examined the particulate reduction potential of rural hawthorn *Crataegus monogyna* hedging in the UK and came up with a very similar $PM_{10}$ reduction potential as Varshney and Mitra (1993): 34%. Whilst hawthorn is the most abundant rural species in the UK, which is the reason they chose to study it, it is unlikely to be dominant in urban areas: their source of information, the Countryside Survey (Barr & Gillespie, 2000), does not include settlements. Earlier work done by Tiwary *et al.* (2005) examined the capture efficiency of three different species used in hedges – hawthorn, yew *Taxus baccata* and holly *Ilex aquifolium* – and showed that the porosity of the hedge to wind-flow was likely to affect particulate capture. A hedge is likely to be less effective at collecting particulates by impaction if airflow is diverted over it than one where air can move through the foliage. However, such turbulent air may result in higher levels of impaction on the lee side of a hedge. Very dense hedges would tend to act as rigid collectors and conifers, especially, will tend to become increasingly so with increasing windspeed; such hedges will tend to send airflow over the hedge, creating a region of turbulent air behind. Less dense hedges, particularly deciduous or evergreen broadleaved species, with less rigid foliage permit airflow through and over; such hedges increase in permeability with increasing windspeed as leaves are displaced (see references in Tiwary *et al.*, 2005). The same study also points out that hedge porosity varies vertically (as anyone who has seen a 'leggy' hedge with little foliage at the base can verify), which will affect collection efficiency, as will the complexity of the internal structure of the hedge. In their study, hawthorn's greatest foliage density was at 0.86 the height of the hedge, holly's 0.88 and yew's 0.5 (e.g. for a 1.5 m high hedge the greatest density would be at 1.29 m above ground for hawthorn, 1.32 m for holly and 0.75 m for yew), but it was clear that the foliage density was much more uniform over the surface of the yew hedge in their study than either that of holly or, especially, their hawthorn hedge. With airflow through a hedge, it is clear that particulates can be captured not only on the foliage-rich surfaces, but also by the internal stem and twig structures, and that the capture efficiency will be affected by their shape, density and resistance to movement by wind (Tiwary *et al.*, 2005). Hedge management will also impact on airflow, as it is likely to affect these parameters, but this was not considered by Tiwary *et al.* (2005). Overall their modelling suggested that capture efficiency of the hedges was hawthorn>holly>yew for 15 μm particles at windspeeds of 2.3 $m.s^{-1}$ at 0.75 hedge height but, of course, this was when hawthorn had foliage. Yew appeared to perform better at capturing very fine particles than hawthorn (below 3.5 μm).

Perhaps one of the most obvious differences between free-standing 'curtain' green walls and those fixed to, growing up, or placed immediately adjacent to walls is that air movements through, over and around the structures will probably differ; partly due to their inherent characteristics, and partly due to their placement. As a result, the same structures (such as hedges) may vary in their ability to capture particulates solely on their proximity to an impenetrable barrier (as it will affect airflow). As yet there is no information available on the impact of placement on particulate capture.

### 5.6.8 Noise pollution

Vegetation can act as a sound barrier, reducing the impact of general urban noise pollution from roads, etc. Buchta *et al.* (1984) and Bastian and Schreiber (1999, cited in Köhler, 2008) suggest that Boston ivy can reduce sound by 5 dB and *Rubus* or *Fallopia* by 2–3 dB.

More recently Wong *et al.* (2010b) tested the eight different green walls (see Table 5.5; Figure 5.25 for descriptions and visuals) set up in Singapore's HortPark (see section 5.6.2) for sound attenuation. Formally, the parameter tested was 'insertion loss' – the reduction in sound caused by an object inserted equidistant between a sound source and the sound receiver (a person or microphone). Because the parameter of interest in these experiments was the sound absorbent effect of the vegetation (and associated infrastructure), what was being assessed was the sound pressure (in decibels) received behind a control wall without vegetation minus the sound received behind walls with added greenery. Four frequency bands were analysed 63–125 Hz, 125–1250 Hz, 1250 Hz to 4 kHz, 4–10 kHz. The growth and cover of foliage, substrate and gaps exposing bare wall surfaces differed between the different green walls and impacted on the sound reduction. Generally speaking, the green walls reduced sound transmission, though not equally, with walls 1, 2, 5, 6 and 7 reducing sound between 5 and 10 dB, in the human hearing range, whilst walls 3 and 4 performed poorly. Performance varied within and between frequency ranges with no impact on the lowest (63–125 Hz) band due to diffraction bending the sound around the relatively narrow walls. The greatest sound reduction took place with two of the living wall systems (7 and 8) and the façade system with planter boxes at the base of the wall (wall 2). In the latter case, it may be the planter box, that was at a similar height to the sound source, that absorbed most of the sound, a reduction which would not be typical of the whole wall. Sound absorption was highest in the 125–1250 Hz (perceptible) range for walls 2 (9.9 dB) and 7 (8.4 dB), and in the 4–10 kHz range for wall 8 (8.8 dB). An air gap behind the planter boxes of green wall 5 was thought to have contributed to a 7.0 dB reduction in the 125–1250 Hz range. Van Renterghem *et al.* (2013) modelled various aspects of greening façades on sound reduction in a courtyard behind a building which fronted onto a road canyon. They found that greening façades could help reduce sound propagation but the extent of the effect depended on the underlying façade material and the placement of the greenery. If the building surfaces (road canyon or courtyard) were made of a relatively soft brick, its absorption of noise would be greater than for a hard, rigid material – and hence the value of greening would be reduced. The placement of greening in the road canyon appeared to have most value when placed on the upper half of building frontages as it reduced sound propagation. Within the courtyard, behind the building, the effect was greatest if greening covered the whole of the vertical surface as it reduces reverberation. The reduction in sound in the courtyard was maximised by greening façades as described and combining it with greening the roofs, as sound propagated over the roof could thereby be reduced. Installation of multiple acoustically absorbing screens placed on rooftops was also shown to assist in sound reduction, and it is possible that free-standing green walls may have a role to play here too.

Ding *et al.* (2013) carried out laboratory studies of the potential sound-reducing action of foliage in three frequency bands: below 250 Hz, 500–2,000 Hz, and 2000–6,500 Hz in a 29 mm diameter impedance tube. They used 25 mm deep cylinders of 'Armafoam Sound 240' and 30 mm deep cylinders of melamine foam to simulate growing material (substrate) with different sound absorption characteristics; the Armafoam had a higher density (and thus lower permeability) than the melamine. They tested the effect of placing discs of leaf material with different densities in front of the model substrates. The plants used, with different densities, thicknesses and leaf surface characteristics, were from Japanese andromeda *Pieris japonica* (leaf density 0.367 kg.m$^{-2}$, 0.41 mm thick), the rhododendron 'scarlet wonder' *Rhododendron forrestii* (0.408 kg.m$^{-2}$, 0.34 mm), primrose *Primula vulgaris* (0.469 kg/m$^{-2}$, 0.74 mm) and Corsican hellebore *Helleborus argutifolius* (0.22 kg.m$^{-2}$, 0.43 mm). The results showed that at low frequencies vegetation has no additional effect on sound absorption above that of the substrate, but that it does increase sound absorption in the 500–2,000 Hz band. This frequency range is that produced by traffic noise, and the value of vegetation in enhancing noise reduction was most effective where the substrate had poorer sound-absorbing qualities. Above 2000 Hz, foliage was shown to reduce sound absorption compared to bare 'substrate' but fortunately this range is not considered relevant in terms of 'environmental' noise. Using the experimental results in a model, they showed that leaf surface density does affect sound absorption, with leaves with lower surface densities increasing sound absorption over a wider frequency range compared with those of higher surface density.

### 5.6.9 Internal air conditioning

Green walls grown on the external surfaces of buildings can also be integrated into the internal air-conditioning system of buildings with air sucked through the plants and growing media into an air space (Figure 3.7). The air is then distributed through the building using ducting. As with active indoor green walls (see Chapter 3), the plants and rooting media act as a biofilter, removing pollutants and moderating the air temperature and humidity (Franco *et al.*, 2012).

### 5.6.10 As components of sustainable urban drainage

Whilst there has been a considerable amount of work done in evaluating the value of green roofs in reducing stormwater runoff, there is very little on the value of green walls, even though they have considerable potential in the interception of rainfall. To investigate this, Ostendorf *et al.* (2011) created 18 circular (2.3 m diameter) dome-like retaining walls made of staggered blockwork at the campus of the Southern Illinois University at Edwardsville in 2007. Three of the 'domes' were left unplanted whilst the remaining 15 were planted up with either one of four different *Sedum* (stonecrop) species or a mixture of all five species (three 'domes' per species or mix). The domes were constructed on an impermeable membrane of concrete blocks with a cavity for growth media and were five 'blocks' high. The growth media for the experimental walls was 80% coal bottom ash mixed with 20% composted pine bark and was used in the growing spaces between blocks and as a 5 cm high filler between courses of block. Water from each dome was collected from the basal membrane and piped to collection buckets. The different planting regimes resulted in different levels of vegetated cover (despite some replanting) with *Sedum kamtchaticum, S. spurium* and the sedum mix having greater than 50% coverage whilst *S. hybridum* 'immergrauch' and *S. (Phedimus) takesimensis* had less than 40%. However, whilst the planted walls tended to have lower runoff and some delay in that runoff compared with the unplanted control, as might be expected, the experiment was inconclusive with no statistically significant differences in runoff. Despite this Ostendorf *et al.* (2011) considered that vegetated retaining walls did have the potential to reduce runoff, but acknowledged that plant characteristics, cover and rainfall appeared to interact in a complex fashion. This is clearly an area where further research is needed.

### 5.6.11 Anti-social behaviour

Whilst green walls can be used to cope with antisocial behaviour such as covering up graffiti and eyesores (Mir, 2011), for example with ivy (Figure 5.27), the law of unintended consequences can come into play. I have been told of the green wall clothing a car park in Birmingham, UK, that had to be removed because dense ivy growth obscured vision into the car park which, unfortunately, increased car-related crime. So careful thought needs to be given to safeguarding; in an attended, limited access or private company car park this would probably not have been an issue.

### 5.6.12 Wildlife

#### Vertebrates

Surprisingly little information is available on the benefits to fauna of green walls, but Johnston and Newton (2004) note that birds such as blackbird *Turdus merula*, house sparrow *Passer domesticus*, song thrush *Turdus philomelos* and wren *Troglodytes troglodytes* use climbers as nesting habitat (Figure 5.28). Köhler *et al.* (1993) reported the use of green walls by house sparrow, greenfinch *Carduelis chloris* and blackbird, including some breeding activity. Chiquet *et al.* (2013) carried out a formal comparison of the use of bare walls and green walls by birds. Twenty-seven bare walls were compared with 27 paired green walls with plants growing directly up the wall. Pairs of walls were in the same locality and were identical in physical characteristics, with the exception of vegetation cover. Plants growing on the walls were mainly evergreens (English ivy, *Hedera colchica* 'sulphur heart', *H. colchica* 'dendrata variegata', the Atlantic or Irish ivy *H. iberica*) but also included deciduous species (*Wisteria, Pyracantha, Jasminum* and *Parthenocissus*: *P. quinquefolia* Virginia creeper, and *P. tricuspidata* Boston ivy or

FIGURE 5.27 Ivy and other climbers can reduce the impact of graffiti. © John Dover

Japanese creeper). Walls were either the outside walls of buildings or boundary walls. Birds were recorded from the walls, the roofs of buildings (or the top of boundary walls) and the vegetation in the immediate locality (enclosed by a semicircle of 10 m radius in front of the wall). Overall they found nine species of bird during the study. Birds were only ever seen on the surfaces of green walls, never on the surface of bare walls; birds were also never seen on the vegetation surrounding bare walls – only on vegetation surrounding green walls. The only place that birds were observed in association with bare walls was on the roof of buildings or on the top of bare walls (Table 5.6).

Unfortunately bird abundance was not high enough to carry out statistical comparisons for individual bird species, but overall there were 4.5 times more birds associated with walls (roof + wall + surrounding vegetation) when the wall was vegetated compared to bare equivalents. The study was carried out in summer (July and August) and winter (January, February and March), and although green walls always had more birds than bare walls, no matter what the season, the difference was greatest in the winter when shelter was scarce. When the data was split between deciduous and evergreen walls, there was no difference between the types of wall in the summer, but in the winter it was clear that the evergreen walls were far superior to both bare walls and to deciduous walls – as might be expected. The work was carried out outside the nesting period so the value of green walls to birds at that time could not be quantified; resources available to birds during the study were considered to be mainly physical ones of sheltered roosting/perch sites, perhaps with a heat subsidy from the walls of buildings, but they are also likely to be a source of invertebrate and plant (berry) food (Jacobs *et al.*, 2010; Köhler *et al.*, 1993). Chiquet *et al.* (2013) stress that theirs was a

FIGURE 5.28 Green walls are used as nesting habitats by some birds, with the evidence often uncovered when the vegetation is cut back. © John Dover

**TABLE 5.6** Birds found in association with green and bare walls, their roofs[†] and surrounding vegetation

| Common name | Scientific name | Wall surface | Roof | Surrounding vegetation |
|---|---|---|---|---|
| Pigeon | Columba livia | ★ | ★ | |
| Collared dove | Streptopelia decaocto | ★ | ★ | |
| Jackdaw | Corvus monedula | | ★ | |
| Rook | Corvus frugilegus | | ★ | |
| Magpie | Pica pica | | ★ | |
| Robin | Erithacus rubecula | ★ | | ★ |
| House sparrow | Passer domesticus | ★ | ★ | ★ |
| Starling | Sturnus vulgaris | | ★ | ★ |
| Blackbird | Turdus merula | ★ | | ★ |

Source: Chiquet et al., 2013.
[†]For boundary walls, the top of the wall is considered the 'roof'.

preliminary study, to establish whether wall vegetation conferred any benefit to vertebrates; in that it was successful, but it is clear that there is much detailed work yet to be done on the value of green walls for birds.

If the work of Chiquet et al. (2013) on birds is somewhat sketchy, then we are probably still sharpening pencils in respect of our knowledge of other vertebrates. Little is known other than snippets such as that shrews and voles use green walls for 'resting and feeding' (Johnston & Newton, 2004) and that bats have also been observed hunting around green walls (Köhler et al., 1993).

## Invertebrates

Surprisingly little information is available on the value of green walls to invertebrates, and most work reported has been done in Germany. Köhler et al. (1993) gave lists of the main beetle, spider and fly species found by Bartfelder and Köhler (1987) (Table 5.7). Köhler et al. (1993) also note that one of the greatest prejudices of the urban dweller against green façades is a fear of pests, but suggest that this is unfounded as other habitat types probably hold higher densities and imply that green façades actually decrease the numbers of invertebrates in buildings. Whilst green façades may act as alternative habitats for invertebrates in urban areas, Köhler et al. (1993) have shown that it can also be the only habitat exploited; as was the case for Agrilus derasofasciatus one of the buprestid jewel beetles found on green walls in Berlin but previously only recorded from vineyards, likewise the small spider Nigma walckenaeri normally found in the Mediterranean. Some ivy specialists were also identified (e.g. the ivy-boring beetle Ochina ptinoides).

Köhler et al. (1993) also gave the results of invertebrates found at five study sites by Hagedoorn and Zucchi (1989) in which young and old plants of grapevine and ivy were compared with bare walls. While some species groups showed little difference between the different climbers and bare wall (e.g. spiders), some groups were clearly more abundant on the vegetation. There were also differences in abundance between the type of vegetation

TABLE 5.7 Invertebrates found by Bartfelder and Köhler (1987) on green walls in Berlin

| Scientific name | Common name |
| --- | --- |
| **Beetles** | |
| *Ochina ptinoides* | Ivy boring beetle |
| *Ptinus fur* | White-marked spider beetle |
| *Meligethes aenus* | Pollen beetle |
| *Mescoelopus niger* | |
| *Ptinus sexpunctatus* | Spider beetle |
| *Otiorhynchus ovatus* | Strawberry root weevil |
| *Dasytes plumbens* | |
| *Propylaea 14-punctata* | 14–spot ladybird |
| *Stegobium paniceum* | Biscuit beetle |
| *Otiorhynchus sulcatus* | Black vine weevil |
| *Anthrenus verbasci* | Varied carpet beetle |
| *Serica brunnea* | Brown chafer |
| *Trogoderma angustum* | Stockholm beetle |
| **Spiders** | |
| *Tegeneria domestica* | Domestic house spider |
| *Theridon melanurum* | |
| *Harpactea rubicunda* | |
| *Harpactea spp.* | |
| *Philodromus spp.* | |
| *Salticus olearii* | |
| *Agyneta rurestris* | |
| *Theridon tinctum* | |
| *Nigma walckenaeri* | |
| **Flies** | |
| *Lycoria modesta* | |
| *Megaselia subgensus* | |
| *Medetera truncorum* | |

Source: Köhler *et al.*, 1993.

with some favouring ivy over vine and vice versa. For example, Acarina (mites) were more abundant on ivy whilst Collembola (springtails) were more abundant on vines, especially young vines, although it might be stretching a point to define Collembola as living on plants: they were found in association with the stems and presumably at the base. There were also differences between age groupings of the same plant with Psocoptera far more abundant on older compared with young vines (1785 and 124 respectively). Dead and old stems were found to be valuable components of older growth, providing habitat for wood-boring and saproxylic species. Hagedoorn and Zucchi (1989) also looked at the food of invertebrates and grouped them into three food-utilising guilds:

- *Saprophages* (Collembola, Cryptonagidae, Endomychidae, Lathridiidae and Psocoptera)
- *Phytophages* (sap sucking: Homoptera & Thysanoptera; leaf-eating: the black vine-weevil *Otiorhynchus sulcatus* and the chrysanthemum flea beetle *Longitarsus succineus*; and flower feeders: the pollen beetle *Meligethes aeneus*, the carpet beetle *Anthrenus fuscus*, Apoidea and Lepidoptera)

- *Predators* (Araneae, Opiliones, the anthocorid bug *Orius majuscules*, the ladybird *Stethorus punctillum*, lacewing (Chrysopidae) and ladybird larvae (Coccinellidae)) (Köhler *et al.*, 1993).

Jacobs *et al.* (2010) studied ivy on hedges rather than green walls on buildings and noted its requirement for insect pollination. Individual species identified as visiting ivy included: bumblebees *Bombus hypnorum*, *B. pascuorum* and *B. terrestris/lucorum*; honeybees *Apis mellifera*; hoverflies *Episyrphus balteatus*, *Eristalis tenax*, *Myathropa florae*, *Syrphus ribesii*, *S. vitripennis*, *Sphaerophoria scripta* and *Syritta pipiens*; and wasps *Vespula vulgaris* and *V. germanica*. Other flower visitors were identified to genus (some hoverflies) or, for bristly flies, to family: the Calliphoridae, Muscidae, Sarcophigae and Tachinidae. Observations of pollinators in September–October of 2005 and 2007 showed that over half the visits (54.7%) were made by wasps, and a third by bristly flies (33.6%) (Figure 5.29). The hymenoptera carried more pollen on their bodies than flies. Of the hymenoptera, wasps were considered to be the most effective pollinators being more abundant, visiting more flowers and being more likely to brush the stigma with their bodies during feeding and so actually pollinating the flower. The attraction of ivy for wasps is perhaps a consideration when greening domestic and office buildings – in that it might be advisable to place ivy on walls with no windows or failing that keeping windows firmly closed during the period that ivy flowers.

Francis and Hoggart (2008, 2009) established that river walls and embankments along the River Thames in London had a mix of terrestrial and riparian plants and were likely to support a mix of terrestrial and freshwater/riparian invertebrates such as the freshwater sandhopper *Orchestia cavimana* (Francis & Hoggart, 2008). Attrill *et al.* (1999) examined a range of wall structures along the Thames estuary foreshore: low brick/boulder walls, high brick walls, concrete walls and metal walls for invertebrates. Perhaps not unexpectedly they found more rugose wall types (brick/boulder) superior to smooth-faced (concrete, metal) as invertebrate habitat, although the relationship was not significant (Table 5.8).

Hoggart *et al.* (2012) also reported the results of a macroinvertebrate survey of 15 vertical river wall sites along the River Thames carried out in 2007; samples were taken 1 m above and 1 m below high tide mark (indicated by algal cover). The survey resulted in the capture of

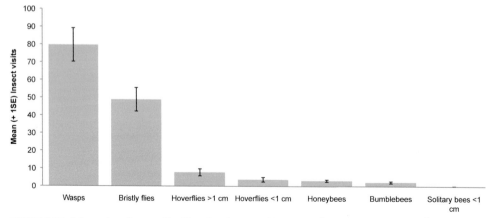

**FIGURE 5.29** Mean abundance of pollinating insects visiting ivy flowers in 18 x 0.5 m$^2$ quadrats. Source: Jacobs *et al.* (2010).

1401 invertebrates from 37 groups mainly from the Isopoda, Araneae, Diptera, Gastropoda, Oligochaeta. Most were not identified to species; those that were included the Isopods (woodlice) *Porcellio scaber*, *Armadillidium vulgare*, the marine amphipod *Orchestia cavimana*, the stripe-legged spider *Harpactea hombergii*, and the freshwater snails *Galba truncatula*, the New Zealand mud snail *Potamopyrgus antipodarium*, and *Radix balthica*. As with the study of Attrill *et al.* (1999), brick surfaces were shown to be better than concrete and greater surface

**TABLE 5.8** Some species found associated with river walls along the River Thames foreshore in London

| Group | Species | Medium complexity* | | Low complexity* | |
|---|---|---|---|---|---|
| | | Low brick/ boulder walls | High brick walls | Concrete walls | Metal walls |
| Araneae | Lepthyphantes tenius | * | | † | |
| Araneae | Philodromus cespitum | | | † | |
| Araneae | Salticus scenicus | | * | | |
| Chilopoda | Lithobius melanops | * | * | | |
| Chilopoda | Lithobius melanops | | * | | |
| Coleoptera | Adalia bipunctata | * | | | |
| Coleoptera | Agonum albipes | * | | | |
| Coleoptera | Amara aenea | * | | | |
| Coleoptera | Bembidion harpaloides | * | * | | |
| Coleoptera | Bembidion maritinum | * | * | | |
| Coleoptera | Coccinella septempunctata | | | † | |
| Coleoptera | Dromius linearis | | | † | |
| Coleoptera | Galerucella nymphaeae | * | | | |
| Coleoptera | Harpalus aeneus | | | † | |
| Coleoptera | Harpalus rufibarbis | | | † | |
| Coleoptera | Micraspisse decempunctata | * | | | |
| Coleoptera | Notiophilus rufipes | | | † | |
| Coleoptera | Thea vigintiduopunctata | | | † | |
| Diplopoda | Blanniulus guttulatus | | | † | |
| Hymenoptera | Lasius flavus | | | † | |
| Hymenoptera | Lasius niger | * | * | * † | |
| Isopoda | Armadillidium vulgare | * | | | |
| Isopoda | Ligia oceanica | * | * | * | |
| Isopoda | Oniscus ocellus | | | † | |
| Isopoda | Philoscia muscorum | * | | † | |
| Isopoda | Porcellio scaber | * | * | | |
| Mollusca | Candidula intersecta | * | | | |
| Mollusca | Cochlicopa lubrica | | | † | |
| Opiliones | Lacinius ephippiatus | * | | † | |
| | Mean species richness | 6.75 | 4.75 | 3.05 (0.75)‡ | 0 |

Source: data provided by M. Attrill from the study by Attrill *et al.*, 1997.

* Complexity = relative habitat complexity of structures.

‡ An atypically high number of invertebrate species was found at Battersea Power station (†) where the concrete wall was topped by a layer of vegetation, removing this data give a more typical species richness for concrete walls.

heterogeneity positively influenced species richness. Additionally, Hoggart *et al.* (2012) also showed that algal cover was a positive influence. When data from above and below the high tide mark were compared, no significant influences were identified for samples taken from above the high tide mark – although the authors suspected a conservative form of analysis may have been over-influential. Below the high tide mark, all the analyses identified brick, concrete, heterogeneity and algal cover as important (as above) but also showed a negative impact of metal surfaces on species richness.

Recent work by Chiquet *et al.* (unpublished) has investigated the invertebrate fauna of green façades, living walls and green screens. The sampling was done using a Vortis® vacuum sampler (which samples all invertebrates) (Figure 5.30) and also through direct visual inspection for snails and spiders (Henschel *et al.*, 1992). A breakdown of the aggregated data (individual species names not given) can be found in Table 5.9 and clearly shows that all forms of green wall, no matter how simple, have value as invertebrate habitat. The data in Table 5.9 should not be used to directly compare between types of green wall as the locations, sample periods, maintenance, etc. differ. Use of green walls by invertebrates will depend on the resources provided, and needed, but may include breeding habitat, food (plant material (living or dead), invertebrate prey, nectar, pollen, etc.), basking surfaces, roosting areas and hibernating/aestivating cover.

FIGURE 5.30 Caroline Chiquet sampling invertebrates from a green façade using a Vortis® vacuum sampler. © John Dover

TABLE 5.9 Breakdown of the invertebrates found on green façades, living walls and green screens in the UK

| | Green wall type | Number of walls sampled | No of orders | No of families | Species richness[†] | Total abundance | Mean inverts/wall |
|---|---|---|---|---|---|---|---|
| **Insects** | Green façade | 29 | 11 | 86 | 208 | 6404 | 221 |
| | Living walls | 22 | 9 | 61 | 137 | 1399 | 64 |
| | Green screens | 4 | 6 | 37 | 59 | 360 | 90 |
| **Spiders** | Green façade | 29 | 1 | 13 | 25 | 2389 | 82 |
| | Living walls | 22 | 1 | 9 | 24 | 1112 | 51 |
| | Green screens | 4 | 1 | 3 | 3 | 67 | 17 |
| **Snails** | Green façade | 27 | 1 | 4 | 7 | 489 | 18 |
| | Living walls | 22 | 1 | 6 | 9 | 59 | 3 |
| | Green screens | 4 | 0 | 0 | 0 | 0 | 0 |

Source: Chiquet et al. (unpublished data), used with permission.
[†]Where species identity could not be established, a morphospecies category was assigned (Oliver & Beattie, 1996); for snails: of 27 green façades only 12 had snails, and of 22 living walls only 13 had snails.

## 5.6.13 Are green walls sustainable?

With increasing sophistication of green walls comes not just increased financial costs (Table 5.1), but also issues of sustainability, or the 'environmental burden' that a technology brings with it. This of course differs from the 'do they work' question and a comparative analysis of establishment and lifetime 'survival' is not available for the different systems – and failures do occur (see Middelie, 2009). Ottelé et al. (2011) carried out a lifecycle analysis of four types of green wall (direct, indirect (façade), living wall (modular) and living wall (felt)). The basic concept for sustainability is that of a cost-benefit analysis: if the environmental benefits outweigh the costs, the technology can be considered sustainable. Based simply on energy, the benefits of the systems tested vary with climate zone, and in the analysis by Ottelé et al. (2011) were greater for a Mediterranean than a temperate climate because of savings on air conditioning. A direct green wall with climbers was considered sustainable because additional resources were not required in its construction; an indirect (façade) system had a higher burden because of the stainless steel support material employed (it could be reduced by using a material with a lower environmental impact); a modular (planter) system, perhaps surprisingly, did not have a high environmental burden due to its construction material having a positive impact as an insulator; a felt-based system, however, was shown to have a high environmental burden – principally because of the short life of its construction materials (ten years compared to perhaps 50 for other systems (though typical guarantee periods tend to be a maximum of 25 years)). Of course, this analysis excluded other potential benefits (e.g. aesthetics, biodiversity, human health, particulate removal, mitigation of the urban climate, protection of building surfaces, etc.) and some costs were location-dependent. Only a small number of systems that are available were studied. A subsequent analysis by Perini and Rosasco (2013) came to similar conclusions in a cost-benefit analysis study of direct and indirect façades, the latter plus planter boxes, and a living wall system: they found that the more complex the system, the less economically sustainable it was. Ultimately there is little surprising in this: improvements in design, materials, and lifecycle analysis will impact on the perceived sustainability of this rapidly evolving technology.

## 5.7 Summary

- Green walls can be completely natural with plants colonising wall surfaces by growing up them or on them, or they can be created with specific intended functions.
- Green walls, as a group, vary in construction from the exceedingly simple to the very sophisticated, incorporating technologically advanced materials and electronic sensors.
- The green wall industry is exceptionally dynamic with new products and concepts constantly being brought to a relatively new market.
- There are real environmental benefits to be gained from using green walls in urban areas for both individuals and society as a whole.
- Whilst there is a considerable body of data existing on some aspects of green walls, e.g. insulation value, this is by no means the case for all attributes (e.g. air pollution mitigation, wildlife habitat) and far more research is needed.
- Values will vary with geographical location, climate and system used.

# 6

# GREEN ROOFS

This chapter will:

- define the different kinds of green roof
- identify the structural issues that may affect green roof choice
- cover retrofitting to existing buildings and incorporation in new developments
- describe green roof components/media as appropriate to type of roof
- cover planting and maintenance
- identify biodiversity and other ecosystem services
- outline their use in development mitigation
- consider the issue of 'whole life' costs.

## 6.1 Introduction

### 6.1.1 Preamble

The roofscape of urban areas can be considerable. Estimates vary between 20 and 40% of the surface area (Akbari *et al.*, 2003; Liptan & Strecker, 2003) and in particularly densely packed areas this may be exceeded; for example, 47% in parts of Sheffield, UK (SCC, 2011). With such high levels of sealed surface come the inevitable consequences of high levels of runoff. The ability of green roofs to act as elements of sustainable urban drainage and reduce the impact of the sealed-surface problem is one important aspect of their appeal to planners, but there are much wider benefits to be had. These include extending the life of roof membranes through reduced exposure to freeze–thaw cycles, reduced mechanical erosion (including from hail), and UV degradation. In addition to the extended lifetime of roof membranes (and hence reduced cost to building owners), an additional benefit accrues to society in reduced material going to landfill (Liu & Baskaran, 2003; Rowe, 2011). Green roofs also contribute to a reduction of the urban heat island effect, reduce energy consumption and thus $CO_2$ emissions, reduce air pollution, provide space for growing food, provide additional recreational and commercial space (e.g. rooftop terrace cafés and gardens), increase property values and create jobs (Peck, 2003). There are a substantial number of books now available on green roofs, pre-eminent is the pioneering volume by Dunnett and Kingsbury (2004) (second edition 2008) but those by Grant (2006b), Snodgrass and Snodgrass (2006), Snodgrass and McIntyre (2010), amongst others, are well worth a look.

### 6.1.2 Introduction: what is a green roof?

At its very simplest, a green roof is simply a roof with some form of vegetation growing on it: naturally colonised or deliberately planted. Grant (2006b) includes a lovely image of a large dock building in Cornwall, UK, whose roof is completely covered in moss that has naturally colonised the asbestos surface. Green roofs have been installed on large commercial buildings, domestic buildings, sheds, garages, bird boxes, small bus shelters and even buses themselves (Dunnett *et al.*, 2011; GRC, 2014; Phytokinetic, 2013) (Figure 6.1).

### 6.1.3 A brief history of green roofs

The history of vegetation on roofs is generally considered to extend back at least to the hanging gardens of Babylon, thought to have been constructed in the reign of King Nebuchadnezzar II (605–562 BC). Documentary evidence is available on its use by the Romans (source Pliny the Elder 23–79 BC), and the Vikings' (C8–11) use of turf on roofs (Figure 6.2) (Peck *et al.*, 1999) which has continued in the Scandinavian countries to this day. Some Icelandic turf houses look more like green mounds than houses, with the eaves of their turf roofs reaching ground level in some cases (Doernach, 1986), and look akin to

FIGURE 6.1 Spanish bus with a green roof created as part of the 'Gardens in Movement' project (Phytokinetic, 2013). © Marc Grañén, PhytoKinetic (www.phytokinetic.net)

FIGURE 6.2 Summer sheiling (hut) at Ulsåk-Stølen, Norway. © John Dover

Bilbo Baggins's home in Hobbiton (except the Icelanders use rectangular doors and windows whereas Hobbits favour round openings) (Tolkien, 1937).

Magill *et al.* (2011) give a nice historical account of green roofs from Babylonian times to the modern day and include the example of sand-and-gravel-topped roofs built in the 1880s in Germany being colonised by plants. Köhler and Poll (2010) give more detail of these roofs and estimated that about 2,000 were built in Berlin with about 50 surviving post-World War II, though by 2008 only three remained. Another example given by Magill *et al.* (2011) was the use of turf on World War II aircraft hangers in the UK to camouflage them (Figure 6.3), an approach sometimes used to help new industrial buildings blend into the landscape. An example of the latter is the 2,382 m² distribution centre for the brewer Adnams in Southwold, Suffolk, UK (Campbell, 2006) (Figure 6.4). In the early 20th century, architects Frank Lloyd Wright (1867–1959) and Charles-Édouard Jenneret-Gris (Le Corbusier, 1887–1965) started to incorporate green roofs in their designs (Peck *et al.*, 1999) and the building designs of Hundertwasser (Friedrich Stowasser, 1928–2000) are magnificent fantasies which incorporate vegetation, including green roofs, into the structures (Figure 6.5).

From the mid- to late 20th century, interest in green roofs picked up, because of their potential environmental benefits; this was primarily in a cluster of countries around Germany, with consequent technical improvements (Peck *et al.*, 1999). The final boost to the adoption of the technology was state grant-aid for the installation of green roofs in Germany (Boivin, 1992, cited in Peck *et al.*, 1999) and Austria (LivingRoofs, 2014), planning policies which required green roofs to be installed (Brenneisen, 2006; LivingRoofs, 2014; Maurer, 2006), and the development of formal technical specifications and standards (the German FLL Guidelines (Forschungsgesellschaft Landschaftsentwicklung Landschaftbau) completed the process (Peck *et al.*, 1999)). Since 2002 the FLL guidelines have been available in English

FIGURE 6.3 Old World War II aircraft hangars at Hullavington, Wiltshire, UK, coated in turf to act as camouflage. Hangar 88, pictured in 2013, is now used as a go-karting centre (UKGK, 2014). © John Dover

FIGURE 6.4 The extensive green roof at the Southwold distribution centre owned by the brewer Adnams is designed to blend into the landscape. © Adnams plc

FIGURE 6.5 Hundertwasser, the Forest Spiral of Darmstadt (1998–2000). Photo: P. Moszden. © 2014 Hundertwasser Archive, Vienna

as the *Guidelines for the Planning, Execution and Upkeep of Green Roof Sites* (FLL, 2008). An outline UK Green Roof Code, based on the German FLL guidelines, has recently been published (Groundwork, 2011). In a first for the UK, the City of Sheffield has recently published supplementary planning guidance (SCC, 2011) which specifies that single buildings in excess of 1,000 m² or a development of ten dwellings or more must have 80% vegetated cover, though this is subject to a viability assessment. In North America, whilst nothing quite like the FLL guidelines exists, there are guidance documents emerging (Dvorak & Volder, 2010) (Table 6.1) and incentives, such as reduced 'drainage' taxes, may be available (Liptan & Strecker, 2003).

The European green roof industry is now quite mature with a range of reliable products and a substantial manufacturing industry behind it, although there are a considerable number of questions relating to environmental performance which still need to be addressed. In the USA the industry is less mature and perhaps less certain of itself. Rosenthal *et al.* (2008) reported personal correspondence that indicated that the US green roof manufacturers would

TABLE 6.1 Green roof guidance documents issued by the American Association of Standards and Testing Materials (ASTM)

| Standard code | Title | Most recent version[†] |
|---|---|---|
| E2396 | *Test Method for Saturated Water Permeability of Granular Drainage Media [Falling-Head Method] for Vegetative (Green) Roof Systems*, to compare one type of media to another | 2011 |
| E2397 | *Practice for Determination of Dead Loads and Live Loads Associated with Vegetative (Green) Roof Systems*, to help assess the building for such considerations as structural design given the weight of the system without and with rain or irrigation | 2011 |
| E2398 | *Test Method for Water Capture and Media Retention of Geocomposite Drain Layers for Vegetative (Green) Roof Systems*, to help assess one system's performance relative to another as well as assess irrigation requirements for system designs | 2011 |
| E2399 | *Test Method for Maximum Media Density for Dead Load Analysis of Vegetative (Green) Roof Systems*, which provides an objective measure of media density for estimating structural loads | 2011 |
| E2400 | *Guide for Selection, Installation and Maintenance of Plants for Green Roof Systems*, includes recommendations regarding choosing, planting and irrigating plants grown on vegetative roofs; and most recently | 2006 |
| E2788 | *Specification for Use of Expanded Shale, Clay and Slate (ESCS) as a Mineral Component in the Growing Media and the Drainage Layer for Vegetative (Green) Roof Systems*, details quality and gradation requirements | 2011 |
| WK25385 | *A New Guide for Vegetative (Green) Roof Systems*. Intended as a comprehensive guide and one that clearly links to an existing building sustainability standard E2432 *Guide for General Principles of Sustainability Relative to Building* | Under development |
| WK28504 | *Selecting Waterproof Membranes for Green Roofs* | Under development |

Source: Enwright (2013) and www.astm.org
† As at 2 April 2014.

FIGURE 6.6 Extensive green roof on a visitor centre near Bewdley, UK. © John Dover

not guarantee membranes covered with green roofs for longer than those on conventional roofs, and yet we know that simple tar-and-paper roofs covered with sand, gravel and plants can last 100 years (Koehler & Poll, 2010)! At least one municipality in the USA has been experimenting with green roofs for some time: Portland, Oregon, started out monitoring a small roof on a garage in 1996 and by 1999 had officially recognised green roof technology as a way of managing runoff in its *Stormwater Management Manual* (Liptan & Strecker, 2003).

In some countries, such as the UK, it has taken some time for the concept of roof greening to take hold. Installations in the UK in the late 20th century were more likely to be on wildlife visitor centres (Figure 6.6) than on residential buildings or office blocks. Typical reasons for the reluctance to fit green roofs, apart from an understandably cautious approach to adopting novel technologies, have centred around fears of water penetration as a result of roots breaking through roof membranes, the weight of the growing medium and plants causing roof collapses, etc.

The UK's reluctance to vegetate roofs is almost amusing when you come across evidence that one of the earliest established examples of a flat green roof still in existence is from England (though the Berlin tar-and-paper roofs considerably pre-date it). This intensive green roof was installed in 1938 on the top of the very fashionable department store of Derry & Toms, located in the centre of London on Kensington High Street! The roof was turned into a pleasure garden and the publicity blurb from an advert (Figure 6.7) in the *Country Life Picture Book of Britain* gushed:

> The Derry Gardens ... overlooking London, one hundred feet in the air, are among the most wonderful in the world, occupying over an acre of roof space, with matured trees and lawns, a stream, waterfalls and bridges, Spanish pergolas and cloistered Tudor

# The Great

# fashion House

## of Derrys

PARTICULARLY modern, with a character altogether its own, and a colour and charm which make the visit a shopping experience of rare pleasure, Derrys is one of the most beautifully appointed Stores in Europe, and has a reputation for fashions of quality and distinction at quite reasonable cost. The amenities of the House include the favourite Cocktail Lounge on the Fifth Floor for the morning visitor, and the Rainbow Restaurant with its unique multi-coloured domes for Lunch and Afternoon Tea . . . nor must we forget the Beauty Salon, on the Third Floor, which offers visitors a delightful and efficient service.

## The Derry Gardens . . .

*overlooking London, one hundred feet in the air, are among the most wonderful in the world, occupying over an acre of roof space, with matured trees and lawns, a stream, waterfalls and bridges, Spanish pergolas and cloistered Tudor walks, woodland scenes, beautifully kept lawns, flower-beds aglow with colour. The Gardens are open to visitors from 9.30 to 5.30 p.m. daily (Saturdays 1 p.m.) during the summer months.*

*'Spanish Gardens.*

*The Sun Pavilion.*

Derry &Toms
Kensington W

Telephone: Western 8181

FIGURE 6.7 A 1950s advert in the *Country Life Picture Book of Britain* for the roof garden on top of the Derry & Toms Department store (now called 'The Roof Gardens') (Anonymous, 1950). Image courtesy of Country Life. Every reasonable effort has been made to trace the copyright holder.

walks, woodland scenes, beautifully kept lawns, flower beds aglow with colour. The gardens are open to visitors from 9.30 to 5.30 p.m. daily (Saturdays 1 p.m.) during the summer months.

*(Anonymous, 1950)*

When the garden was renovated some 50 years later the roof membrane was found to be in excellent condition, remarkable when you consider that a normal roof membrane of the period would have been expected to last 10–15 years (Johnston & Newton, 2004) – and

FIGURE 6.8 A domestic, turfed, green roof in Stoke-on-Trent (with snowdrops). © John Dover

certainly shows that a carefully constructed green roof can preserve rather than damage roof membranes! Indeed, the US Environmental Protection Agency estimate membrane life could be extended by 20+ years using green roofs (EPA, 2000).

The Derry Gardens live on, and since 1981 have been owned by Richard Branson's Virgin Group and are now called 'The Roof Gardens' (Velazquez, 2004). They are open to the public free of charge and there is a restaurant that can be hired for weddings and events. The gardens, frankly, look gorgeous (VHG, 2014). Not all green roofs are so exotic. I came across a lovely example on a pair of semi-detached houses in Stoke-on-Trent in the UK, which is essentially a turf roof (mown once a year – access via a skylight), and when I visited snowdrops (*Galanthus* sp.) were flowering on it (Figure 6.8).

## 6.2 Types of green roof

### 6.2.1 Classification

As with most types of green infrastructure, there is a multiplicity of types, definitions and categorisations that are not always clear, or their terminology helpful. At least seven categories are used by the construction industry (lightweight extensive, super lightweight, extensive, semi-intensive, intensive, roof gardens/podium decks, and biodiverse/wildlife (extensive) (see Groundwork, 2014)). The major terms likely to be encountered are as follows:

- *Intensive green roofs* – here intensive means it needs a lot of management. Such green roofs look like parks and gardens, examples are such exotica as The Roof Gardens or the roofs of underground car parks. The growing media depths are substantial so trees and shrubs can be included in the planting designs.
- *Semi-intensive green roofs* – typically have reasonable depths of growing media so they can retain enough water to grow a wide range of plant species, small shrubs are probably their limit though. Such green roofs are often called 'biodiverse' but a recent UK guidance document (Groundwork, 2011) (based on the FLL guidelines) recognises this as a different category.
- *Biodiverse green roof* – this category attempts to specifically replicate more natural plant communities and is aimed at creating wildlife habitat (flora and fauna); it would typically be planted up but could be left for natural colonisation. There is a variant of this type of roof which is often called a 'brown roof'.
- *Brown roofs* – attempt to recreate elements of brownfield conditions (otherwise, and clumsily, known as 'Open Mosaic Habitats on Previously Developed Land' (JNCC, 2010)). Brownfield land is that which has typically been used for industrial or commercial activity (could be domestic) but where buildings have been demolished and abandoned, for varying periods of time, leaving a low-nutrient status substrate with a lot of building rubble. Such areas, when sparsely colonised by vegetation, make excellent habitat for many quite rare species (Gibson, 1998).
- *Extensive-green roofs* – in this usage of the word it does not mean they are very big, but that they have low maintenance requirements, including for water. The growing medium is quite shallow and as a result only plants that are able to cope with dry conditions do well. These are typically sedum-covered roofs.

## 6.2.2 Influence of growing medium (substrate) and depth

These different roof types will be explored in the following sections, but Dunnett and Kingsbury (2004) gave a handy rule of thumb based on substrate depth (Table 6.2). More recent guidance suggests that 80 mm should be the minimum depth unless using pre-grown sedum mats (Groundwork, 2011).

The depth of the substrate is important as it affects the kind of plants that can be grown on the roof, the amount of water that can be retained by the roof, and crucially the weight of the roof. As is obvious, the three factors are interrelated and are also affected by the design purpose – what is the specific function of the roof (e.g. stormwater control (a SUDS component), biodiversity, aesthetics, development mitigation). The design weight of a roof is a combination of the weight of all the roof components (including the dry weight of green roof components and growing media) and the more variable components such as the maximum amount of water that is held in the roof components and media following rainfall (obviously more with a green roof than a concrete roof), and also the weight of snow, the effect of wind, and there also has to be an allowance for the weight of, for example, maintenance crews and their equipment (Magill et al., 2011).

Depth of substrate clearly has an impact on weight, but different growth media vary considerably in weight, composition and characteristics (Kolb & Schwarz, 1986). Ampim et al. (2010) review the different types of green roof growing media which differ little between 'types' of roof (e.g. extensive vs intensive) but see Table 6.3; they also highlight a recent trend for media depths to be variable on any given roof rather than uniform. This latter point is important, because such an approach allows elements of extensive and semi-intensive flora to be combined on the same roof (Bates et al., 2013), providing better biodiversity benefits (Brenneisen, 2006) and also allowing for very creative visual designs (Figure 6.9).

Materials often used in making up green roof substrates are given in Table 6.4. The specific mix for any given roof will depend on whether the mix is a commercially available one or bespoke to the site; the availability of on-site recycled materials having the advantage of no transport costs whereas commercial media have known properties (Ampim et al., 2010). Media mixes are primarily of inorganic materials (typically 80–100%) and the most popular are the expanded minerals, primarily because of their light weight, water holding and aeration characteristics. The remaining 1–20% is made up of organic materials and, if required, some slow-release fertiliser. Nagase and Dunnett (2011) in a greenhouse experiment with four

**TABLE 6.2** Definition of green roof type based on substrate depth and indication of supportable vegetation

| Roof type | Typical plants used | Depth of growing medium |
|---|---|---|
| **Extensive** | Small sedums + moss | 20–30 mm |
| | Taller sedums + grasses, drought-tolerant sp., alpines, small bulbs | 50–80 mm |
| **Semi-intensive** | Low-medium dry habitat perennials, grasses, annuals, small shrubs, turf grass | 100–200 mm |
| **Intensive** | Medium-sized shrubs, vegetables, generalist perennials, grasses | 200–500 mm |
| | Small deciduous and conifer trees | 500+ mm |

Sources: Dunnett & Kingsbury, 2004; see also Groundwork, 2011.

**TABLE 6.3** Wet weights of 1 m³ of proprietary green roof soil media designed for a range of green roof types

| Media type | Content | Wet mass (kg.m⁻³) |
|---|---|---|
| 'Zincolit' | Recycled crushed brick 4–15 mm dia. | 1250 |
| Zinco intensive soil | Zincolit, compost, sandy loam | 1450 |
| Zinco semi-intensive soil | Zincolit, mature compost | 1400 |
| Zinco extensive soil | Zincolit and fines★ | 1250 |

Source: Alumasc, 2004.
★ Fines are particles that can pass through a #200 sieve (0.075 mm).

**FIGURE 6.9** An extensive green roof, designed for biodiversity, on a new office building in Basel, Switzerland. Note the varying substrate depths and mounds of stones. © Gary Grant/Green Roof Consultancy

extensive green roof plants (chives *Allium schoenoprasum*, sea lavender *Limonium latifolium*, hairy melic *Melica ciliate* and grey or Faasen's catmint *Nepeta x faasenii*) and dry and wet conditions concluded that 10% organic matter (green waste compost) was optimum for plant establishment. All media have to be lightweight; maintain air gaps (to stop anoxic conditions developing and killing the plants); drain well, yet retain sufficient water and nutrients to support plant growth, and not be subject to compaction; not blow away, be sucked off by the wind, or be eroded by water; carry no weeds, pests or diseases (Ampim *et al.*, 2010; Groundwork, 2011). Fire breaks (areas without vegetation cover made of inert material such as stones or concrete) may have to be incorporated in the design and the maximum 20% organic matter is also a fire prevention feature (Groundwork, 2011). The balance of compost in the mix is important: too little and plant growth may be inhibited, too much and nutrient leaching may take place which is undesirable as Moran *et al.* (2004) found in their study of extensive green roofs that incorporated 15% compost, although they suggested that total nitrogen and phosphorus levels in runoff may decline over time as the roof ages.

Other materials used include dried mats of turf grass sod and amended furnace bottom ash (FBA – a by-product of coal-fired electricity generation) primarily composed of iron, silica and aluminium. FBA is an attractive material: partly because of its low weight and porosity, and partly because it is cheap, and incorporation in green roofs is a way of recycling an industrial waste product. Unfortunately, furnace bottom ash drains too rapidly and would need additional organic material and fertiliser to be useful (Kanechi *et al.*, 2014).

Groundwork (2011) give approximate values for porosity, water-holding capacity, air content, water permeability, pH ranges and organic content for intensive and extensive green roofs. Media weights are clearly important and the FLL guidelines suggest 'dry bulk density' of media to lie within the range 600 to 1200 kg.m$^{-3}$, which when saturated should lie within 1000 to 1800 kg.m$^{-3}$ (Ampim *et al.*, 2010). Alumasc Roofing Systems supply proprietary soil mixtures for use with their green roof components and give wet weights for different types of green roof which fall within this range (Table 6.3) (Alumasc, 2004).

Of course, recommendations of substrate depth suitability and makeup must take into account local climates and, as Kotsiris *et al.* (2013) note, the FLL guidelines were created for a northern European climate – and specifically for Germany – and are unlikely to be appropriate for all situations. It will thus be more challenging to create green roofs

TABLE 6.4 Materials used in the formulation of green roof growing media

| Natural minerals | Artificial/modified minerals | Recycled | Organic |
| --- | --- | --- | --- |
| Sand | Perlite | Crushed clay bricks, tiles, brick rubble | Peat |
| Clay | Vermiculite | Crushed concrete | Coir fibre dust |
| Lava (scoria) | Expanded shale, clay, slate | Aerated concrete | Compost: bark, sawdust, poultry litter, yard waste |
| Pumice | Rockwool | Subsoil | Worm castings |
| Gravel | | Styrofoam | |
| | | Urea-formaldehyde resin foam | |

Source: Ampim *et al.*, 2010.

in countries with more elevated temperatures and reduced rainfall (relative to northern Europe). The simple classification in Table 6.2 may therefore be meaningless in, for example, the Mediterranean context (Kotsiris *et al.*, 2013). Attempts to adapt green roof technologies to local conditions are most likely to succeed if they take into account the realities of the need for irrigation, use local materials (factors include cost, availability, sustainability of supply and carbon costs, e.g. transport), and use plants adapted to local conditions. Kotsiris *et al.* (2013), in Greece, used this approach in their experiments using the drought-tolerant, evergreen shrubs/trees Japanese pittosporium *Pittosporium tobira* (one year old) and olive *Olea europea* (two-years old). All growth substrates were of local origin apart from peat, which acted as a control organic input, and were dominated by volcanic material (pumice or perlite; 65% by volume) mixed with 5% zeolite (an aluminosilicate, also of volcanic origin), and either 30% of peat or compost (with the pumice) or sandy loam soil (with the perlite). Two media depths were used (300 mm and 400 mm) and automatic trickle irrigation. After the first year, both plants were found to have grown better, and to have higher chlorophyll levels, in the compost-containing medium. In the second year the 400 mm depth marginally improved the growth of olive trees, but the results demonstrated that 300 mm was quite adequate. The carbon footprint of the perlite mixture was close to double that of the pumice (due to the energy required to expand perlite from the native material – increasing its bulk relative to its density and increasing its water-holding capacity). The peat:pumice mix also had a higher footprint than the compost:pumice mix.

## 6.3 Structure of green roofs

### 6.3.1 Roof shape

Green roofs can be fitted to a range of roof types: from curved, to pitched (as long as not too steep a pitch) to flat. Pitches greater than 20° require additional cross-battens (horizontal projections across the roof) to prevent slippage or 'shear'; with a 20° slope battens can be fixed at 10 m centres getting progressively closer the steeper the slope: at 25° battens at 8 m centres and at 30° battens at 5 m centres (Alumasc, 2004; Groundwork, 2011). The exact structural requirements will depend on the limitations imposed by the specific roof type (Figure 6.10).

A typical conventional flat roof on a concrete deck would have a vapour control membrane above the upper roof deck, some form of additional insulation, and above that a waterproof membrane. Green roofs can be fitted to buildings with metal decks as well. In addition to these conventional features, a green roof would have a number of additional layers. The essentials – root barrier, drainage layer, filter and growing medium – have been known for some time (Kolb & Schwarz, 1986) but the following is that as recommended by the company Alumasc Roofing Systems (Alumasc, 2010) (see Figure 6.11). The layers, as built up from the roof deck, are given below in reverse order:

- *A root barrier* – ensures that roots do not penetrate the roof membrane – this would not be required if the existing roof membrane already installed was resistant to root penetration.
- *A moisture mat* – made of rot-proof fibre, it provides additional protection to layers below it, but the primary function is to retain water.

**FIGURE 6.10** A green roof above Sidmouth, Devon, UK, showing both curved and flat roof surfaces. © John Dover

- *A drainage layer* – this both retains water and allows excess to runoff. It has holes to ensure aeration of the root zone.
- *A filter mat* – this prevents fine material washing out and impeding drainage.
- *A fall net* – this is an optional layer which spreads the load and provides added safety for maintenance teams.
- *Growing media* – the planting substrate, at whatever depth is required, is spread above the fall net/filter layer and the appropriate flora planted into it.

### 6.3.2 Roof irrigation

In addition to the above components, irrigation systems may need to be incorporated, depending on the roof design. As with most planted systems watering will be needed at first, but once plants have established the extensive systems and many semi-intensive roofs may not need irrigation at all (except for during exceptional drought events). Intensive green roofs, by their nature, with taller vegetation, perhaps including tall shrubs and even trees, are likely to require a permanent irrigation system (Groundwork, 2011). Water retention on green roofs is related to a range of interacting factors including the slope, exposure, provision of drainage channels, the water retention capacity of the growing media due to its organic/inorganic nature, porosity of the latter, and depth. Of course many of these become less important in terms of plant survival if irrigation is always available or if plants are resistant to desiccation (Kanechi *et al.*, 2014).

Rainwater from green roofs can also be harvested and stored for subsequent non-potable purposes such as toilet flushing, and of course for irrigation. This approach was used in the

**FIGURE 6.11** The typical components of a green roof, bottom to top: a root barrier membrane (RB), a water retention and protection mat (WM), a perforated drainage layer (DL), a filter sheet (FS) and a 'fall' net (FN). The growing medium sits above the fall net. Material supplied by Alumasc Roofing Systems. Image © John Dover

building of Daimler City (DaimlerChrysler Areal) at the Potsdamer Platz in Berlin designed by Renzo Piano and Christoph Kohlbecher and completed in 1998. The project included the use of extensive and intensive green roofs on the 19 buildings of the development (37.5% of roof area greened = approximately 16,690 m²) with the runoff collected in three underground cisterns with a joint capacity of 2,550 m³ (or 12% of annual rainfall) and a specially constructed lake ('Piano Lake') with a surface area of 13,000 m² (11% of annual precipitation) and a biofiltration bed of some 1,900 m² (Berlin.de, undated; Overstrom, 2010; Schmidt, 2009).

Green roof designs can also incorporate rainwater harvesting systems that process a building's grey water (from baths, basins and showers). In such systems, the grey water would typically be piped on to the roof and trickled through the growing media via perforated pipes and subsequently blended with rainwater, collected, and stored for later use (Shirley-Smith, 2002).

### 6.3.3 Installing green roofs

The ease and cost of fitting a green roof depends very much on the starting point: is the green roof to be incorporated in a new build or is it a retrofit? If the former, then the costs can be extremely low – estimated at 0.5% of the total build costs (Stifter, 1997, cited in Peck et al., 1999) – if the latter, it depends on the capability of the roof to take additional weight and the kind of green roof that is desired. Liptan and Strecker (2003) estimated that a new-build green roof would cost between $5 and $12 per 0.0929 m² compared with between $2 and $10 for a conventional roof; for retrofits the equivalent figures were between $7 and $20 and $4 and $15 respectively. If the existing roof has a heavy material such as concrete slabs over the roof membrane, then their removal and replacement with a lightweight sedum green roof may actually reduce the loading; for example, 50 mm paving slabs on asphalt = 185 kg.m$^{-2}$, a lightweight extensive sedum roof weighs 60–90 kg (Wickham & Hallam, 2006). The latter authors recommend that all existing roofs should be checked by structural engineers as there is no golden rule as to which type of roof will support a retrofitted green roof and which will not. Having said that, Stovin (2010) suggests that, in the UK at least, many existing medium-height office blocks with concrete decking roofs should be able to be retrofitted with green roofs without structural modification. Buildings constructed prior to 1980 may be particularly suitable as roof load calculations are likely to have been over-specified compared with current practice. Stovin (2010) is less sanguine about roofs with metal structural beams and steel decking, and even less so for timber roofs. Whatever the structure, provided it is safe to fit a green roof, the expectation is that the most likely retrofit would be an extensive one because of its low weight necessitating no, or relatively modest, changes to roof structure (Oberndorfer et al., 2007).

The process of installing a green roof is clearly one for trained contractors, unless it is quite a small area such as on a garden shed or garage where water leakage is, whilst not desirable, at least less important than in a domestic, commercial or industrial situation. Material handling in large installations is also a consideration, along with health and safety aspects and adherence to all building codes, planning permissions, etc. Large installations will almost certainly need crane access for shifting the green roof components and media, although media can be pumped into place. A typical installation can be seen in Figure 6.12. Whilst the installation of the green roof structure clearly needs contractors or at least

FIGURE 6.12 Installation of a green roof at Parliament Hill School, London: a) preparation of refurbished roof deck, b) rolling out the drainage layer with filter mat, c) installation of a fall net, d) growing medium spread from crane-hoisted bags; e) spreading of growth medium to desired depth, f) planting-up – showing detail of drain/fire breaks around edges, drains, etc. Photographs © Alumasc Roofing Systems

professional expertise, the growing and planting of the roof could be a community effort provided it is safe enough. For example, the installation of the green roof on the Agriculture Building at the Southern Illinois University Carbondale campus involved over 150 people (Magill *et al.*, 2011).

### 6.3.4 Planting-up a green roof

A rooftop is a harsh environment for plants (e.g. exposed conditions resulting in extremes of temperature and wind), naturally limiting the range of species that can be grown, and the thin media depths used in extensive green roofs make plants particularly subject to desiccation (Ampim *et al.*, 2010). Kolb and Schwarz (1986) considered that the optimum time for planting in their climate (Würzburg, Germany) was April–June and cautioned against autumn plantings. With extensive roofs, the type of plant that can be used is pretty limited and typically consists of low-growing plants like grasses, sedums and mosses. Although Kolb and Schwarz (1986) were working with a test palette of 150 species, many would probably not have been appropriate for extensive plantings; they listed 33 species which they felt would grow well in shallow media depths and this naturally included a fair few grasses and sedums. Some plants have an adaptation to drought conditions whereby they reverse the 'normal' pattern of stomatal opening (Borland *et al.*, 2011; Getter & Rowe, 2008). In most plants the stomata, air pores allowing gas exchange whose opening and closing is under the control of 'guard cells', open during the day (to take in $CO_2$) and close at night. In plants adapted to drought conditions, the stomata open at night, when it is cooler, and keep closed during the day, when temperatures are higher, to reduce water loss. This approach is known as Crassulacean Acid Metabolism (CAM) after the stonecrop plant family Crassulaceae, in which it was first discovered – and which contains the sedums. CAM uses a modification of what is known as the C4 rather than the more widespread C3 photosynthetic pathway. The adaptation is not simply the use of a different metabolic pathway to most plants, but reflects the way $CO_2$ is initially stored at night, in a four-carbon compound (malic acid), prior to use in photosynthesis during the day (Borland *et al.*, 2011).

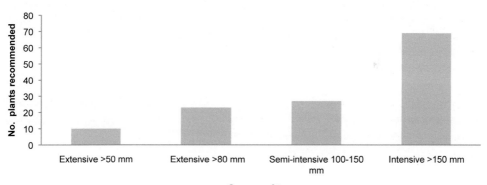

FIGURE 6.13 Media depth affects the number of plant species or varieties recommended for use. This example is taken from the list recommended by Alumasc Roofing Systems for their green roof systems and includes different colour varieties of the same species – e.g. for the extensive system, with the shallowest depth, of the ten plants recommended three are different colour varieties of *Sedum spurium* (Alumasc, 2004).

TABLE 6.5 Suitability of plants grown on 50 and 100 mm deep extensive green roofs at Wayne County Community College (Goldsboro) and the Neuseway Nature Center (Kinston), North Carolina, and the impact of depth on percentage cover of three species after one year

| | | Green roof location | | | |
| | | Goldsboro | Kinston | Suitability | Impact of depth[†] |
|---|---|---|---|---|---|
| *Delosperma cooperi* | Trailing iceplant (pink) | · | | – | |
| *Delosperma nubigenum* | Yellow iceplant | · | | + | x2.3 |
| *Sedum acre* | Jelly bean | ? | ? | – | |
| *Sedum album* | White stonecrop | · | · | + | x3.6 |
| *Sedum album chloriticum* | White stonecrop | · | | – | |
| *Sedum album murale* | White stonecrop | · | · | + | |
| *Sedum floriferum* | Orange stonecrop | | · | + | |
| *Sedum grisebachil* | | · | | – | |
| *Sedum reflexum* | Reflexed stonecrop | · | · | + | x2.4 |
| *Sedum sexangulare* | Tasteless stonecrop | · | · | + | |
| *Sedum spurium fuldaglut* | Two-row stonecrop | · | | + | |

Source: Moran *et al.*, 2004.

? = location of roof not given in source reference; [†] multiple of percentage cover with increasing substrate depth.

Planting lists typically expand with increasing substrate depth (Figure 6.13) (Ampim *et al.*, 2010) and stress tolerance/drought adaptations are characteristic of plants suitable for extensive green roofs (Oberndorfer *et al.*, 2007).

The actual approach to planting up a green roof depends on the type of the roof and therefore substrate depth. It also depends on whether an instant effect is required, cost, whether the intention is to introduce vegetation (as seed or growing plants) (Groundwork, 2011) or let it develop naturally. The latter is of course the cheapest but will take some time to establish and is not under the control of the installer or client except in that maintenance visits can remove undesirable species. Even where roofs are planted up, the final vegetation community may not reflect the starting point. Hutchinson *et al.* (2003) describe how the east and west sides of the roof of the Hamilton Apartments Building in Portland, Oregon, USA, were planted up in an identical fashion with over 75 different species of plant. Many of the original plants died, new ones colonised (particularly grasses), and the two halves of the roof had distinctly different plant communities only three years after establishment.

Moran *et al.* (2004) in a study of 11 plant species (Table 6.5) found that plant growth was better on 100 mm compared with 50 mm substrate depths, although some species showed little growth and were not recommended for use on green roofs in North Carolina.

## 6.4 Plants for green roofs

### 6.4.1 Impact of climate on plant choice

Much useful advice on plants for green roofs can be found in Snodgrass and Snodgrass (2006). Roofs will inevitably vary in their exact roof design, and on the basis of substrate depth, but the roof environment will also vary on the basis of their climate zone/geography – it is thus important that plants are selected for local conditions (Dvorak & Volder, 2010; Getter & Rowe, 2008; Kotsiris *et al.*, 2013; Wolf & Lundholm, 2008). For example, Benvenuti and Bacci (2010), based in Pisa, Italy, trialled 20 species over one growing season for their values as green roof plants: partly to try to improve the range of plants available for dry roofs and partly to use species native to the Mediterranean. Seeds of xerophytes (plants adapted to low-water environments) were gathered from high-water stress environments including old quarries, rocky areas, sand dunes, road verges and 'sea rocks'. Material was grown-on and tested in simulated (ground level) green roof plots of 150 and 200 mm depth – but with no added water. They found that all species germinated well (due to wet weather) and tolerated extreme temperatures (about 50°C) and drought conditions (no rain from early June to early September). Eighteen of the species flowered well in their first season (though not the euphorbs); cover and biomass was improved for a small majority of species with greater substrate depth, and flowering was earlier (Table 6.6). As rooftop conditions would be harsher than ground level, and in areas with substantial drought periods, the authors concluded that irrigation would probably have to be built in to roofs with media depths of less than 200 mm. Where irrigation was not possible, slow-growing, resistant species such as *S. rupestre*

**TABLE 6.6** Species used in trials of Mediterranean climate zone green roof plants

| Original Biotope | Species | Common name | Impact of lower media depth on | |
| --- | --- | --- | --- | --- |
| | | | % cover | Biomass |
| Dunes | *Anthemis maritima* | Corn chamomile | | |
| | *Glaucium flavum* | Yellow horned poppy | | < |
| | *Helichrysum stoechas* | Curry plant | < | < |
| | *Otanthus maritimus* | Cottonweed | < | |
| Quarries | *Helichrysum italicum* | Curry plant | < | < |
| | *Satureja montana* | Winter savory | | |
| | *Sedum rupestre* | Reflexed stonecrop | | |
| | *Calamintha nepeta* | Lesser calamint | < | < |
| | *Centranthus ruber* | Red valerian | < | < |
| | *Dianthus carthusianorum* | Carthusian pink | | |
| | *Euphorbia characias* | Mediterranean spurge | < | < |
| | *Leontodon tuberosus* | Tuberous hawkbit | | |
| Rocky areas | *Lavandula stoechas* | French lavender | < | < |
| | *Scrophularia canina* | French figwort | < | |
| | *Verbascum thrapsus* | Great/common mullein | < | < |
| Roadside | *Scabiosa columbaria* | Small scabious | < | < |
| Sea rocks | *Armeria pungens* | Spiny thrift | < | < |
| | *Euphorbia pithyusa* | Little fir spurge | < | < |
| | *Helichrysum italicum* ssp. *microphyllum* | Curry plant | | |
| | *Crithmum maritimum* | Rock samphire | | |

Source: Benvenuti & Bacci, 2010.

< = significant reductions in percentage cover or biomass with 150 mm media depth compared with 200 mm media depth.

or *Helichrysum italicum* ssp. *microphyllum* were recommended. Köhler and Poll (2010), in a meta-analysis of several studies, showed a significant linear relationship (r = 0.49) between percentage plant cover of green roofs and the logarithm of the available water capacity of the substrate (in l.m$^{-2}$).

MacIvor and Lundholm (2011) examined the performance of 15 native species drawn from the Atlantic coast of Canada on 60 mm deep extensive green roof plots (0.36 x 0.36 m each); the work was carried out between May and October 2009. The plants were derived from seeds and cuttings and planted up as plugs; six species were graminoids, three were tall forbs, one creeping forb, and the remaining five were creeping shrubs. The plants grew very well with 12 of the 15 species having 100% survival and three 80%, although one failed completely (low-bush blueberry *Vaccinium angustifolium*). Cover values were in excess of 90% for ten of the species – the grasses achieving the highest cover values. Dvorak and Volder (2010) carried out a literature review of the use of vegetation on green roofs in North America and suggested that the existing data supported the idea that plant survival varied depending on ecoregion. The same review appeared to support the idea that more diverse vegetation resulted in both improved plant condition and delivery of ecosystem services. There is some evidence from the UK, for example, to support the concept that a more diverse plant assemblage would be more resistant to drought than a monoculture or species-poor community (Nagase & Dunnett, 2010), and this may be another approach that could be used to climate-proof green roofs in water-scarce areas or to avoid the expense of installing an irrigation system. Caution should probably be exercised, however, as the plants used in the research were rather smaller than is normally the case and plants specific to the climate region should be evaluated for use. The grass and forb species that Nagase and Dunnett (2010) considered of being able to recover well from drought stress were: thrift *Armeria maritima*, self-heal *Prunella vulgaris*, sea campion *Silene uniflora*, crested hair-grass *Koeleria macrantha* and yellow oat-grass *Trisetum flavescens*.

## 6.4.2 Sedums and related drought-tolerant species

For extensive green roofs, installation of a pre-grown sedum mat is popular. In this situation vegetation is pre-grown on a fleece-type base which may be biodegradeable or have reinforcement and geotextile characteristics depending on the manufacturer. The mats are typically delivered as multiple flat sections (typically 1 x 1.2 m) on a pallet with enough to cover about 50 m$^2$ (SedumDirect, undated), or as rolls of various sizes – typically 1 m wide by 2 m long, although some manufacturers supply them in rolls up to 10 m in length (Bauder, 2010) (Figure 6.14).

The alternative approach to vegetation mats is to use plug plants, small plants rooted in growing medium which, because of their small size, are relatively cheap and should establish well, provided watering is not neglected (as with all newly introduced species). Sedum plugs can be used on extensive roofs and have been trialled in a number of studies such as that by Getter and Rowe (2008) for the Midwestern USA at Michigan State University. In that study 12 *Sedum* species were grown in green roof substrate of 40, 70 and 100 mm depths and cover recorded bi-weekly from first planting in early June 2005 to the onset of the first autumn frosts – the aim was to determine which plants established most rapidly. Unsurprisingly, growth was generally better on the 70 and 100 mm substrate depths than on the 40 mm; 70 mm was considered the minimum depth for rapid establishment of most

FIGURE 6.14 Left: sedum mats are pre-grown in specialist nurseries and rolled up like turfgrass for delivery. Right: once on-site, the sedum mats are carefully unrolled directly into their final position. © Bauder Ltd

species. Stringy stonecrop *Sedum sarmentosum* performed best of all those tested in terms of cover development – but was considered probably too aggressive for use with other species and likely to result in a monoculture. Three species were considered the best for the deeper media (Kamtschatka stonecrop *Sedum floriferum*, the 'stonecrop' *S. stefco* and creeping sedum *S. spurium* 'John Creech') and two species suitable for the shallow (40 mm) media (the aggressive *S. sarmentosum* and *S. stefco*). Five species were considered to be poor performers: crooked stonecrop *S. rupestre* 'Angelina', the 'stonecrops' *S. ewersii* and *S. cauticola* 'Lidakense', European stonecrop *S. ochroleucum* and the crooked stonecrop *S. reflexum* 'Blue Spruce' (syn. *rupestre*).

Wolf and Lundholm (2008) examined the effect of different watering regimes over a two-month period on 14 plant species, all but four native to their area (Nova Scotia, Canada), in simulated green roof 'microcosms' filled with 70 mm of substrate. Five species were succulents, four herbaceous, three grass and two woody; unplanted controls were also included in the experiment. Watering was every 4, 11 or 24 days; the most frequent watering simulated unrestricted water conditions and the least frequent drought. Only succulents (the non-native sedums: *Sedum acre*; the jelly bean plant or 'pork-and-beans' *S. x rubrotinctum*, *S. spurium* and the native golden root *Rhodiola rosea*) survived the drought treatment.

It is possible to simply use a seed mix on a green roof, but commercially available mixes usually come in a range of formulations such as with hydroseeding (Groundwork, 2011) or mixed with a bulking agent, an adhesive to prevent blow-off, fertiliser and also mycorrhizal fungi (Bauder, 2012b). The latter have been found to be essential to the establishment of some plant species by forming a symbiotic relationship and can even improve the visitation rate of pollinating insects as a result of their influence on plant traits such as number of flowers or flower size (Gange & Smith, 2005). Mycorrhizal fungi invade plant roots and from this gain the products of the plants' photosynthesis (sugars); in return, the fungi improve the plants' uptake of mineral nutrients and water – mycorrhizae can also link several plants together

**TABLE 6.7** Trials carried out (2004–2006) by Nagase and Dunnett (2013c) of 26 geophyte species in experimental green roof plots with two media depths (50 and 100 mm)

| Family and species | Common name★★ | Better emergence with | | | Better growth‡ with | | | Better flowering‡ with | | | Better reproduction | |
| --- | --- | --- | --- | --- | --- | --- | --- | --- | --- | --- | --- | --- |
| | | Sedum cover | 100 mm substrate | Poor emergence | Sedum cover | 100 mm substrate | Poor Growth | Sedum cover | 100 mm substrate | No flowering | Sedum cover | 100 mm substrate |
| **LILIACEAE** | | | | | | | | | | | | |
| *Allium flavum* | Yellow-flowered garlic | + | | | | | | | | | | |
| *Allium karataviense* 'Ivory Queen' | Turkestan onion | | | • | | | • | | | | | |
| *Allium ostrowskianum* (syn. *oreophilum*) | Pink lily leek | + | | | | | | | | | | |
| *Allium unifolium* | American onion | | + | | | + | | | | | | + |
| *Puschkinia libanotica* | Striped squill | | | | | | | | | | | |
| *Scilla siberica* | Siberian squill | | + | • | | + | | | + | | | + |
| *Tulipa bakeri* 'Lilac Wonder'★ | Canada tulip | | | | | | | | | | | |
| ***Tulipa clusiana* var. *chrysantha*** | **Tubergen's gem** | | | | | | | | | | | |
| *Tulipa hageri* 'Splendens' | Splendens tulip | | | • | | | • | | | | | |
| **Tulipa humilis** | **Persian pearl** | | | | | | | | | | | |
| *Tulipa kolpakowskiana* | Kolpakowsky's tulip | | + | | + | + | | + | | | | |
| *Tulipa linifolia* | Flax-leaved tulip | - | | | | + | | | | | | |
| *Tulipa saxatilis*★ | Canada tulip | | | | | | | | | • | | |
| **Tulipa tarda** | **Late tulip** | | | | | | | | | | | |
| **Tulipa turkestanica** | **Turkestan tulip** | + | | | + | + | | + | | | + | |
| **Tulipa urumiensis** | **Iran tulip** | + | | | | + | | | | | | + |

**IRIDACEAE**

| Species | Common name | | | | | | | |
|---|---|---|---|---|---|---|---|---|
| *Crocus sieberi* 'Tricolor' | Three-coloured Sieber's crocus | | | • | | | • | |
| *Crocus tommasinianus* | Early crocus | | | • | | | • | |
| *Crocus vernus* 'Vanguard' | Dutch crocus | | | • | | | | |
| **Iris bucharica** | **Bokhara iris** | | + | | | + | | |
| *Iris danfordiae* | Bulbous iris | | | • | | | • | |
| *Iris reticulata* | Early bulbous iris | + | | | + | | | |
| **Muscari azureum** | **Azure grape hyacinth** | | + | | + | + | | |
| *Sparaxis tricolor* | Wandflower/ Harlequin flower | + | - | | | | • | + |

**AMARYLLIDACEAE**

| Species | Common name | | | | | | | |
|---|---|---|---|---|---|---|---|---|
| *Ixioliron pallasii* | Blue mountain lily | | | | + | | | |
| **Narcissus cyclamineus** **'February Gold'** | **Cyclamen-flowered daffodil** | + | | | + | | | + |

Species names in bold indicates that plants are suitable for extensive green roofs (Nagase & Dunnett, 2013c); grey shading indicates species are considered to be poor in at least one of the following parameters: emergence after winter, growth of foliage, flowering.

Key: ⋆ listed as synonyms by the Royal Horticultural Society (RHS, 2011); ⋆⋆ common names from a range of sources; + = significantly better with deeper substrate or Sedum cover; = = significantly worse with Sedum cover; † measured as maximum leaf length; ‡ enhanced length of flowering period.

including different species forming a complex sub-surface ecology (Harrison, 2005; Selosse *et al.*, 2006). Hydroseeding solutions, useful for seeding large areas for green roofs, involve the spreading of sedum shoots (Optigreen suggest 50 g.m$^{-2}$) over the growing substrate followed by spraying a growing medium mixed with seeds and a cellulose mulch, which also reduces erosion (Optigrün, undated).

### 6.4.3 Geophytes

Nagase and Dunnett (2013c) reasoned that low-growing (dwarf) geophytes would be ideal extensive green roof plants as they are typically found in the wild in extreme environments (e.g. dry, rocky, mountainous areas such as parts of the Mediterranean, Asia, South Africa). Geophytes are perennial plants whose above-ground foliage dies back every year and which survive overwinter as a storage organ such as a corm, rhizome, bulb or tuber. Part of the aim was to extend the species and flowering range of vegetation on extensive green roofs which are typically *Sedum* dominated and can be visually uninteresting. They trialled the performance of 26 species (Table 6.7) grown with or without a cover of sedum (*Sedum album*) sown as seeds in two green roof substrate depths (50 mm and 100 mm from Alumasc) neither of which was irrigated. The experimental plots were planted up in December 2004 and measurements taken in 2005 and 2006. Overall, emergence was better in 2006 with the deeper substrate – no difference was evident in 2005. The summarised results can be found in Table 6.7 and show that, in general, an increase in substrate depth and a covering of *Sedum* had a positive value probably as a result of improved over-winter survival (reduced extreme temperatures) and making it more difficult for birds to dig them up (but see results for *Tulipa lineola* and *Sparaxis tricolor*). Not all of the geophytes performed well though, with eight of the 26 being poor in at least one of the four characteristics reported – the Iridaceae seem to be least useful as extensive green roof species including all the *Crocus* species. One of the aims was to extend the flowering period on extensive green roofs and this was successful. Collectively the geophytes that did flower gave continuous flowering from March to June, of benefit both visually and to pollinators. In total, eight species were considered to have high potential for use on extensive green roofs, two of which were recommended only where depth was 100 mm (*Narcissus cyclamineus* and *Tulipa urumiensis*) (Table 6.7).

### 6.4.4 Perennial wildflowers

For the semi-intensive/biodiverse roofs some manufacturers have also extended the sedum mat technology to include native wildflowers; for example, the 'Xero Flor XF 118' from Bauder contains 24 species of native plant with a combined, predicted flowering period from March (daisy *Bellis perrenis*) to October (e.g. maiden pink *Dianthus deltoides*). The mat requires the minimum typical 'semi-intensive' substrate depth of 100 mm rather than the shallower depths suitable for the sedum mats (Bauder, 2012a, 2012b) and also requires a longer minimum 'watering-in' period than the sedum mats (four weeks vs ten weeks (Bauder, 2010, 2012a)).

A wide range of wildflower, herb and grass species for roofs with deeper media than the extensive type is widely available (e.g. Bauder, 2012b). Technical data usually supplied includes the expected height, the flower colours, whether suitable for sunny, shady or

TABLE 6.8 Annual species showing promise in trials on an extensive green roof (70 mm depth) in Sheffield, UK, in 2006

| Family | Scientific name | Common name[†] | Native to UK? |
|---|---|---|---|
| Brassicaceae | *Alyssum maritimum* | (sweet) Alyssum | No |
| Brassicaceae | *Iberis amara* | Rocket, wild or bitter candytuft | Yes |
| Brassicaceae | *Iberis umbellata* 'Fairy' | Dwarf fairy candytuft | Cultivated variety |
| Boraginaceae | *Echium plantagineum* 'Blue Bedder' | Viper's bugloss 'Blue Bedder' | Cultivated variety |
| Caryophyllaceae | *Gypsophila muralis* | Baby's breath | No |
| Scrophulariaceae | *Linaria elegans* | | No |
| Scrophulariaceae | *Linaria maroccana* | Moroccan toadflax | No |

Source: Nagase and Dunnett (2013b).
† From a range of sources, no obvious common name found for *L. elegans*. Note: *Linaria* is now in the Plantaginaceae following recent phylogenetic analyses.

partially shaded conditions, and the minimum substrate depth. As well as the standard plants available, producers will also work with the client to produce custom-grown plugs for specific requirements – the lead time for the production and delivery of such items is clearly greater (typically a full season) than for standard product lines (Bauder, 2012b). More mature plants can be used rather than plugs in the semi-intensive/intensive roof categories; these will be more expensive and have higher watering requirements, but can be an important consideration where a faster cover and impact is required. The density of planting is likely to have an effect on plant growth (crowding reducing growth of plants) as is the choice of plants and with stronger competitive interactions in high-density plantings (Nagase *et al.*, 2010).

### 6.4.5 Annual species

The approach to planting up a green roof has generally involved the use of pre-grown plants, typically perennials, although seeding is used (see section 6.3.4). Nagase and Dunnett (2013b) investigated the possibility of using annual plant species on green roofs in 2006. The use of annuals was something of a departure from the norm for green roofs but potentially could result in an aesthetically pleasing sward. They introduced 22 species on an experimental roof, with a 70 mm substrate depth, as seed at two rates: 2 $g.m^{-2}$ and 4 $g.m^{-2}$. They also imposed two watering regimes: no irrigation or irrigation, the latter was done when substrate moisture content was below 15%. Water was applied with a fine-spray hose until water started to runoff the experimental plot; irrigation was only used four times between July and August. Of the 22 species used, seven showed promise as green roof plants (Table 6.8) due to their good germination, growth and flowering characteristics. The original seed mix contained 12 non-native species, two commercial cultivars and eight native species, but only one of the seven best-performing species was native.

Without irrigation, good cover was best achieved with the 4 $g.m^{-2}$ sowing rate, but where water was provided the lower rate of 2 $g.m^{-2}$ resulted in better-grown plants. Seeds were sown in mid-June; the first plants flowered about a month later and flowering continued until late

October. The establishment of the roof was very cheap compared with the use of pre-grown material, but would probably need at least some additional re-sowing each year to maintain its quality.

## 6.4.6 Vegetation development

Long-term studies of vegetation development are rare as they do not tend to fit into the pattern of grant funding (usually three years); the problem is especially acute when the technology, such as green roofs, is relatively new. When vegetation development is through natural colonisation, vegetation dynamics (change/turnover of species) are expected; there is, perhaps, an expectation that a 'planted up' system will be more stable, but this is unlikely to be the case. Rowe *et al.* (2012) followed the fate of 25 succulent species over a seven-year period on different depths of extensive green roof media (25, 50 and 75 mm) and showed that whilst most species (22) survived after one year, this had declined to seven species after five years with two species – false stonecrop *Phedimus spurius* and Chinese mountain stonecrop *Sedum middendorffianum* – dominating on the 50 mm and 75 mm deep media. Different species dominated on the shallowest (25 mm) media depth: biting stonecrop *Sedum acre* and white stonecrop *Sedum album*. The following sections cover some of the few longer-term studies from which data is available.

### Long-term work on extensive roofs in Berlin, Germany

The work of Köhler (2006) is of especial value covering as it does a 20-year evaluation of extensive (100 mm depth) green roofs in Berlin from 1985 to 2005 and which included different designs from flat (actually about 2°, to facilitate drainage) to pitched (47°). The first installation (in 1985) was at Kreuzberg in Berlin and was an inner-city residential development called the Paul-Lincke-Ufer (PLU) project. The development was actually two buildings which had ten 'sub-roofs', with a range of aspects and pitches rather than uniform single-decked flat roofs; the total area was 650 m². Köhler and Poll (2010) demonstrate graphically how aspect and slope of a roof determines the amount of sunlight experienced by plants. They also demonstrated how, on the PLU building, plant coverage on sub-roofs with a 45° pitch decreases with increasing levels of sunshine in the order N>W>E>S and that the dominant plants changed from chives *A. schoenoprasum* with grasses on the north, to *A. schoenoprasum* on the west, and with sedums on the east and south. Floral quality (an index combining plant coverage and species richness) started to decline with sun exposure greater than 70%. Returning to the study of Köhler (2006): the growing medium was a mixture of expanded clay, sand and organic matter; vegetation was applied as mats (in themselves an innovation at the time) containing ten pre-grown species. The results, given for one of the sub-roofs (No. 1), showed strong vegetation dynamics at work. Initially annuals in the seed-bank of the growing medium of the PLU roofs germinated adding to the community but were lost in later years; some new species colonised and some were subsequently lost, a lichen *Cladonia coniocera* appeared after 11 years and became widely established. The total richness of the roof over the 20-year period was 55 ranging from 7 to 25 in any one year. Of the ten species introduced in the mat, five persisted throughout the study: three grasses (flattened meadow grass *Poa compressa*, sheep's fescue *Festuca ovina* and drooping brome *Bromus tectorum*), chives *A. schoenoprasum* and *S. acre*. For all sub-roofs combined, the total species count was 110 and richness was greater in years when summers were 'wet'; the roof pitch did not appear to affect species richness.

The other installation studied was in the suburban Templehof area on the Ufa-Fabrik (UF) cultural centre and consisted of six roofs (over 2,000 m² of green roof) built to the same depth as on the PLU buildings but the growing medium was mainly composed of garden soil with 10% expanded clay. The roofs were planted with material collected as seed from alpine meadows and were not all constructed at the same time. Results were given for the concert hall roof which had a cumulative richness of 91 vascular plant species and was planted in 1986; 17 species were known to have been introduced in the seed mix (there were probably others) and in any one year the species richness fell between a minimum of 22 and a maximum of 64. Irrigation was used at different levels of saturation in the earlier years but was terminated in 1997 and subsequently *S. acre* started to dominate the roof.

The species richness of the UF roofs was greater than on the PLU; the former was characterised by the presence of sedum, the latter by chives. The differences in richness and dynamics were considered to result from: locality of roof (with consequent differences in seed sources for colonisation), the UF roofs being shaded by trees (and thus providing a wider range of microclimates), irrigation and establishment (in the early years a gardener looked after the UF roof and removed tree seedlings).

### Brown roofs in Birmingham, UK

Bates *et al.* (2013) studied two brown roofs in Birmingham, UK, that had a range of aggregate sizes (coarse, fine, very fine), topography (depths 40–100 mm on one, 60–120 on the other) and a compost-based mulch over a four-year period (2007–2011). The mulch was spread thinly (2–3 mm) on one of the roofs, on the other a swirled design was created with the mulch applied at a greater depth (10 mm). Growing media were derived from recycled bricks, concrete and sand. The use of different aggregates, depths and mulch, together with piles of sand, logs and chunks of green oak, created a range of growing conditions and microclimates depending on aggregate size and presence/absence of mulch and other material. The vegetation was 'started off' by applying a seed mix (1.6 g.m⁻²) containing 25 native species including *Sedum acre*. As with all extensive roofs, water and organic matter levels were low.

Not all seeded plants germinated (three to four per roof) and some (two to three) failed over the study period depending on the roof. Plants with seeds already in the media or colonised by other means (e.g. wind, bird) dominated the roof lists, which totalled 59 on one roof and 52 on the other (totals include successful seeded species). Annual species typical of disturbance regimes such as poppy *Papaver rhoeas* did well in the first year, as might be expected, but then declined and only showed resurgence following drought conditions. Plant richness was highest in the second year (47 on one, 36 on the other) and then declined. Whilst the different aggregate sizes varied on a year-to-year basis in terms of species richness, by the end of the study period the areas of the roofs with coarse aggregates in the growing media had the richest communities. The species that eventually dominated the flora, especially on the finer aggregates, were mosses. The sedum included in the initial mix ended up with the highest non-moss cover – predictable given its drought-resistant, succulent nature. Forbs were generally in low abundance and did best in areas with coarser media; species that were obviously more abundant than others included ox-eye daisy *Leucanthemum vulgare*, bird's-foot trefoil *Trifolium corniculatus* and haresfoot clover *Trifolium arvense*, the latter two being able to fix nitrogen. The general conclusions were that the coarser and deeper media were able to retain water better than the finer and shallower ones, allowing plants in the former to survive drought periods (2+ weeks without rain), also that shelter from wind and some shading helped plants avoid desiccation. It was also clear that even related species can differ in drought tolerance (e.g. *T. corniculatus* fared less well than *T. arvense*). Ironically, though the presence of organic material can promote plant growth and conserve water, it appeared that under drought conditions the lush top growth became a liability, reducing survival. Optimal organic content was tentatively suggested as ranging from 0 to 10%, but with a nod to potential interactions with water availability.

## 6.4.7 Salt spray

The impact of de-icing salt on green roof plants is something which may seem an irrelevance as they are elevated some way above ground level. However, wind-blown spray can carry considerable distances, up to 100 m from the roadside, although little is known of the height to which such spray can reach and damage may only occur on roofs near elevated highways or where public access to roofs requires de-icing salt to be put down on walkways (Whittinghill & Rowe, 2011). Nevertheless, it appears that some green roofs have been damaged by salt (Whittinghill, 2011).

Whittinghill and Rowe (2011) compared the tolerance to salt spray of seven species used on green roofs: five sedums and two alliums (Table 6.9). The salt levels applied varied from none (control) increasing in 10 g.l$^{-1}$ steps to 50 g.l$^{-1}$ and applied as 4.5 ml of spray per plant on five occasions starting on 11 January 2009 and separated by three weeks between applications; five plants of each species were exposed to each experimental treatment. The results of measurements in May 2009 showed that one sedum (*S. reflexum*) established poorly and suffered substantial mortality even with no salt applied; as might be expected from such inauspicious results, it did not fare well, with high levels of mortality with even light salt levels. All other species were tolerant to spray in the 10–20 g.l$^{-1}$ range, but *S. floriferum* and *S. kamtschaticum* suffered some mortality at higher salt levels and *S. spurium* substantial mortality. A turfgrass comparison (primarily of *Festuca arudinacea* varieties with 10% *Poa*

TABLE 6.9 Mortality of common green roof plants with different levels of exposure to salt applied as a spray

| | *Percentage mortality* | | |
|---|---|---|---|
| *Plant* | *Control (no salt)* | *10–20 g.l$^{-1}$ salt* | *30–50 g.l$^{-1}$ salt* |
| *Allium cerunuum* | 0 | 0 | 0 |
| *Allium senescens* ssp. Montanum | 0 | 0 | 0 |
| *Sedum ellecombianum* | 0 | 0 | 0 |
| *Sedum floriferum* 'Weihenstephaner Gold' | 0 | 0 | 0–40 |
| *Sedum kamtschaticum* | 0 | 0 | 0–20 |
| *Sedum reflexum* | 40 | 60–100 | 60–100 |
| *Sedum spurium* 'Dragon's Blood' | 0 | 0 | 60–80 |
| Turfgrass | 0 | 0 | 0 |

Source: Whittinghill and Rowe, 2011.

*pratensis*) suffered no mortality whatever the regime (Table 6.9). Plant health and volume (size) generally declined with increasing salt concentrations. Moderate volume declines were evident in the 20–30 g.l$^{-1}$ range, becoming increasingly severe over the subsequent levels and reaching a mean decrease in volume (over all species) from just over 200 cm$^3$ to below 100 cm$^3$ (Whittinghill & Rowe, 2011).

The same study also compared the effect of watering other plants with 50 ml of the different salt levels using the same application frequency and intervals. The general pattern of tolerance by species was similar to that of the salt spray but the impacts were evident at lower rates and were far stronger; for example, complete mortality at 10 g.l$^{-1}$ and higher for *S. reflexum* (although survival without salt was low at 20%), no survival for *S. kamtschaticum* at 30 g.l$^{-1}$ and 80% mortality from 40–50 g.l$^{-1}$, no survival of *S. floriferum* at concentrations of 50 g.l$^{-1}$ and strong impacts even at 20 g.l$^{-1}$, strong effects at 10 g.l$^{-1}$ and complete mortality from 20 g.l$^{-1}$. An additional cohort of plants was placed at three different distances (19, 38 and 58 m) from the side of the road Interstate 96. Distance from the road did not change survival rates – although those at greatest distance were healthier. The low survival rates of *S. reflexum,* and to a lesser extent *S. spurium,* in the experiments was partly attributed to the smaller size of the plants used – with smaller plants known to tolerate winter conditions less well than larger plants, so the relative tolerance status of these plants needs confirmation (Whittinghill & Rowe, 2011).

The general conclusions were that soil salt concentrations were more problematic than salt levels delivered as spray to the foliage; that differences in salt tolerance were evident in the groups studied, with the two alliums and *Sedum ellecombianum* considered tolerant and *S. floriferum, S. kamtschaticum* and *S. spurium* being moderately tolerant. The planting of salt-resistant species in such areas likely to receive spray is therefore worth considering, and especially where roof walkways are likely to be accessed during the winter and will thus be subject to increased soil salt levels.

## 6.4.8 Weeding

Removal of tree seedlings (Figure 6.15) and aggressive invaders such as the butterfly bush *Budleja davidii* (Bates *et al.*, 2013) is essential unless the design specifically caters for self-seeded tree species. The latter is unlikely as even intensive roofs, which have the appropriate media depth to support trees, are typically designed as gardens where self-seeding would be inappropriate. Nagase *et al.* (2013) showed that high-density planting of a semi-extensive green roof in Rotherham, UK, reduced the ability of weeds to colonise as did a 25 mm depth of gravel mulch; high species diversity plantings did not reduce weed colonisation. The more rapid the initial establishment of green roof vegetation, the less chance there is for weed colonisation (Getter & Rowe, 2008). Weeding is not necessarily a high-cost activity: Kolb and Schwarz (1986) estimated that weeding their extensive roof experiments took about 8 minutes.m$^{-2}$.year$^{-1}$.

## 6.5 Ecosystem services (or benefits) delivered by green roofs

As with all green infrastructure components, green roofs have a range of environmental benefits claimed for them including improving aesthetics, improving building insulation and associated energy consumption reductions, reducing the urban heat island, reducing

FIGURE 6.15 Green roofs are not maintenance free, and removal of tree seedlings is essential – for obvious reasons. © John Dover

stormwater runoff and attenuating peak flows as sustainable urban drainage components, and providing wildlife habitat. However, Simmons *et al.* (2008) demonstrated how green roofs supplied by different manufacturers exhibited strongly different performance levels in attribute tests. They concluded that it was desirable to identify the particular performance requirements for a specific green roof installation at an early stage in project development so that they could be explicitly 'designed in' rather than assuming that the desired attributes would be met by generic green roof designs.

## 6.5.1 Aesthetics

Whilst green roofs are, in general, considered to be more aesthetically pleasing than unvegetated roofs (see, for instance, Dunnett and Kingsbury, 2004), Kanechi *et al.* (2014) considered the red colouration of water-stressed *S. album*, a plant typical of many extensive green roofs, to be unattractive (Figure 6.16). Sendo *et al.* (2010) and Kanechi *et al.* (2014) explored the use of substitutes in Kobe City, Japan, including the plant blue daze *Evolvulus pilosus*. The aim was to identify plants that would retain their green foliage colouration throughout the season as well as providing flowers. Nagase *et al.* (2010) working in Rotherham, UK, indicated that planting up 'semi-extensive' roofs with 100 mm and 200 mm depths with a higher species richness (six species in their work) could achieve a longer continuous flowering period over a season compared with species-poor plantings (two or three species).

**FIGURE 6.16** Sedum roof in the Netherlands showing red colouration due to water stress during the summer. © John Dover

## 6.5.2 Building insulation and the urban heat island effect

As with all other green infrastructure elements, green roofs have a role to play in reducing the urban heat island effect by improving energy consumption in buildings for air conditioning. For example, less heat is emitted from buildings, less $CO_2$ is generated, and evapotranspiration conditions the local climate (Liu & Baskaran, 2003; Oberndorfer *et al.*, 2007). Vegetation also has a higher albedo (reflectivity) compared to a black conventional roof (Susca *et al.*, 2011). 'Sustainable South Bronx', for example, includes the use of green roofs as one way of helping to combat the urban heat island effect (Rosenthal *et al.*, 2008). Schmidt (2009) considered greening buildings to be second only to having open green spaces in a list of priority measures to tackle the urban heat island effect. Perini and Magliocco (2014) modelled the effects of ground-level and rooftop-level vegetation on typical and hottest summer temperatures in three Italian cities: Milan, Genoa and Rome. They showed that whilst green areas at ground level were more effective at reducing thermal effects than at 1.6 m above ground (street) level (their reference point), vegetation on roofs did contribute to lower temperatures by reducing the 'cooling load' of buildings (the need for air conditioning). Bass *et al.* (undated) estimated that if 5% of the land area in Toronto, Canada, were covered in irrigated green roofs, then it would result in a 1–2°C reduction in air temperature at 1pm. Jim and Peng (2012), working in the humid sub-tropical climate of Hong Kong, China, found that a green roof could reduce the average daily maximum surface temperature of a roof by 5.2°C and the air temperature above the vegetation by 0.7°C, although differences in air temperature could not be detected 1.6 m above the roof.

**TABLE 6.10** Comparative energy budget in (Wh.m$^{-2}$) of an asphalt (black) roof compared with an extensive green roof

|  | Asphalt roof | Green roof | % change |
|---|---|---|---|
| Total incoming radiation from the sun | 5355 | 5354 | 0.0 |
| Long-wave radiation from roof | 2923 | 2494 | –17.2 |
| Reflection from roof surface | 482 | 803 | 40.0 |
| Evaporation/Evapotranspiration | 123 | 1185 | 89.6 |
| Sensible heat | 1827 | 872 | –109.5 |

Source: Schmidt, 2009.
Note: evaporation from roof membrane of conventional asphalt roof, evapotranspiration from green roof with plants. Net radiation received by the roof = latent heat of evaporation + sensible heat.

Some of the earliest modern observations of the thermal values of green roofs were made by Darius and Drepper (1984, cited in Stülpnagel et al., 1990), who noted that self-established vegetation on tenement roofs in Berlin could moderate building temperatures. Kolb and Schwarz (1986) in Würzburg demonstrated reduced temperature variation under green roofs compared with those topped with gravel and suggested that this would lead to energy savings and improved membrane life. Liu and Baskaran (2003) carried out work on the thermal properties of extensive green roofs and found that whilst a conventional roof membrane with a light grey surface could reach 70°C on a summer afternoon, the membrane under a green roof with 150 mm substrate depth only reached 25°C. The role of the green roof in dampening spring and summer temperature variations was substantial: the conventional roof temperature had a median daily range of 45°C whilst the green roof membrane fluctuated by a mere 6°C. Over the study period of November 2000 to September 2002, this translated into a reduction of 75% in energy required for heating or cooling. Schmidt (2009) compared the summer energy budgets of conventional asphalt roofs and extensive green roofs in Berlin (Table 6.10) and estimated that 57.6% of the net radiation received by green roofs was consumed by evapotranspiration as well as increased reflection through having a higher albedo. Latent heat is heat used in a process that does not result in a change in the temperature of an object – so heat from the sun is used in the process of evapotranspiration and does not increase the roof temperature – whereas sensible heat is heat that directly changes the temperature of an object – in this case the roof.

Oberndorfer et al. (2007) also noted the value of green roofs in acting as a thermal insulator: retaining heat in winter and preventing it from entering in summer as well as cooling via evapotranspiration and the shading action of plants. The spring and summer temperature of a 279 m$^2$ green roof on the Fencing Academy of Philadelphia with a 70 mm thick vegetation and drainage layer varied by only 10°C, whereas the surface of a 'black tar' roof in the vicinity varied by as much as 50°C. Vegetative cover also increases solar reflection (compared to bare soil) and thus keeps the ground cooler. In a study by MacIvor and Lundholm (2011), the best-performing species of 15 natives tested in Canada increased the roof albedo by 22% compared with the bare substrate and over 200% compared with a conventional flat roof deck; the average roof temperature was reduced by 3.4°C. Species complexity might be expected to deliver better ecosystem services than monocultures and Lundholm et al. (2010) compared the effect of species monocultures, uniform structural types (creeping shrubs, creeping herbaceous species (forbs), grasses, succulents or tall forbs), and three and five mixtures of the structural forms on green roof surface temperatures using green roof modules. The results were mixed, with generally the three and five growth-form plantings, which included tall forbs, reducing surface temperature more than monocultures or single structural types, but not always. Monocultures of biting stonecrop *Sedum acre* (a low-growing succulent) and silver-rod *Solidago bicolor* (a tall forb) were shown to be as good as the best multiform mixtures, with the latter being best of all (temperature with no plants 26.56 ± 0.39°C, with *S. acre* 23.36 ± 0.58°C, with *S. bicolour* 21.77 ± 0.49°C, best structural type (tall forb) 22.71 ± 0.81°C, best mixture of structural types 22.71 ± 0.34°C). The take-home message being that it is best not to rely on assumptions, but on observed performance! Niachou *et al.* (2001) used infra-red (IR) photography on the green roof of a hotel in the Lotraki area of Greece, and showed heterogeneous surface temperatures due to differences in vegetation and the presence of bare substrate. The IR image was taken on a day in late June 2000 with an air temperature of 28°C. Thick, dark green foliage gave the lowest surface temperatures in the range 26–29°C; sparse red vegetation 36–38°C, and almost 40°C on bare substrate. In comparison, a nearby building without a green roof (but which did have insulation) had white vertical painted surfaces registering 27°C but, as with the unvegetated substrate of the green roof, the bare horizontal roof surface registered 40°C. They considered that the improved insulation of a green roof saved only about 2% of the annual energy requirement of a building with a well-insulated roof, rising to 4% in a moderately insulated roof, but was a substantial 37% saving in a building without insulation. If additional night-time cooling in the summer, based on ten air changes per hour (ACH), was included in the analysis, then the annual savings for the uninsulated roof increased to 48% and the moderately insulated roof to 7%. On a student residence building in Portland, Oregon, USA (the Broadway Building), the green roof had a 13% reduced heat flux in winter compared with a conventional roof, and 72% in the summer, giving substantial energy savings (Spolek, 2008).

Simmons *et al.* (2008) compared the thermal performance of six different green roof designs in a replicated experiment (three of each type) – the substrate depth and planting design was the same in each case (16 species, 100 mm depth) but the specifics of the design varied on the basis of proprietary media design and the materials involved in the various drainage/membrane layers. The plots were relatively small (2.0 x 1.7 m) but the experimental design did include a conventional black-coloured roof and a high albedo white roof for comparison. Temperature sensors were placed below the roof membranes and below the green roof platform (steel decking) to measure 'inside' temperatures. In the subtropical environment of Austin, Texas, irrigation was essential to maintain plant health, and plots were watered (equivalent to a weekly rainfall of 20 mm/week) when there was inadequate rainfall during the active growing season. Whilst there were minor variations at the membrane level, the green roofs all out-performed the conventional and high albedo roofs, having substantially cooler membrane temperatures. Example data were given for a mid-afternoon with an ambient air temperature of 33°C: the green roofs ranged in membrane temperature between 31 and 38°C (depending on design), the black roof reached 68°C and the white roof 42°C. Under the roofs, the green roofs' temperatures ranged from 36 to 38°C, the black roof reached 54°C and the white roof 50°C. On slightly cooler days (ambient = 27°C), the differences were still evident, though less pronounced. On cool days (ambient = 5°C), black and white roofs were cooler than green roofs at the membrane level. Also in Texas, Dvorak and Volder (2013) showed that whilst the surface of an extensive green roof was cooler (on average by 18 °C) than a similar roof without vegetative cover, the greatest temperature reductions were seen underneath the modules containing the growing media (average 27.5°C; peak 37.3°C). The daytime temperature range recorded on a conventional roof surface was 50°C but under the green roof the temperature range was only 17°C and this reduced thermal stress is probably one of the main reasons why roof membranes have greater longevity under green roofs.

Jim and Peng (2012), in Hong Kong, compared the impact of weather conditions on green roof cooling. They used data taken from a roof with a conventional surface in August–September 2008 and compared it with data from the same roof, and same months, in 2009 after it had been given a 484 m² green roof (a control, ungreened roof was also included as an additional comparison). During windy periods, evapotranspiration, and hence cooling, was increased on the greened roof, but the wind also cooled ungreened roofs. On hot days, the green roof experienced a lower temperature range (28.2–30.3°C) compared with the same roof before greening (25.3–39.4°C), with the biggest difference occurring at noon. At night the ungreened roof was 1–2°C cooler than the green roof, as the green roof stored heat and released it more slowly. The cooling effect of the green roof was still evident on cloudy days but the difference was less than on sunny days, as the unvegetated roofs were naturally cooler than on sunny days. On rainy days the impact of a green roof was quite different to that on both sunny and cloudy days: the temperature of the roof surface of the unvegetated roof dropped sharply with the onset of rain, whilst the surface of the green roof (that is the upper roof deck surface, below the vegetation and green roof drainage layers) remained warmer with a slow reduction in temperature. The authors concluded that the level of solar radiation and humidity were the main driving factors behind the cooling effect of green roofs, but that soil moisture was also critical in that it interacted with both solar radiation and humidity. The annual reduction in electricity needed for summer air conditioning was calculated as $2.80 \times 10^4$ kWh. In the rather cooler conditions of Ottawa, Canada, the main value of a green roof was shown to be in spring and summer (see above). As temperatures dropped in late autumn/early winter, the growing medium started to freeze and its value as an insulator declined; when full snow cover was established in winter, the green roof and conventional roof had similar insulation properties.

Alexandri and Jones (2008) carried out a modelling study using data from nine cities (listed in section 5.6.4) and showed similar results to other studies – that a green roof reduces the heat flux through a building's roof and that the range of the fluctuations in heat flux is much lower when compared to a conventional roof. The 24 h convective flux density for Montreal was estimated as: conventional roof = $345.1$–$128.6$ $W.m^{-2}$; green roof $51.3$–$99.9$ $W.m^{-2}$. Green roofs were considered to contribute to the cooling of the urban canyons below (i.e. the street level) because cooler air would flow down from the rooftop to displace warm air from the urban canyon. Roof surface temperatures were shown to be much lower with green roofs, though the benefit was expected to be greatest for Mumbai with a maximum temperature decrease of 26.1°C, whilst it was 'only' 19.3°C for London. In terms of average daily roof surface temperature reductions, Riyadh was predicted to have the best decrease at 12.8°C and Moscow the least pronounced reduction at 9.1°C. Overall, the study concluded that the greatest benefit in temperature reduction would be experienced when both roofs and canyon walls were greened (best case: Riyadh air temperature in the street canyon decreases by a peak of 11.3°C and a daily average of 9.1°C).

Martens *et al.* (2008) carried out simulation modelling and showed that the value of a green roof in reducing the overall cooling requirements of a building decreases as the number of storeys increases – so a green roof is of most benefit in a single-storey building. Likewise, the greater the roof area relative to the wall area, the greater the energy saving. They compared the energy saving of a green roof on a 250 x 250 m building with one, two, or three floors and a 50,000 watt internal cooling requirement. The results showed that a green roof could save 73% of the energy requirement for a one-storey building, 29% for the two-storey and 18% for the three-storey building.

### 6.5.3 As components of sustainable urban drainage

The water retention ability of green roofs is of interest not simply because of the need to keep roof plants alive but also because of their value as part of sustainable urban drainage systems (SUDS), helping to reduce the impact of sealed surfaces on local hydrology (Carter & Butler, 2008). Oberndorfer *et al.* (2007) cite the value of green roofs as SUDS components, an approach which has gained much traction (Rosenthal *et al.*, 2008). Liptan and Strecker (2003) reported on monitoring projects in the city of Portland, Oregon, USA, where several 'ecoroofs' had been installed as part of a wider sustainable urban drainage programme; Portland had been suffering from stream destabilisation through erosion, and increasing pollutant problems from runoff as a result of high levels of sealed surface. Their choice of roof term is interesting; they point out that because of their climate their roofs are not green throughout the year, but deliver both 'eco'-system services and 'eco'-nomic benefits.

In terms of sustainable urban drainage, it is the ability of roofs to intercept water as it falls and store it *in situ*, remove it via evapotranspiration, clean it up and delay its onward movement to the sewer/drains or storage areas that is of interest. The retention ability depends on the roof design (Simmons *et al.*, 2008) and especially the depth and nature of the growing media (Mentens *et al.*, 2006), whether it also incorporates a drainage/storage level (see above) together with the roof angle (slope), and the predominant wind direction (Köhler

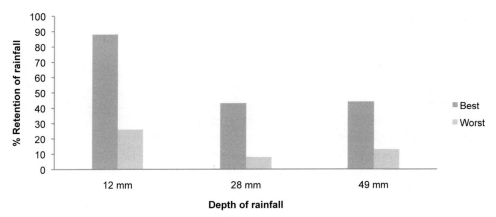

FIGURE 6.17 Best and worst levels of % rainfall retention from a study of six different green roof designs with different precipitation levels. All designs had the same substrate depth and planting design, the work was carried out in the subtropical climate of Austin, Texas, US. Source: Simmons *et al.* (2008).

& Poll, 2010). Simmons *et al.* (2008) in their experimental study of six green roof designs demonstrated how different manufacturers' designs could strongly affect water retention. All designs could fully retain a 10 mm rainfall, but retention levels varied considerably with larger rainfall depths (Figure 6.17). For some of the most poorly performing green roof designs, although retention was greater than conventional roof designs, the results were not always significantly different. The work clearly highlighted the differences that materials in green roof design could have on water retention performance with some performing exceptionally well – and some underperforming.

Liptan and Strecker (2003) estimated that the ecoroofs they studied in Portland could retain 10–35% of rainfall during their wet season and 65–100% during the dry months and were effective in attenuating peak flows (see also Hutchinson *et al.*, 2003). Moran *et al.* (2004), in a study of two extensive green roofs in North Carolina, USA, estimated that just over 60% of rainfall was retained over the period April–December 2003. In the cooler seasons, with more rainfall, the substrate retains water for longer because plants are not actively transpiring and cooler weather inhibits evaporation, thus there is less capacity to store water than in warmer seasons (Mentens *et al.*, 2006). Carter and Rasmussen (2006) at the University of Georgia, Athens, USA, compared the stormwater retention of an experimental green roof section (76 mm depth of 55% Stalite, 30% USGA sand and 15% worm-cast compost planted up with four *Sedum* species and two *Delosperma* species) with a conventional roof section. The roof sections, 42.6 m$^2$ each, were monitored from November 2003 to November 2004 during 31 rain events (2.8 to 84.3 mm in depth) and water collected and measured, and the time taken to drain recorded. As expected, the green roof retention declined with increasing intensity of rainfall: with rain of less than 25.4 mm almost 90% of water was retained; as the substrate became saturated with larger events (in excess of 76.2 mm), the roof held just less than 50% of water deposited. Runoff from the conventional roof, on average, followed 17.0 minutes after the onset of rain, whereas this was 34.9 min with the green roof – a delay of a further 17.9 minutes. Moran *et al.* (2004) reported mean peak flow reductions of 78% and 87% for their North Carolina study roofs, and even longer retention times. They gave an example of a 23 mm rainfall event starting at about 07.55 and lasting until about 14.10 on 7 April 2003; 75% of water was retained, with a 90% reduction in peak flow, and runoff was not evident until about four hours after the onset of rain. Spolek (2008), also working in Portland, examined the value of two retrofit roofs on the Multnomah County building (one planted with wildflowers, the other with grasses; both 100–150 mm in depth) and also a designed-in green roof on the Broadway Building (a student residence) at Portland State University – this was 150 mm deep and was planted with sedums, periwinkles and blue fescue grass. On the Multnomah roofs, summer retention rates typically ranged from 7 to 85% (mean 42%) and winter from 0 to 52% (mean 12%). Over a 28-month study period, the annual mean retention rate was 18% for Multnomah, and over 20 months for the Broadway building 29%. These values were rather lower than reported in other studies, but there was no obvious explanation as to why that should be. Stovin (2010), in Sheffield, UK, used a small 3 x 1 m green roof testbed (sedums growing on 80 mm of substrate) and found that mean retention during the spring of 2006 with 11 rain events was 34% with mean peak attenuation of 57%. In May, June, July and August the mean retention was 79%, 33%, 45% and 100%, which reflected the exceptionally heavy rainfall in June. Attenuation of individual rainfall events depends on the volume and intensity of rainfall, and crucially on the 'antecedent dry weather period' (ADWP). This

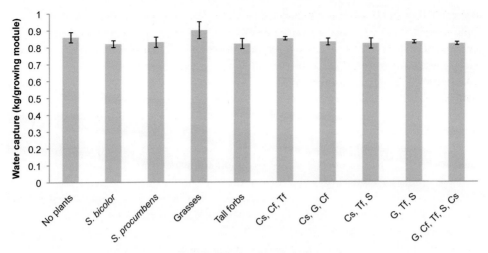

**FIGURE 6.18** The ten best-performing combinations of growth form for water capture from 30 combinations tested (bare growing substrate, 13 species monocultures, 5 structural combinations (Cs, Cf, G, S, Tf), 10 x 3-growth form mixes, 1 x 5-growth form mixture). Source: Lundholm *et al.* (2010). *S. bicolor = Solidago bicolor* (tall forb), *Sagina procumbens* (creeping forb), Cs = creeping shrub, Cf = creeping forb, G = grasses, S = succulents, Tf = tall forbs.

latter is essentially the amount of time since the last rainfall – which allows the substrate to drain/evaporate off water and thus increase its holding capacity for new rain events. The ADWP will, of course, be affected by the season and general weather patterns, with wet weather slowing the capacity recharge and warm weather facilitating it.

The FLL guidelines suggest that the minimum water capacity of growing media should be 35% for extensive and 45% for intensive roofs, although there is some debate about this (Ampim *et al.*, 2010). Alumasc Ltd give a range of retention data depending on substrate depth and plant type for their green roof systems: with extensive roofs of 20–40 mm and moss and sedum cover the retention rate was given as 40%, rising to better than 90% for intensive roofs with in excess of 500 mm substrate depth and a cover of lawns, shrubs, bushes and trees (Alumasc, 2004).

Vegetation would be expected to improve the water retention capability of green roofs, but the work of MacIvor and Lundholm (2011) was less encouraging in that respect. The majority of the 15 species they tested were not significantly different from the control in water retention, and only three of them retained more water, on average. The best species in their trial, hay sedge *Carex argyrantha*, retained over 75% of the water applied to it and was significantly better than the control, by some 4%; wire grass *Danthonia spicata* retained significantly less than the control (6.9%). In their experiments with plant functional groups (above), Lundholm *et al.* (2010) included an assessment of the value of mixtures of plant growth form on water capture and also found variation in response to this attribute. In general, mixtures outperformed monocultures, but modules with bare substrate were better than all the mixtures and a monoculture of grasses was superior to all other combinations (Figure 6.18). More recently Nagase and Dunnett (2013a) demonstrated, with simulated extensive green roofs and 12 test species, that water retention for the three major classes of plant typically used on green roofs was in the order grasses>forbs>sedums. Unsurprisingly, the features that contributed to these findings were related to physical structure and size, with taller species with greater bulk and biomass (above and below ground) being better at reducing water runoff. Bare substrate was used as a control in these experiments and the four sedums used all had water runoff greater than from soil alone – possibly because of their small leaves retaining water less well on the plant surfaces and their growth forms impeding uptake by the growing medium.

Mentens *et al.* (2006) used data from 18 publications and 628 records (primarily German language in origin) to develop a model of runoff from green roofs. They then used this model to estimate the effect of having 10% of the roofs in the Brussels city-region converted to extensive green roofs with 100 mm of substrate. Their findings were that, as a whole, the result would be to reduce runoff in the region by 2.7% which seems modest, but represents a lot of water ($1.7 \times 10^6$ litres). In the more densely built-up areas, including the city centre, the runoff reduction would be higher at 3.5% and of course less in the less densely populated suburbs. The value of roof greening was clearer at the individual building level where runoff reduction was shown to be closer to 54%. With conventional roofs about 90% of all rain ends up as runoff, with green roofs this was estimated to be reduced to 42%.

### 6.5.4 Green roof runoff water quality

Water quality from green roofs is expected to be good, partly because of the ability of plants to take up pollutants, but that will depend on the levels of pollutants in the substrates used. As Moran *et al.* (2004) found, nutrients can leach from compost in green roof media and these nutrients then become pollutants in runoff; however, leaching was expected to decline over time. Likewise, Hutchinson *et al.* (2003) found total phosphorus and ortho-phosphorus levels in green roof runoff to be above water quality criteria. Aitkinhead-Peterson *et al.* (2011) working in south-central Texas found that runoff from a green roof planted up with *Sedum kamtschaticum, Delosperma cooperi* and *Talinum calycinum* leached carbon, nitrogen, potassium, calcium and magnesium from the substrate over the first six months post-installation. Some plants (*D. cooperi* and *T. calycinum*) were shown to use nitrate for plant growth, reducing leachate levels. Green roofs in Texas need irrigation to survive during the summer, and sodium levels were shown to increase in the growth medium from the irrigation water. They suggested that decaying plant material could also contribute to eutrophication of the runoff but, as with other authors, suggested that over time green roofs were likely to remove nutrients from rainwater, but that both testing of roof media and long-term studies of installed roofs was desirable. To a certain extent, that data gap has recently been filled by Köhler and Poll (2010), who showed that the carbon/nitrogen ratio of green roofs declines rapidly from relatively high values of about 25 to a more acceptable level (@13), reflective of natural substrates after ten years with little subsequent variation – although their analysis used rather few data points.

Alsup *et al.* (2010), in pot trials, showed that a range of materials included in some green roof substrates (Axis, Arklyte, coal bottom ash, Haydite, Lassenite, lava rock and composted pine bark) had similar concentrations of metals in them as 'typical' soils. Surprisingly, the pine bark compost had relatively high levels of copper, iron, zinc, cadmium and manganese given it is an organic material; this was thought to have arisen as a result of contamination by rain and dust from nearby industrial processes during storage and indicates that pollutant loads in green roof media need to be tested and not assumed. Some of the metals in the media leached at concentrations that exceeded water quality criteria. Manganese from Lassenite was identified at levels that would affect the taste and staining qualities of drinking water; lead and cadmium from all the media tested had at least one test sample that exceeded quality levels. The leaching of the metals over time was complex, depending on the particular substrate and the presence of plants (*Sedum hybridum* 'immergrauch'). Generally, lead exhibited less leaching with plants (the exception being with Lassenite), whilst cadmium had increased leaching when plants were present. Of course these pot-based studies had far less water running through them than would be expected of green roofs *in situ* and so the potential for contamination of runoff revealed by the experiment may be potential rather than real. Despite the majority of records from study roofs in Portland, Oregon, having copper levels below quality thresholds, Hutchinson *et al.* (2003) found 21% of samples exceeded acceptable levels. Potential sources could have been treated timber used as edging material, natural topsoil, green roof media, etc. – the conclusion was that care needs to be taken in the selection of green roof materials. Hutchinson *et al.* (2003) noted that in southern California there are restrictions on copper and zinc in green roofs.

### 6.5.5 Air pollution

Currie and Bass (2008) used the Urban Forest Effects (UFORE) model to investigate the effect of green roofs on air pollution in the Midtown area of Toronto, Canada. There were a number of assumptions built into the model: one important simplification being that data on grass (leaf area index, evapotranspiration rates) could be substituted for the various plants used on extensive green roofs and that data on shrubs could be used to represent intensive green roofs (the latter, in particular, seems to be rather unrealistic and is not covered here). Scenario 1, the baseline, represented the pollution removal ability of the existing vegetation composition (trees and shrubs) of Midtown, the baseline was augmented with 20% of flat Midtown roofs being covered with grass in Scenario 6 and all flat roofs were assumed to be

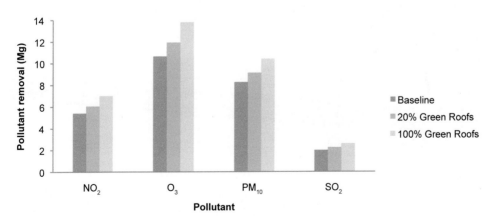

**FIGURE 6.19** The impact of green roofs on a range of air pollutants in Midtown, Toronto, Canada. The baseline represents the pollutant removal provided by the existing tree and shrub vegetation prior to modelling the impact of adding green roofs. Source: Currie and Bass (2008).

grass covered in Scenario 7. Overall, the findings showed that substantial levels of pollutants were removed by the existing vegetation and that the two green roof scenarios could also potentially capture useful, though rather lower, levels of pollutants (Figure 6.19).

Yang *et al.* (2008) estimated the levels of different pollutants removed by green roofs in Chicago, Illinois, USA. Chicago was reported to have about 300 green roofs, and data on the kind of green roof, total roof area, area of each vegetation type on the roof, building parameters, etc. was gathered on 71 (19.8 ha) using high (160 mm) resolution aerial photography. This study used a 'big-leaf resistance' dry deposition model and used a full year's hourly data on nitrogen dioxide, sulphur dioxide, ozone and $PM_{10}$ concentrations from the centre of Chicago itself over the period August 2006 to July 2007. As with Currie and Bass (2008), extensive green roofs were modelled as short grasses but intensive roofs were considered to consist of short grass, tall herbaceous species (crop data was used), and small trees. In addition to the situation at the time of the study, three further green roof scenarios were used to examine their impact on pollution removal: planting all roofs in Chicago with the same proportion of extensive to intensive roofs, planting up all remaining roofs only as extensive roofs, and planting all remaining roofs as intensives (tall herb species, shrubs and small trees). Most roofs were on commercial and office buildings compared with only 4% on residential developments. Despite the dominance of extensive green roofs in the literature, over 67% of the Chicago green roofs were semi-intensive/intensive compared with nearly 33% extensive. The green roofs studied were estimated to have removed 1,675 kg of pollutants (871 kg $O_3$, 452 kg $NO_2$, 235 kg $PM_{10}$ and 117 kg $SO_2$) (Figure 6.20); there were seasonal differences, with more removal in May and least in February. If remaining roofs in Chicago were planted up with the same extensive/intensive ratio, then pollutant removal could reach 1,835 tonnes/year, if only as extensive 1,405 tonnes, but in the extreme case of all the remaining roofs being planted up as intensives, the annual pollution removal was estimated at 2,047 tonnes – but there would be an eye-watering one-off installation cost of $35 billion. As with all modelling studies simplifications, assumptions and missing parameters mean that the pollutant removal figures cannot be taken as exact values but rather as working approximations. The costs of installation are large when considered purely for air pollution control, but do not take into

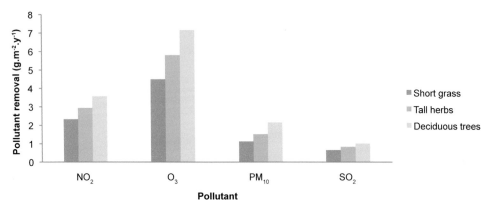

FIGURE 6.20 Air pollutants estimated to have been removed by different vegetation types on 19.8 ha of predominantly semi-intensive/intensive green roofs in Chicago, Illinois, 2006–2007 using a dry deposition modelling approach. Source: Yang *et al.* (2008).

account other ecosystem services delivered by green roofs, and further opportunities for optimisation exist (e.g. of plant species), and reduced installation costs are likely as green roofs become more popular.

Air pollution reduction by plants is difficult to measure directly with real green roof installations and this probably explains the paucity of studies reporting such work. It is far easier to do such work under more controlled conditions in laboratory fumigation chambers as Morikawa *et al.* (1998) did when screening some 217 plant species in Japan for nitrogen dioxide uptake. Morikawa *et al.* (1998) did not do their work specifically to screen green roof plants for their ability to remove nitrogen dioxide. They used 50 common herbaceous species of roadsides, 60 cultivated herbaceous and 107 cultivated woody species. Some of the findings from their work suggest a green roof plant screening programme is urgently needed. For example, they found a 657-fold difference in the uptake of nitrogen dioxide from their worst (the non-rosette-forming Bromeliads *Tillandsia ionantha* and *T. caput-medusae)* to best species (drooping or manna gum *Eucalyptus viminalis*). Only one plant from the same Family as the popular extensive green roof sedums was assessed, flaming Katy *Kalanchöe blossfeldiana*, and it was not a particularly impressive performer. Takahashi *et al.* (2005) found that a sedum they tested (species name not given) was also poor at $NO_2$ uptake. Morikawa *et al.* (1998) also suggested that there were $NO_2$-philic plants that had a metabolic pathway that could more effectively use $NO_2$ (which is known to have a range of toxic effects in plants including the inhibition of photosynthesis (Segschneider, 1995)). Unfortunately, monocotyledonous plants – which include the grasses – do not appear to be in this group. Morikawa *et al.* (2003) subsequently identified plants that can grow with their only source of nitrogen coming from nitrogen dioxide. Given that the modelling approaches of Currie and Bass (2008) and Yang *et al.* (2008) used data from grasses to simulate extensive green roofs, it is obvious that there is an important data need here. Identifying changes in gaseous air pollution concentrations above green roofs compared with control roofs would be exceptionally difficult, and uptake by the plants of the pollutants would effectively be by inference; for example, enhanced growth of plants on roofs in response to nitrogen dioxide (Adam *et al.*, 2008) in polluted compared to non-polluted locations. However, Tan and Sia (2005) in Singapore attempted to evaluate the impact on sulphur dioxide and nitrogen dioxide levels of installing green roofs above a multistorey car park by comparing concentrations in the air before and after installation. Pre-installation assessments were made in June/July 2003 with the installation later that year; the post-installation assessments were made in February/March 2004. The results suggested a decrease in sulphur dioxide by 37% and nitrogen dioxide by 21%. Whilst this study is interesting, the failure to monitor un-greened control roofs over the pre- and post-installation data collection period, and the differing time periods compared (June/July vs February/March), means it is not really possible to be sure that the reduction in pollutants claimed was not due to some other influence (e.g. of seasonality, changes in traffic flows along adjacent roads).

Whilst there are great difficulties in directly measuring uptake of gaseous pollutants on green roofs, measuring the removal of particulates ($PM_{10}$ and smaller) can be more straight-forward: particulates can be counted and size-ranged on plant surfaces using environmental scanning electron microscopes or can 'simply' be washed off and weighed; a more techno-logical approach by assessing the magnetic metal content of the particles is also possible. Speak *et al.* (2012) used this latter approach in their comparison of the uptake of particulates on four species of plants placed, in trays, on two roofs in the centre of Manchester, UK. The plants used were two grasses – creeping bentgrass *Agrostis stolonifera* and red fescue *Festuca rubra* – and two herb plants – ribwort plantain *Plantago lanceolata* and white stonecrop *Sedum album*, a succulent commonly used on extensive green roofs. In this study the grasses were more effective than the herbaceous or succulent species at particulate capture and the order of capture efficiency (on an equivalent area basis) was creeping bentgrass>red fescue>ribwort plantain>white stonecrop. The differences in capture efficiency were considered to be at least partly related to the surface microstructure of the leaves, with the ribbed grooves of the two grasses being particularly effective whilst the polygonal grooves of ribwort plantain and white stonecrop appeared to be less effective, the latter possibly partly because of its waxy surface. An estimate of the annual particulate removal from the 326 ha of Manchester city centre and the associated Oxford Road area was made assuming all flat roofs were covered in sedum. Total emissions for the area were given as 9.18 tonnes/year and the sedum capture as 0.21 ± 0.01 tonnes/year or 2.3% of the total emissions; with the most effective grass species this would be expected to achieve a removal of 1.61 ± 0.05 tonnes/year or 17.5% of total emissions.

## 6.5.6 Noise pollution

Van Renterghem and Botteldooren (2008) examined the noise reduction value of green roofs. They were particularly interested in whether placing a green roof on a building would reduce the propagation of noise from the road canyon on one side of a building to the canyon on the other side of the building. In other words, would green roofs help in the creation of quieter spaces for urban dwellers (even just by having quieter rooms at the rear of the house) and thus reduce stress. They used a modelling approach and did find that green roofs could reduce sound propagation in the octave bands with 500 and 1000 Hz centres. The best attenuation for extensive roofs (of about 10 dB) was with a substrate depth of between 150 and 200 mm, with the effectiveness reducing with shallower substrates; for intensive green roofs, with media depths above 200 mm, increasing depth did not increase noise reduction. There was also a linear relationship between the proportion of the roof that was 'greened' and the reduction in noise. They subsequently tested their predictions with real roofs by measuring noise levels before and after the installation of five green roofs (Van Renterghem & Botteldooren, 2011). The interpretation of the results was complex, dealing with different frequency ranges, different building and green roof design layouts, and even different levels of sound diffraction, and in some frequency ranges and situations there were negative effects with more sound transmission. However, essentially the answer was 'yes', green roofs have the potential to reduce sound diffraction by up to 10 dB. Further modelling (Van Renterghem *et al.*, 2013) showed how multiple sound-absorbing screens on rooftops could also contribute to sound reduction in a courtyard behind a building and that greened pitched roofs reduce sound better than green flat ones.

## 6.5.7 Food

It is also possible to grow food on green roofs and Liptan and Strecker (2003) claimed that one Portlander 'harvests hundreds of pounds of tomatoes each week' on a so-called 'Ag-roof'. Whittinghill *et al.* (2013) list over 20 plants (including rice) that have been grown on roofs, but they point out that these are usually in intensive roof substrate depth levels or in containers. They explored the potential of growing vegetables (cucumber *Cucumis sativus*, green bean *Phaseolus vulgaris*, pepper *Capsicum annuum*, tomato *Solanum lycopersicum*) and herbs (basil *Ocimum basilicum*, and chives *A. schoenoprasum*) on shallow green roof substrates over a three-year period (2009–2011) at Michigan State University, East Lansing, USA. The trial was relatively small with only two plants of each type grown in four replicates. There were three different sets of growing conditions: in the ground, in a green roof, and on a 'green roof platform' – the latter two had substrate depths of 105 mm. The green roof platforms might also be termed 'raised beds' as they were supported on blocks (see image in VanWoert *et al.*, 2005). Overall the work showed that as long as irrigation and some fertiliser were provided, it was possible to grow crops with reasonable yields at roof level. There was scope for improvement in production technology, which it was considered would raise production levels to the equivalent of ground-level production. One supermarket of the Budgen's chain in Crouch End, London, is part of a project called 'Food from the Sky' – volunteers grow vegetables on the roof in 300 planters made from recycling boxes donated by Haringey Council and then sell it in the supermarket below on Fridays. The plants get a heat subsidy from the supermarket as the roof is warmed by its heating and lighting, helping to reduce the effect of frost (Barnett, 2011).

## 6.5.8 Development mitigation

Development mitigation can take many forms, and is often used to replace habitat lost through building or infrastructure works. However, this is not always the case, and Badeja (1986) reports the planning stipulation that the roof of the Crédit Suisse administrative building have a green roof so that it would not ruin the view from the Uetliberg mountain near Zurich, Switzerland. The roof was installed in 1980. The approach used was interesting in that different substrate depths were used: 300 mm to support low hedges (on banks 2 m wide) and 80 mm for planted extensive vegetation; additionally, about a quarter of the roof was left bare and covered with 100 mm gravel to allow for natural colonisation from the planted areas. The hedges were used to create mini-fields to provide shelter and warmer microclimates for the lower-growing herbs and grasses; native shrubs were used in the hedging. The roof design was carried out in 1978–1979 and very little guidance was available at the time, so a rather rough-and-ready approach was used with shallow wooden vegetable and fruit crates used as planting modules, filled with growing media, planted up and placed on the roof! The use of wood was considered acceptable as it would hold water and would eventually rot down. The bare gravel areas had been colonised by 40 plant and grass species from adjacent plantings and from natural sources two years after installation.

Brenneisen (2006) reported the use of green roofs as compensation for land-use change, which would include development mitigation, but noted that substrate depth was a crucial factor in limiting their value. He also found that invertebrate colonisation was better if 'natural soils' and other media (e.g. from riverbanks) were used – probably due to their structure being most appropriate for the colonising species, though they may have also carried individuals of some species with them during translocation. In Basel, Switzerland, this use of natural soils from the locality has been built into the planning Acts and is usually obligatory for roofs over 500 m$^2$ and the depths of the growing media must vary (Brenneisen, 2006).

Olly *et al.* (2011) were interested in whether brown roofs, a subset of the semi-intensive 'biodiverse' type of roof, truly recreated the habitat of brownfield land. They did this by comparing the growth of a seed mix (and media seed bank) in experimental arenas which replicated the brown roof structure of aggregate, drainage layers and impermeable membranes and compared them with controls composed only of a layer of aggregate but in direct contact with the ground (and therefore soil moisture). Essentially they made experimental 'ground-level' brown roofs – there were two roof treatments: 100 mm aggregate depth and 150 mm depth. The results showed that real differences emerged in plant growth and survival in the 'brown roof' and 'brownfield' treatments and that the differences were related to depth of growing media and therefore water regimes. They concluded that brown roofs (even with careful design) could not act as direct replacements for brownfield habitat. However, periodic drought on brown roofs appears to help 're-set' succession (Bates *et al.*, 2013) which may extend their useful life in recreating early brownfield conditions (Olly *et al.*, 2011).

### 6.5.9 Biodiversity value of green roofs

Grant *et al.* (2003a) collated information on the biodiversity value of green roofs for England's nature conservation agency English Nature and noted that, at that time, relatively few green roofs had been specifically designed with wildlife in mind. This is beginning to change as green roofs are considered viable ways of mitigating for development losses and new guidance documents are produced for conurbations, such as Toronto (Torrance *et al.*, 2013), and for species groups, such as invertebrates (Gedge *et al.*, undated). Green roofs can never completely replace ground-level habitats, if for no other reasons than the limited range of habitats that can be created on the shallow substrate depths of most roofs (Grant *et al.*, 2003a), the relatively small size (area) of individual roofs, some species have extreme difficulty getting up to them, and for some soil-living species, such as earthworms, the growing media are not deep enough (Brenneisen, 2006). They can, however, make a substantial contribution to biodiversity conservation and may improve connectivity as part of a green corridor or stepping-stone network.

Butler *et al.* (2012) was interested in the promotion of native species plantings on green roofs and, from a literature survey, found that it was related to the discipline of the writer with 'engineers' less likely to specify the use of native plantings (only 2%) than 'architects' (49%), 'landscape architects' (38%) or 'biologists' (30%) (other/unknown made up the remaining 25%). They identified the four main reasons for natives being recommended as follows:

1. For aesthetics (which they considered rather subjective).
2. Because natives are expected to be better adapted to local conditions and require fewer inputs (even though environmental conditions at ground level are typically less harsh than on rooftops).
3. Because native flora might be expected to prove a more viable habitat for wildlife (although the resource provided depends on the plant species – grasses do not provide nectar, but may act as food plants for some herbivores).
4. Because native species were considered to be less likely to become invasive (though few non-natives become invasive).

Papers were identified that addressed the 'adaptation issue', though they felt that claims did not bear much scrutiny as many roofs needed irrigation (and therefore an 'input' that would not have been required at ground level). They found no papers directly comparing native vs non-native plantings in terms of habitat or invasiveness propensity. Their review questioned the use of the term 'native' – what it actually means in context – and found that 42% of research and conference papers that used the term did not actually define what they meant by it, and that usage varied in scope. The review effectively highlighted assumptions, suggested the blanket 'native is best' mantra to be over-simplistic, and pointed to the need for more studies, whilst accepting that evaluating a range of native species for use on green roofs was a sensible approach, allowing greater flexibility in planting schemes.

## Flora

The majority of green roofs are planted up in some way, although some may be left to be colonised naturally fully or partially. Floral biodiversity is thus, generally, under the control of the roof designer – at least at first, as the initial planting may not be the same as that which ultimately develops (see above and Grant, 2006a). Nevertheless, there are some data on the natural colonisation of roofs by vegetation: Payne (2000) produced a small booklet detailing his eight-year survey of the flora found naturally colonising 1,629 roofs and associated features, primarily in East Anglia, UK, and supplemented with data from a review of old county floras. The dataset contained information collected directly from 639 roofs. The nature of the surfaces studied was wide and varied from the top of concrete wartime pillboxes, to garden sheds, bus shelters, conventional and thatched roofs, etc.; he also delved into drainage gutters, though fortunately these ephemeral water bodies were treated separately. He called roof and gutter flora 'tecticolous' from the Latin for 'a roof' following the advice of a colleague. In total 135 species of vascular plant were identified from roofs, but only five species were found on 10% or more. The ten commonest species listed by Payne (2000), perhaps unsurprisingly, included three sedums, and the most abundant plant of all was golden stonecrop *Sedum acre* (Figure 6.21). The top ten included a tree, sycamore *Acer pseudoplatanus*, which is not a species desirable on roofs.

Another example of 'natural' colonisation of a green roof is the Wollishofen water plant in Zurich, in this case presumably from the seedbank of the soil used in its construction. This

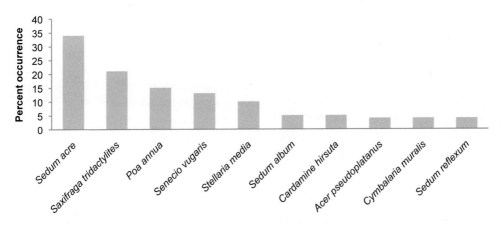

**FIGURE 6.21** Percentage occurrence of the ten species found most frequently in a survey of 639 roofs in East Anglia, UK, and elsewhere. Source: Payne (2000).

facility has four green roofs that were built in 1914 to keep the buildings cool. These roofs are still in existence and host 175 plant species (including nine orchids, three of which are endangered) – ironically the roof flora has a rich vegetation community whilst the surrounding area is now depauperate due to agricultural intensification (Brenneisen, 2006). The roof is poorly drained (150 mm topsoil on top of 50 mm of gravel), something that is designed out of modern roofs – and yet it is this additional wetness that allows for a wider range of wet grassland species to persist than would otherwise be the case.

## Invertebrates

Buglife (the British nature conservation charity) produced a best practice guide for creating invertebrate habitat on green roofs (Gedge *et al.*, undated) which provides a range of case studies of roofs designed for invertebrates. The guide highlights the value of varying substrate depths to maximise opportunities for a wide range of flora, and the provision of a range of microclimates. Varying substrate types (such as sand and shingle) are also promoted as they provide opportunities for different flora as well as conditions suitable for a wider range of invertebrate; for example, bee banks provide nesting habitat for solitary bees and wasps. Including piles of wood is also beneficial providing overwintering shelter for many species and food for saproxylics (those that feed on dead wood). A green roof's environment can be augmented with invertebrate nesting boxes, or 'bug hotels', etc. Advice on flora includes provision of nectar and pollen sources over the whole season and being sure to make provision for specialist species that require specific types of flower because their mouthpart structure affects their ability to access nectar; for example, long-tongued bumble bees. There is surprisingly little, however, on provision of host plants for herbivores to consume and lay eggs on. Some roofs may have the ability to provide wetland habitat in the form of ponds, which substantially increases the species that can breed on them. Some local planning authorities codify invertebrate requirements in planning law. In Basel, Switzerland, for new flat roofs the Building and Construction Law specifies the use of local soils, minimum substrate depths (100 mm), native plant species local to the area, and the construction of substrate mounds 300 mm high and 3 m wide specifically to provide invertebrate habitat. In addition there is also a statutory requirement for the local authority's green roof expert to be involved in the design and construction phase for roofs over 1000 m² in extent (Kazmierczak & Carter, 2010).

Ladybugs (ladybirds, beetles) were found on green roofs in Portland, Oregon, (Liptan & Strecker, 2003). Grant *et al.* (2003a) reported that the Derry & Toms roof gardens have had a range of invertebrates on them including slugs and snails, aphids, bees, wasps and butterflies though no details are given of which species. Jones (2002) surveyed eight green roofs in London using a vacuum sampler and found 136 species of tecticolous invertebrate (*sensu* Payne, 2000), a number he considered rather low for the sampling effort. He put this paucity of species down to the uniform nature of the roofs: most had similar substrate, had mostly low-growing mat-forming species such as sedums, and had a limited range of flora. The list of invertebrates was dominated by beetles (51 species, 38%) followed by spiders (31 species, 23%) and bugs (25, 18%). Whilst the majority of the 136 species were relatively common, ten species of conservation interest were identified, seven of which have specific conservation designations (Table 6.11). Some of the species found on green roofs may have been imported with the plants or sedum mats used rather than having colonised naturally, particularly interesting when vegetation is imported from other countries! Jones (2002) mused on a contradiction: that none of the species identified by Grant *et al.* (2003a) from a survey of biodiversity action plans that included brownfields and that were considered to potentially benefit from green roofs were found in his survey, or that of Kadas (2002). His interpretation was that the green roofs he studied, although providing harsh conditions and bare substrate, were not actually very typical of brownfields, that is, they did not contain soil mixed with concrete and brick rubble, and green roofs would have superior drainage to 'natural' brownfield sites. Ultimately, he seemed to view green roofs as a unique, new habitat type, which favoured species of harsh xerophillic conditions.

Kadas (2006) compared the spider, beetle and hymenopteran fauna of four brownfield habitats with two types of green roof (three extensive sedum roofs and two 'brown' or 'biodiverse' roofs) in London, UK. Whilst Jones (2002) used an insect vacuum sampler, Kadas (2006) used pitfall traps (ten on each roof); sampling was continuous from May to October 2004 and traps, which were filled with a preservative, were emptied every three weeks. In terms of abundance, two of the sedum green roof sites had the greatest invertebrate abundance of all sites monitored, with the brown roof sites having the least. These latter sites were only one year old at the time of sampling and could not be considered to have reached any kind of maturity; they would have experienced only limited colonisation in the time since construction. Also, over half the invertebrates on the sedum roofs were snails; these were probably introduced to the roofs with the sedums from the plant nursery and their numbers subsequently unchecked by significant predation. Species richness data were not given for the different roof types, but the Shannon–Weiner diversity index indicated that the brownfield sites were more diverse than the extensive or brown/biodiverse roofs. As might be expected from their youth, the latter sites had the lowest diversity. Kadas (2006) like Jones (2002) found species of conservation value on all the roofs studied with numbers of species on the sedum roofs comparable to those on the brownfield sites. Overall more than 10% of the species found on the roofs were of nature conservation interest and the roofs yielded records for five

**TABLE 6.11** Species of conservation interest found in a survey of eight green roofs in London in 2002

| Order | Species | Family/description | Conservation status |
|---|---|---|---|
| Coleoptera | *Anthicus angustatus* | Anthicidae (ant beetle) | Nationally scarce (Notable B) |
| | *Helophorus nubilis* | Hydrophilidae (crawling water beetle) | Very local |
| | *Olibrus flavicornis* | Phalacridae (flower beetle) | RDB K |
| | *Oxypoda lurida* | Staphylinidae (rove beetle) | Nationally notable |
| | *Tachys parvulus* | Carabidae (ground beetle) | Nationally scarce (Notable B) |
| Hemiptera | *Chlamydatus evanescens* | Miridae (leaf bug) | RDB3 |
| | *Chlamydatus saltitans* | Miridae (leaf bug) | Local |
| Aranaea | *Erigone aletris* | Linyphiidae (money spider) | Naturalised, very local |
| | *Pardosa agrestis* | Lycosidae (wolf spider) | Nationally scarce (Notable B) |
| | *Philodromus albidus* | Thomiscidae (crab spider) | Nationally scarce (Notable B) |

Source: Jones, 2002.
Note: IUCN Categories RDB K = rare, but insufficiently known; RDB3 = rare species with small populations and at risk (found in 15 or fewer Ordnance Survey (OS) 10 x 10 km grid squares). UK designations: nationally scarce (Notable B) = very local, found in 31–100 OS grid squares; nationally scarce: found in 16–100 OS grid squares

species of spider new to London (*Pardosa agrestis*, *P. arctosa*, *Steatoda phalerata*, *Silometopus reussi*, and *Erigone aletris* and a very rare beetle (*Microlestes minutus*)). None of the species found by Kadas (2006) were on the list compiled by Grant *et al.* (2003a). Considerably more detail of Kadas's PhD research is given in Kadas (2010).

Brenneisen (2006) reported surveys of green roofs in Basel, Switzerland, capturing 12,500 individuals. The biodiverse vegetation on the green roof on the Rhypark building (constructed in 1987 and owned by the City of Basel (GreenRoofs, 2013)) had 79 beetle and 40 spider species recorded from it; of these 13 and 7 respectively were considered endangered. Over a three-year period, roofs designed for wildlife had continual increases in species richness, whereas it appears that the species richness of extensive roofs, with thin media, remained roughly static.

Earlier work on invertebrates on green roofs in the literature was typically carried out on relatively few sites, with obvious limitations in being able to uncover important influences on the biota. As a result, Madre *et al.* (2013) embarked on an ambitious 115 roof study of invertebrates on green roofs in northern France; sampling was carried out between April and June 2011. They used a slightly different classification system to other studies, preferring to emphasise the vegetation structure/complexity of roofs rather than substrate depths (though it comes to much the same thing). Their typology was as follows (with the number sampled in their study in parentheses):

- *Muscinal* (M-type) dominated by low-growing bryophytes and vascular plants (e.g. sedums) with no more than 80% of the flora consisting of herbaceous species. This equates to the term 'extensive'. (45)
- *Herbaceous* (H-type) contains M-type plants as an understorey, but has more than 20% of the vegetation composed of non-woody herbaceous species including grasses. This definition equates roughly to the semi-intensive/biodiverse green roofs. (38)
- *Arbustive* (A-type) has both M- and H-type vegetation, but has more than 20% of its area covered in woody shrubs. This roughly equates to the intensive type of green roof. (32)

Sampling was by ten minutes visual search of a 2 m wide x 20 m long transect. This is slightly curious as increasing complexity of vegetation would imply a decreased level of searching efficiency resulting in possible underestimation of both richness and abundance on the more complex roof vegetation types. Invertebrates were captured by hand using pill boxes; sampling was restricted to Araneae (spiders), the Heteroptera (true bugs), Coleoptera and Hymenoptera. The surveys found, in total, 66 species: 21 spiders, 11 beetles, 10 bugs and 24 hymenoptera; species richness and overall abundance was low (total 290: 50 spiders, 51 beetles, 26 bugs, 163 hymenoptera) probably due to the visual searching method employed. The majority of species (88%) were capable of flight, 44% were generalist species, 30% were thermophiles and 26% were specialists of hot and dry conditions. The abundance of hymenoptera was dominated by ants (63%), of which most were the black ant *Lasius niger* (49%); the red-tailed bumblebee *Bombus lapidarius* made up 12% of hymenoptera; beetles were dominated by the seven-spot ladybird *Coccinella septempunctata* (45%), and bugs by *Pyrrhocoris apterus* (62%). Total and hymenopteran species richness was shown to be significantly lower on M-roofs compared with the other two roof types. Spider and beetle richness was significantly lower on M-roofs compared with A-type roofs (but not H-type roofs).

Increasing building height also reduced spider richness; bugs were positively affected by extent of vegetation cover on the green roof; hymenoptera responded positively to increasing roof surface area. Total abundance of invertebrates, and beetle and hymenoptera abundance were all significantly related to roof type in the order A>H>M-type; abundance also grew with increasing green roof area. Bug abundance was significantly related to extent of plant cover on the green roofs. Community analysis showed that spiders exhibited different groups of species depending on roof type (and was probably related to prey capture strategies, i.e. hunting vs web spinning); beetle make-up differed with building height; for hymenoptera, ants were found on lower roofs and bumblebees on higher roofs. The wider context, that is, the make-up of the landscape around the green roof (the percentage of habitat within a 2 km radius of each roof), was only shown to have an effect on the hymenoptera. Tonietto et al. (2011), investigating bees on green roofs, parks and prairies in the Chicago area, also found that bee abundance and richness was correlated with increasing proportions of green space in the locality – except where the green space was mostly amenity grass.

Bees (and butterflies): green roofs on high-rise buildings might seem to be unlikely places for wildlife, but Johnston and Newton (2004) cited a report of butterflies and bees using nectar on intensive roofs 23 storeys high in the USA and bumblebees seem quite capable of exploiting resources on tall buildings (Madre et al., 2013). Gong (2007) reported the use of green roofs by six species of bumblebee in a study that took place between July and September 2007, in Sheffield, UK. She found six species using green roofs: the common garden *Bombus pascuorum*, the red-tailed bumblebee *B. lapidarius*, the early nesting bumblebee *B. pratorum*, the buff-tailed bumblebee *B. terrestris*, the white-tailed bumblebee *B. lucorum* and the garden bumblebee *B. hortorum*. Roofs with wildflowers were favoured over those with extensive vegetation, which were largely ignored. The most visited plant species on the wildflower roofs were lesser calamint *Calamintha nepeta,* oregano *Origanum laevigatum*, and lamb's tongue or lamb's ear *Stachys byzantina*. Colla et al. (2009) compared bee abundance and richness on two extensive green roofs in Toronto (one seeded with alpine grasses and wildflowers, the other naturally colonised by grasses and herbs – mostly vetches and clovers) with four ground-level sites between 2004 and 2006. Sampling was via coloured pan traps containing soapy water. They captured 337 individuals of 43 species on the seeded roof and 96 individuals of 22 species on the unseeded roof (data from their Table 1), and considered that the green roof and ground-level bee communities were similar. Matteson and Langellotto (2011), in their study of urban green space, considered that provision of green roofs with floral resources might be a way of improving resource availability for bees and butterflies. Their analyses revealed that floral area and sunlight availability were the main determinants of species richness; rooftops, they reasoned, were less limited in sunlight availability than ground-level green spaces. Nagase and Dunnett (2013b) reported that bees, especially bumblebees, visited the cultivar of viper's bugloss and the *Linaria* sp. used in their experiments with annual plants (Table 6.8).

Ksiazek et al. (2012) compared the pollinator communities at four ground-level sites and on four green roofs in the Chicago area using pan (water) traps and found that the green roofs had a lower species richness and abundance (green roof: 133; ground: 281) of bees compared to ground-level plants, a result that mirrored that of Tonietto et al. (2011). Whilst bee communities on roofs and at ground level were dominated by small bee species, green roofs had more 'medium-sized' bees than the ground level and fewer large bees. However, despite the differences in pollinator community and abundance, overall, the percentage mean seed set of nine native prairie plants was actually superior on green roofs than at ground level. Of the six species that had greater seed set on the roofs (nodding onion *Allium cernuum*, lead plant *Amorpha canescens*, red columbine *Aquilegia canadensis*, purple prairie clover *Dalea purpurea*, wild bergamot/bee balm *Monarda fistulosa*, and golden Alexander *Zizia aurea*) the difference was only statistically significant for *D. purpurea*. Of the two species where seed set was higher at ground level, foxglove penstemon *Penstemon digitalis* and blue wild indigo *Baptisia australis,* the difference was only significant for the latter. No analysis was possible for the remaining plant – wild white indigo *Baptisia alba* (syn. *leucantha*) – because of statistical difficulties. Their conclusion was that growing native species on green roofs would not be limited by pollinator services.

## Vertebrates

Whilst a fox *Vulpes vulpes* has been recorded on a green roof on a two-storey house in Lewisham, London (Grant *et al.*, 2003a), and squirrels on the 19th floor of a high-rise in New York (Johnston & Newton, 2004), the majority of records of vertebrates on green roofs are for birds (Table 6.12). Green roofs have been recorded as nesting habitat for a range of ground-nesting species including little ringed plover *Charadrius dubius*, northern lapwing *Vanellus vanellus,* skylark *Alauda arvensis* (Baumann, 2006; Brenneisen, 2006) and many others (Ohlsson, 2004). In Portland, Oregon, 'hummingbirds, blue jays, crows, swallows, pigeons, sparrows and signs of hawks or owls' have been recorded on green roofs (Liptan & Strecker, 2003). Whilst New York, along with the aforementioned squirrels, is said to host woodpeckers, goldfinches and blue jays on the 19th floor of a building (Johnston & Newton, 2004). Grant *et al.* (2003a) give details of a development in Deptford Creek in south-east London on a brownfield site potentially threatening the black redstart *Phoenicurus ochruros*. Green roofs were planned in mitigation which were to include nesting habitat for the black redstart, and also for sand martins *Riparia riparia* and kingfisher *Alcedo atthis* (see also Grant, 2006a). Many green roofs are now designed specifically for black redstarts (Grant, 2006a). A recent review by Fernandez-Canero and Gonzalez-Redondo (2010) provides an overview of birds' use of green roofs and the resources available to them (breeding habitat, food, cover, water (where ponds are provided on intensive roofs), space).

## 6.6 Whole-life cost analysis of installing green roofs

There is no getting around the issue that green roofs cost more than standard roofs to construct – and this is one of the issues that has slowed the adoption of the technology. It is pretty obvious that the use of regulations overcomes developers' reluctance, of course, but it is a shame that coercion is required. Unfounded perceptions about the impact of vegetation on the longevity of roof membranes should by now have been dispelled – though it is surprising how many conversations often start with this gambit – and positive data on a wide range of ecosystem services are now available. Part of the problem probably lies in the distinction between the cost of a building to sell and the cost to operate. A developer will wish to stick to well-known approaches, avoid frills, build quickly and rapidly, and sell fast. The user of a building, not always the same as the purchaser, wants a building that does not cost much to run. There is a disconnect between the start of the chain and the end. An initially higher building cost, due to the incorporation of green technologies such as a green roof, is likely

**TABLE 6.12** Birds observed breeding or otherwise present on green roofs

| Common name | Scientific name | Breeding | Present |
|---|---|---|---|
| Sparrowhawk | *Accipiter nisus* | | **+** |
| Skylark | *Alauda arvensis* | + | |
| Mallard | *Anas platyrhynchos* | + | |
| Meadow pipit | *Anthus pratensis* | + | |
| Swift | *Apus apus* | + | |
| Grey heron | *Ardea cinerea* | | + |
| Canada goose | *Branta canadensis* | + | |
| Goldfinch | *Carduelis carduelis* | | + |
| Greenfinch | *Carduelis chloris* | + | |
| Little ringed plover | *Charadrius dubius* | + | |
| Ringed plover | *Charadrius hiaticula* | + | |
| Killdeer | *Charadrius vociferus* | + | |
| City dove | *Columba livia* | + | |
| Wood pigeon | *Columba palumbus* | + | |
| Olive-sided fly catcher | *Contopus cooperi* | | + |
| Carrion crow | *Corvus corone* | + | |
| Rook | *Corvus frugilegus* | | + |
| Jackdaw | *Corvus monedula* | | + |
| Blue jays | *Cyanocitta cristata* | | + |
| House martin | *Delichon urbicum* | | + |
| Robin | *Erithacus rubecula* | | + |
| Falcon | *Falco sp.* | + | |
| Chaffinch | *Fringilla coelebs* | + | |
| Crested lark | *Galerida cristata* | + | |
| Oystercatcher | *Haematopus ostralegus* | + | |
| Swallow | *Hirundo rustica* | | + |
| Herring gull | *Larus argentatus* | | + |
| Common gull | *Larus canus* | + | |
| Black-headed gull | *Larus ridibundus* | | + |
| White wagtail | *Motacilla alba* | | + |
| Grey wagtail | *Motacilla cinerea* | | + |
| Spotted flycatcher | *Muscipapa striata* | + | |
| Wheatear | *Oenanthe oenanthe* | + | |
| Blue tit | *Parus caeruleus* | + | |
| Great tit | *Parus major* | + | |
| House sparrow | *Passer domesticus* | + | |
| Tree sparrow | *Passer montanus* | + | |
| Black redstart | *Phoenicurus ochruros* | + | |
| Willow warbler | *Phylloscopus trochilus* | + | |
| Magpie | *Pica pica* | + | |
| Whinchat | *Saxicola rubetra* | | + |
| Serin | *Serinus serinus* | | + |
| Common tern | *Sterna hirudo* | + | |
| Collared dove | *Streptopelia decaocto* | | + |
| Starling | *Sturnus vulgaris* | | + |
| Blackcap | *Sylvia atricapilla* | | + |
| Blackbird | *Turdus merula* | + | |
| Fieldfare | *Turdus pilaris* | | + |
| Lapwing | *Vanellus vanellus* | + | |
| Mourning doves | *Zenaida macroura* | | + |

Sources: compiled from data in Johnston and Newton (2004), Ohlsson (2004), and Fernandez-Canero and Gonzalez-Redondo (2010).

TABLE 6.13 The Whole–Life Cost Analysis for the Springboard Centre in Bridgwater, Somerset, UK, using different roof types, including Net Present Value (NPV), Total Whole–Life Cost (TWLC) and Annualised Whole–Life Cost AWLC (TWLC/roof life)

| | Capital costs ($£.m^{-2}$) | Annual maintenance costs ($£.y^{-1}$) | Annual repair costs (yrs 1 & 2 only) ($£.y^{-1}$) | Energy savings ($£.y^{-1}$) | Use of foundation material on roof (one-off saving $£$) | NPV | TWLC | AWLC | Best option 1 = best |
|---|---|---|---|---|---|---|---|---|---|
| | | | | *Assumptions used in the comparisons* | | | | | |
| **Conventional roof** | 47 | 150 | 0 | 0 | 0 | -49,160 | 51,500 | 1,716 | **3** |
| **Sedum mat roof** | 93 | 600 | 2,500 | 5,200 | 0 | -21,268 | -132,000 | -2,640 | **2** |
| **Biodiverse roof** | 79 | 150 | 1,250 | 5,200 | 4,800 | 7,453 | -175,800 | -3,516 | **1** |

Source: Bamfield, 2005.

For NPV -ve values = higher costs; for TWLC -ve values = net gain so most negative is best; AWLC = annual cost, -ve values = net gain, so most negative best. Assumption figures based on a 1000 m² roof extent.

to be strongly discounted over the whole-life cost of running the building. Bamfield (2005) addressed this question in a report for the cost of a green roof on the Springboard Centre in Bridgwater, UK. Three options were considered: a conventional exposed roof, a sedum-covered roof and a biodiverse roof. Factors used included life expectancy of roof materials (30 years for the conventional roof, a conservative 50 years for the green roofs), the capital costs, the annual maintenance and repair costs, energy cost reductions from use of green roofs, and recycling of materials in construction. The results (Table 6.13) show an unequivocally better return using green roofs compared with conventional roofing.

Carter & Keeler (2008) working in Georgia, USA, estimated that an extensive green roof would cost something of the order of 10–14% more in terms of NPV. They carried out a sensitivity analysis to examine which costs would most strongly affect their assumptions in the construction of their estimates, and found that the most important factors were likely to be construction costs (which would probably decline as the market became more efficient – the USA has higher unit costs compared with Germany due to the relative novelty of the systems in the USA) and energy costs (which were expected to increase with greater levels of regulation and higher prices). These factors would eventually make it cheaper to install green roofs compared with conventional roofs on the basis of costs over the whole life cycle. One stunning conclusion was that it was the ability of green roofs to extend roof membrane longevity that made the biggest contribution to reducing the overall costs – without that they estimated that it would cost 85% more to build green roofs!

## 6.7 Summary

- Green roofs have a long history demonstrating viability of the concept and with increasing levels of innovation and development since the late 20th century.
- It is cheaper to incorporate green roofs in new builds than to retrofit; retrofits are likely to be restricted to 'extensive' green roofs due to structural loading limitations.
- Green roofs are not the preserve of DIY enthusiasts, but are being increasingly incorporated into both public and commercial developments.
- Green roofs have been shown to have real economic benefits to building owners and operators, and also to society.
- A wide range of ecosystem services are delivered by green roofs including helping to reduce the impact of climate change.
- The specific functions desired for a specific green roof need to be identified at the design stage.
- The local climate will affect the design parameters and the need for features such as irrigation.
- Despite the maturity of the technology and the development of a support industry, we still have much to learn about how to optimise green roof functions.

# 7

# STREET TREES

This chapter will examine the role trees play in greening streets including:

*   health and safety issues and the impact of the street environment on trees
*   the species of trees used in urban areas and the selection criteria used for street tree planting
*   the value of street trees in terms of the ecosystem services they deliver.

## 7.1 Street trees and city greening

Street trees are perhaps one of the most obvious features of city greening (Figure 7.1) and have probably been part of cityscapes since the beginning of urbanisation; records exist from the 16th century (Forrest & Konijnendijk, 2005). Urban centres with high population density are typically negatively correlated with vegetation, but this generalisation breaks down with street trees (Pham *et al.*, 2013). As a single entity, as opposed to being in combination with other structures, trees embody perhaps more multifunctional benefits than any other aspect of green infrastructure, contributing significantly to local economies through ecosystem services and increasing the value of adjacent properties (Maco & McPherson, 2003; Nowak, 1999; Speirs, 2003). Citywide, Peper *et al.* (2008) estimated that street trees enhanced property values in Indianapolis by some US$2,848,008 annually. Trees have also been identified as the most important feature in improving streetscapes (Antupit *et al.*, 1996), providing structural greenery outside park environments (Zhou & Kim, 2013) and contributing to the likelihood that older age groups would use them for strolling and hence contribute to their longevity (Takano *et al.*, 2002). Despite the obvious benefits of street trees, they are also considered by some as a nuisance (see refs in Flannigan, 2005), expensive to maintain, a health and safety risk, and a causal agent of damage to buildings (Dandy, 2010) and sewers (Randrup *et al.*, 2001).

Moll (1989) bemoaned the paucity of street trees in American cities claiming that in streets with trees, street tree numbers could be doubled, and that in cities with fewer than one million citizens the majority (89%) needed far more as only 60% of streets had trees. The mayor of London, Boris Johnson, made political capital out of his commitment to invest more

FIGURE 7.1 Top left: a street without trees in Stoke-on-Trent, UK; top right: a parallel street with trees. Bottom left: street trees in Warwick, UK; bottom right: manicured trees in the Loire valley, France. © John Dover

in street trees and planted the 10,000th street tree of his administration (a field maple *Acer campestre*) in February 2012 (GLA, 2012). Despite the known value of street tree plantings in the urban environment (including timber production (Kuchelmeister, 2000)), many local authorities do not always appear to appreciate their value (Treeconomics, 2011), relegating them to an 'afterthought' (Pauleit, 2003). This is infinitely depressing, given that many of the values we now recognise and are seeking to exploit were being articulated at the latter end of the 19th century: Smith (1977) quoted Smith (1899) who recognised the value of shade trees in ameliorating the summer heat of urban areas, the cooling effect of evapotranspiration, and the ability of trees to absorb pollutants. Dandy *et al.* (2012) note that street trees are under-represented in planning policy documents. Nevertheless, the perception of street trees as vehicles for delivering ecosystem services rather than as an aesthetic add-on with substantial costs does appear to have been increasingly gaining legitimacy in local government arenas (Seamans, 2013) and Flannigan (2005), in a survey of residents in north Somerset and Torbay in England, found the majority liked street trees, with benefits rated higher than disbenefits (e.g. the need to clear fallen leaves in autumn). A later paper (Schroeder *et al.*, 2006) compared the attitudes of residents in UK with those in the USA and found broad agreement in the positive attitude to trees, although there were some differences in views on some aspects: UK residents were more concerned with the disbenefits, valued shade less, and considered slow growing, smaller forms as optimum tree characteristics. McPherson and Simpson (2002), in a study of urban trees in two cities in California, estimated that every $1 spent per year on tree management resulted in $1.85-worth of societal benefits in Modesto and $1.52 in Santa Monica; the majority of trees were street trees (83 and 85% respectively). The study demonstrated that the specifics are important with costs and benefits differing between the two cities: they differed in climate, in house prices, in population densities (Santa Monica is more urbanised), the tree stocks were different in terms of physical size (similar-aged trees were bigger in Modesto), the ratio of deciduous to evergreens differed, and management policies were less aggressive in Santa Monica. McPherson et al. (2006) considered that trees in the Piedmont region of the USA delivered environmental benefits up to three times the cost of tree care. Peper et al. (2008) estimated the benefits of the 117,525 street trees (177 species) in Indianapolis to be $6.09 for every $1 spent on tree management. Systems such as CAVAT (Capital Asset Value for Amenity Trees) may help amenity managers appreciate the value of their stock (Neilan, 2010) and gain compensation for damage, but it is rather limited in that it does not value ecosystem services delivered by trees, unlike the more comprehensive approach of McPherson et al. (2006) and McPherson and Kotow (2013). CAVAT can be used as part of the i-Tree approach by substituting it for the structural value from the CTLA trunk formula method used in the USA (as part of the i-Tree methodology) (Neilan, undated).

Pauleit (2003), reporting a study of street tree plantings in European countries, demonstrated the shortcomings of the UK in spending on trees, tree quality and site preparation. Nevertheless, in the UK Britt and Johnson (2008a) found 70% of all urban trees to be in good condition. Across Europe average costs of planting per tree varied enormously from a low of 13€ in Spain to 1670€ in Finland; the UK spent on average 120€/tree. Britt and Johnson (2008a) found that local authorities in England spent £1.38 on trees per head of population in 2003/2004. Britt and Johnson (2008a) reported on a survey of urban trees in 147 towns and cities in England and found considerable variation in densities across the country (1.0 ha$^{-1}$ to 886.5 ha$^{-1}$); as might be expected, city centres and high-density residential areas had least trees and lowest canopy cover. Only 12% of trees in the survey could be considered 'street or highway' trees. In high-density residential areas, street trees provided 22% of the total canopy area.

Work in Montreal, Canada, on social demographics and street tree density suggests that there can be inequalities in street tree plantings, with the presence of recent immigrants correlating negatively with street tree density, but positively with university degree holders (Pham et al., 2013). Likewise, Zhou and Kim (2013) showed differences in neighbourhood canopy cover on the basis of race and ethnicity in cities in Illinois, USA.

## 7.2 Trees and health and safety

Trees may be considered a problem, with public safety/'health and safety' worries (Britt & Johnson, 2008a; Dandy, 2010) and consequently financial liabilities. Watt and Ball (2009) quantified the risk of death caused by trees over the period 1999–2008 and concluded that, in the UK, there were about 6.4 deaths per year which equated to a very small risk of 1 in 10 million per year (compared to, say, the risk of death from cancer of 1 in 387 or 1 in 16,800 for a road accident). Ball (2009) found the public to be sanguine about the risk of death from trees, considering it to be 'one of the ordinary, everyday risks of life'. Of course health and safety is usually used in a negative sense in relation to street trees, but they deliver a wide range of positive health benefits that far outweigh the minimal risks from physical injury; for example, providing cooling and shading in the summer, reducing UV exposure, reducing wind speeds in winter, reducing air and noise pollution and creating a less stressful environment with concomitant mental and physical health benefits (Sarajevs, 2011). Street trees can reduce reverberation in urban canyons, but as Herrington (1974) pointed out: the perceived reduction in noise due to trees in urban areas is probably partly psychological and partly physical with the psychological component being more dominant.

## 7.3 The street environment

The urban street environment is inherently challenging for tree growth and longevity (Bassuk *et al.*, 2009; Sieghardt *et al.*, 2005). Part of the problem, of course, is the complexity of underground infrastructure (sewers, electricity, gas, cable) and the limitations this places on available space for tree roots (Kelly, 2012). Street trees are known to contain less biomass than the same species grown in non-urban areas (Nowak, 1994b) and root growth and ultimate tree size is related to available soil volume (DeepRoot, 2011). The carbon storage capacity and annual carbon sequestration of large trees can be 1,000 times and 90 times greater (respectively) in large trees (>76 cm dbh) than small trees (<8 cm dbh) (Nowak, 1994b) and a surprisingly high proportion (about 22%) of tree biomass is held in the root system (see references in Nowak, 1994b).

Sieghardt *et al.* (2005) observed that 'urban trees have a hard life' comparing the likely 60-year lifespan of a street lime to the 1,000 years potential of one grown in a forest. Moll (1989) suggested that urban centre trees had an average lifespan of seven years compared to the average 32 years of suburban trees, but a recent meta-analysis suggests that 19–28 years is closer to the mark (Roman & Scatena, 2011). Negative factors include water stress (drought), constrained growing conditions, damage, lack of root oxygen, inadequate hardening before frost and pollutants (McCarthy & Pataki, 2010; Sæbø *et al.*, 2003, 2005; Sieghardt *et al.*, 2005; Sjöman *et al.*, 2010) (Table 7.1). As Cekstere and Osvalde (2013) found in Riga, Latvia, de-icing salt (sodium chloride NaCl) is a particular hazard to street trees, especially those nearest the road, causing post-flushing die-back, foliage discolouration, leaf necrosis and crown damage. They found salt in considerably higher concentrations (Na x24 and Cl x22) in the soil around street trees compared with those around trees in the Viestura Gardens, a local park. They also found elevated levels of magnesium (Mg) and lower levels of potassium (K), iron (Fe), copper (Cu) and boron (B) as well as suboptimal ratios of Mg:K, Ca:Na and Mg:Na in areas where street trees showed most damage. Whilst salt applications immediately make many think of de-icing roads, in arid regions of North America magnesium chloride and calcium chloride are used on roads to act as spring/summer dust suppressants. Experimental applications of magnesium chloride to the soil surrounding Douglas fir *Pseudotsuga menziesii*, lodgepole *Pinus contorta*, ponderosa *Pinus ponderosa*, limber pines *Pinus flexilis* and aspen *Populus tremula* were carried out and shown to have negative effects on roadside trees including causing leaf necrosis, leaf loss, crown loss and even tree death with high concentrations (Goodrich & Jacobi, 2012). Day and Dickinson (2008) note that whilst the optimum pH for trees is 5–6.5, urban conditions are more likely to be more basic at 7.5–8.5.

TABLE 7.1 Environmental stresses affecting street trees

| Stress factor | Major effect |
| --- | --- |
| **Climate** | If the tree is not adapted to the planting site's climate, it may be susceptible to frost damage (either in the spring due to earlier bud-burst than local trees, or in the autumn if leaf fall is not early enough to harden the trees). May result in greater susceptibility to disease. |
| **Water stress** | In high temperatures trees need lots of water: this may not be available due to constricted root space and the sealed-surface preventing rain penetration to roots. Urban centres are warmer than the countryside. |
| **Oxygen starvation** | In some circumstances, e.g. with highly compacted ground, waterlogging, etc., trees may be killed by starving the roots of oxygen. |
| **Light** | Trees that are not shade adapted may suffer from low photosynthetic activity in narrow streets; the effects are exacerbated when combined with high temperatures and thus high transpiration. Street lights may cause delayed leaf fall/winter hardening leading to frost damage. |
| **Soil** | Soil conditions need to be appropriate for the species/cultivar and should be determined before planting and soil modified or a different species used if necessary. |
| **De-icing salt** | Some trees are tolerant, but as tolerance appears to be a genetic trait, planting the correct genotype is important. Minimise roadwater runoff to tree roots by infrastructure design. |
| **Air pollution** | Trees are susceptible to damage by a range of pollutants, e.g. particulates, reducing photosynthesis, gaseous pollutants taken in via stomata; direct tissue damage. Some species are more tolerant than others (deciduous more than conifer). Trees valuable as pollution-removing agents. |
| **Heavy pruning** | Heavy pruning of trees to limit crown spread can leave trees open to disease attack; heavy management can lead to decline of mature specimens leading to premature felling and replacement. |

Source: from Sæbø et al., 2003

Development pressure also impacts on tree stocks through removal and poor/reduced growth conditions for remaining trees (Jim, 2004). Such stresses make street trees more vulnerable to attack by pests and diseases than park trees (Sjöman & Nielsen, 2010) and also limit the choice of tree species and cultivar that can be grown (Bassuk et al., 2009). Many of the stress factors can be ameliorated by careful design and pre-planting preparation; for example, use of drainage pipes, soil media specifically designed for street trees, the use of a rock matrix to create 'structural' or 'sketetal' soils (Bartens et al., 2009; 2008) whereby rocks or other materials are included in the soil mix. The rocks in structural soils create a compaction-resistant framework to hold the soil and thereby maintain drainage, reduce anoxia and provide channels for root growth (Sieghardt et al., 2005). The pH of structural soils will differ according to the material from which they are composed and a mismatch between soil pH and the tolerance range for the tree species can result in poor tree growth and chlorotic leaves (Day & Dickinson, 2008) (for a brief history of structural soils, see Day and Dickinson, 2008). A more recent development has been the design of suspended pavements using systems such as Silva cells: interlocking modules composed of a glass/polypropylene compound reinforced with galvanised steel tubing (Figure 7.2). Silva cells (1.2 m long x 0.6 m wide and 0.4 m high) create a platform above which can be laid concrete, paving slabs and asphalt; the

FIGURE 7.2 A Silva cell showing the upper platform for pavement slabs/tarmac/concrete, the weight-transmitting pillars, and void for uncompacted soil and tree roots. An A4 piece of paper top left gives the scale. © John Dover

weight load is transferred by vertical columns to the base of the module and non-compacted soil is held between the upper and lower surfaces (92% of the volume of a Silva cell contains growing medium compared with 20% for structural soils (Deeproot, 2012)).

Suspended pavement systems appear to have considerable benefits in terms of tree growth. Smiley *et al.* (2006), in North Carolina, compared growth of snowgoose cherry *Prunus serrulata* and bosque lacebark elm *Ulmus parviflora* for 14 months using five experimental soil treatments (80% gravel/20% soil; 80% Stalite (a lightweight aggregate)/20% soil; 100% Stalite; compacted sandy clay loam; non-compacted soil under suspended pavement) and showed that trees in the suspended pavement treatment were bigger, faster growing, had more chlorophyll and better root growth than trees in the other systems; the superiority of the suspended pavement system has been maintained over time (Figure 7.3).

The experimental approach of Smiley *et al.* (2006) could be extended further to help local authorities in their decision-making by developing the street tree equivalent of the 'landscape laboratory' approach. Landscape laboratories have been used to evaluate different management and planting approaches for urban woodlands (Tyrväinen *et al.*, 2006). Regional street tree laboratories could be established so that the performance of trees recommended for use in particular regions/climate zones could be evaluated under different growing and planting conditions. Seeing is also believing, and street tree laboratories would not only allow the collection of comparative data on tree performance but also allow local authority staff (foresters and budget holders), political leaders and stakeholders to view trees *in situ*. A somewhat different perspective can be gained by viewing real trees as opposed to computer visualisation of plantings (Sjöman & Nielsen, 2010) or simply reading about experimental

FIGURE 7.3 Top: tree growth at 14 months after planting using five soil treatments; bottom: the same experiment nine years after planting. © Tom Smiley (see Smiley *et al.*, 2006)

results and can be valuable in gaining approval for action as part of a 'participatory planning' approach (Tyrväinen *et al.*, 2006).

Improved root growth also improves tree stability. The experiment set up by Smiley *et al.* (2006) was used by Bartens *et al.* (2010) to test trees 3.5 years after planting for deflection resistance in three of the soil treatments: compacted soil; 80% gravel/20% soil; 80% Stalite/20% soil. Deflection resistance is a measure considered to reflect resistance to uprooting by extreme weather (Smiley, 2008). *U. parviflora* stability was unaffected by soil type but *P. serrulata* showed better stability in the gravel/soil mixture and this was considered to be due to enhanced root growth. Data for trees planted into uncompacted soil under a suspended pavement (Smiley *et al.*, 2006) was included as a comparison, but could not be incorporated into the statistical analysis because of the initial trial layout. Nevertheless, the figures provided suggest that trees planted into suspended pavements with uncompacted soil would be at least as good, if not better, in terms of deflection resistance than in the gravel/soil mix. The inevitable caveat, given in the study by Bartens *et al.* (2010), is that the stability tests on young trees (as per their work) may not reflect that of mature trees.

There is also the question of 'which tree where?', not simply because the street environment is hazardous, but also because not all streets, or even parts of the same street, are the same: residential streets tend to be narrower and overhead infrastructure closer together, sunlight may vary with one side more shady than the other, commercial streets may have more traffic, shopping centres may be pedestrianised. Such considerations may impact on the choice of species or cultivar with respect to the growth form (the height and spread of trees), the degree of management (trimming, pollarding), and whether they are coniferous or deciduous (Nagendra & Gopal, 2010). The width of roads may also affect the density of tree planting, with densities (trees/km) declining in line with reducing road widths; an effect which is enhanced if trees are planted in strips between the carriageways of very wide roads (Nagendra & Gopal, 2010). Bassuk *et al.* (2009) gives guidance on site assessment and the selection of trees for stress tolerance in the USA. Despite the hostility of the environment, some settlements may retain quite old 'heritage' trees such as in Guangzhou, China, where over 300 trees were identified as being over 100 years old (Jim, 2004) and many may be street trees. In Guangzhou the Chinese banyan *Ficus microcarpa*, camphor tree *Cinnamomum camphora* and Indonesian cinnamon *C. burmanii* were the main heritage trees found along roadsides. Many of these trees were found in the Liwan district – the second oldest area in the city with a development history stretching back some 1,500 years. Nagendra and Gopal (2010) found that the largest trees in Bangalore, India, were located in the widest roads and that such roads were targeted for further road widening; as a result the largest trees in the city were under threat.

## 7.4 Street tree inventories

The quality of information on urban/street trees in a given urban area may be complete, out of date, or simply not available; in Stoke-on-Trent where I work, at the time of writing, an out-of-date GIS layer existed for street trees, but no information was held on off-street plantings (pers. comm.). However, recent work by Britt and Johnson (2008b, cited in Dandy, 2010) places Stoke-on-Trent in the group of 20 towns with the lowest canopy cover in the UK. Some cities recognised early on that urban trees were important in climate control; Chicago, for instance, initiated a project in the early 1990s to quantify the urban tree resource and recognised the need to drill down to the location, species composition and abundance, leaf-area index, etc. (Nowak, 1994c). The results showed that although Chicago's street trees made up only 10% of the 4.1 million tree population, the leaf surface area of those trees made up 24% of the area's total (Nowak, 1994c; Nowak et al., 1994). The proportion of street trees in Guangzhou City, China, was lower than that of Chicago; from a total of 1,794,455 urban trees only 3.9% were street trees (roughly, 67,000) (Jim & Chen, 2008). According to McPherson and Simpson (2002), Modesto in California had a street tree population of 75,649 (83% of the street and park tree population) made up of 184 species and Santa Monica, in the same state, had a street tree population of 21,698 (85% of the total); unfortunately they did not give a figure just for street tree species richness. In a study of ten Nordic cities comprising 190,682 trees, Sjöman et al. (2012c) found Malmö had the most tree species growing in the streets (113) and Gothenberg the lowest (24) (Table 7.2). The street scenes were dominated by just seven species, although the species dominant in any given city varied (Sjöman et al., 2012c) as did their origin. Sæbø et al. (2005) identified the commonest tree species grown in Europe, and showed that the range of trees used, as one might expect, increases from north to south along with the warmer climate (Table 7.3). They also comment on the rather restricted range of genera (*Acer, Aesculus, Platanus* and *Tilia*) that dominate the central and northern European street environment. Several authors consider that high species and genera-level richness of urban trees is a safeguard against loss due to pest and disease impacts (which are likely to increase in frequency with climate change) (Sjöman et al., 2012c), although there is the view that tried and tested species at least demonstrate their tolerance to urban conditions and that the trend to diversity in street tree plantings, without information on tolerance, may not be wise (Richards, 1983). However, many countries, including the UK, have experienced a number of alien pest and disease invasions that threaten street tree populations (Tello et al., 2005; Tubby, 2006). Recent examples include ash dieback *Chalara fraxinea* in the UK in 2012 (FC, 2012a; Helmschlager & Kirisits, 2008; McVeigh & Layton, 2012), and Barber et al. (2013) found nine species of *Phytophthora* associated with dying vegetation in Perth and part of Western Australia, including street trees. Dominance by a very small number of tree species is looking like an increasingly untenable strategy. Various suggestions have been made as to the maximum proportion of a particular species or genus in the tree population of urban areas; for example, no species to make up more than 10% of the tree composition and preferably no more than 5% (see references and discussion in Sjöman et al., 2012c). If we apply this rule selectively to just the street tree population, then of the ten Nordic cities studied by Sjöman et al. (2012c), none met the 10% rule, with Helsinki far exceeding it with 44.4% of the trees being *Tilia cordata* (Table 7.3). Nevertheless, it is possible for cities to achieve the 10% standard, as applied to street trees, as Nagendra and Gopal (2010) demonstrated for Bangalore, India; where, of the 108 species found (in a survey of 2,339 trees), the most abundant was the rain tree *Albizia saman* which made up 9% of the street tree population (Table 7.3). Unfortunately McPherson and Kotow (2013) consider the 10% rule to be outdated given the recent problems with pests in the USA that are not species-specific, such as the emerald ash borer beetle *Agrilus planipennis*, gypsy moth *Lymantria dispar*, and various diseases. McPherson and Kotow (2013) recommend that tools are used that highlight commonalities of pest and disease vulnerabilities between species (such as the Pest Vulnerability Matrix (PVM) which was designed for the conditions of North Carolina, but whose principles can be adapted for whatever location is required) and modify the species mix appropriately – including elimination if the risk is very high. They also highlight the use of tools to map the spread of invasives – in the USA these are the 'Early Detection and Distribution Mapping System' (EDDMapS) and iMapInvasives. In the UK the equivalent would be the 'Recording Invasive Species Counts' (RISC) scheme hosted by the Department for the Environment, Farming and Rural Affair's (Defra) 'GB Non-native Species Secretariat Website' (Defra, 2011a) with information on plant diseases from the Food and Environment Research Agency (FERA, 2013) and general mapping of species at the National Biodiversity Network's NBN Gateway (NBN, 2012). By linking the PVM with i-Tree Streets, McPherson and Kotow (2013) were able to identify various risk parameters: over-reliance on a few abundant species, unbalanced age structure, degree of pest and disease threat, potential asset loss (loss of ecosystem services due to pest disease susceptibility), and aggregate the scores to a final summarising risk grade. For California, McPherson and Kotow (2013) analysed data for 29 cities with up-to-date inventories of trees (street and park) and identified the species of tree most at risk. Whilst the work was done for California, it is a little alarming that sycamore/plane came up as the species at highest risk in nine cities and second most at risk in two other cities – especially as its widespread planting was probably due to its resilience to urban conditions. The recommendation was to eliminate the tree in all 11 cities as well as modify its occurrence in a further 11.

Tree stocks in urban areas, as elsewhere, are dynamic: trees age and die, new trees are planted, existing trees removed, etc. (Nowak & Greenfield, 2012). Stocks vary greatly: New York has about 500,000 street trees (Brown et al., 2005), Greater London's streets also have about 500,000 trees, about 7% of the total number of trees in the area; Davis, in California, a much smaller city with 58,000 inhabitants compared with London's 8 million, had about 24,000 (Hill, 2012; Maco & McPherson, 2003). In the five years prior to 2007, the stock of street trees declined in about a third of London's boroughs, severely in some cases, though overall there was a net gain of 1.7% or 8,205 trees (GLA, 2007). Harrow, for example, with a starting stock of 18,000 trees, lost 5,000 and replanted 2,000 so the net loss was 3,000 trees, some 16% of the total tree stock (GLA, 2007). Such losses are shocking, especially as the 2,000 replanted trees may take some time to deliver the environmental benefits of the mature trees they replaced, and may have been replaced with smaller varieties to minimise maintenance and to reduce the impact of climate change (GLA, 2007) – on the tree stock

**TABLE 7.2** Street tree species richness and abundance in ten Nordic cities

| Urban area | Country | No. Species | Species origin | | Tree abundance | |
| | | | Native | Non-native | No. trees | % native |
| --- | --- | --- | --- | --- | --- | --- |
| Gothenberg | Sweden | 24 | 12 | 12 | 1628 | 86.4 |
| Tampere | Finland | 27 | 18 | 9 | 18594 | 97.1 |
| Oslo | Norway | 29 | 16 | 13 | 1411 | 72.4 |
| Helsinki | Finland | 48 | 23 | 25 | 21089 | 95.5 |
| Espoo | Finland | 54 | 23 | 31 | 6019 | 94.9 |
| Stockholm | Sweden | 54 | 26 | 28 | 7572 | 91.6 |
| Aarhus | Denmark | 57 | 22 | 35 | 12036 | 62.5 |
| Copenhagen | Denmark | 66 | 22 | 44 | 16636 | 66.5 |
| Turku | Finland | 74 | 27 | 47 | 13257 | 84.1 |
| Malmö | Sweden | 113 | 27 | 86 | 20493 | 61.2 |

Source: Sjöman et al., 2012c.

not the local environment. In Bangalore, the oldest (and largest) species of tree in the streetscape, *A. saman, Peltophorum pterocarpum, Markhamia lutea* and *Millingtonia hortensis*, are slowly being replaced by smaller species which require less management and are less likely to be damaged by rainy season storms; whilst understandable, it is clearly changing the city aesthetic (Nagendra & Gopal, 2010).

In London, many trees were apparently removed following storm damage for 'Health and Safety' reasons, but 5% were removed as a result of insurance claims related to subsidence (GLA, 2007). The latter is interesting as some suspect individual trees are being targeted for removal without any real evidence that they are causal agents of subsidence (Barkham, 2007; GLA, 2005, 2007), although this is contested by the insurance industry (GLA, 2007). Certainly Randrup *et al.* (2001) in reviewing the extent of damage to sewers by tree roots identified that they were involved in more than 50% of all blockages. Östberg *et al.* (2012) examined the issue of sewer intrusion by roots of 52 tree and shrub species in the cities of Malmö and Skövde in Sweden, and showed broadleaved trees (especially *Malus floribunda* and *Populus canadensis* 'Robusta') as the main culprits, although conifers (including *Juniperus* and *Thuja* species) and some shrubs (including *Ligustrum, Spirea* and *Syringa*) were also involved. Careful selection of species for planting close to pipes was recommended.

## 7.5 Which tree species is best? Selection criteria

Which tree species are most appropriate for tree planting in streets (or urban areas in general)? The question is important as Hirons and Percival (2012) cite a figure of 30% mortality for landscape trees in general. Obviously avoiding species that are most susceptible to pests and disease is a starting point (Sæbø et al., 2003), but the question is harder to resolve than it might at first seem. There are street-related stresses, geographical/climate issues, and some social pressures which need to be taken into account (Sæbø et al., 2003, 2005) (Table 7.1). For example, in Beijing, China, the most common trees in street plantings are the pagoda tree *Sophora japonica* and Chinese white poplar *Populus tomentosa* because they are resistant to compacted soil conditions, air pollution and drought (Yang *et al.*, 2005). Perhaps the most difficult question to answer, in relation to the 'which is best' issue, is that related to climate change: how will existing street tree species and cultivars perform in the future under a different/changing climate regime (e.g. Leuzinger *et al.*, 2010)? Climate clearly shapes the nature of street trees; in a study of 22 cities, McPherson and Rowntree (1989) showed that deciduous broadleaved trees dominated in temperate areas and broadleaved evergreens in subtropical ones. Climate change will not simply intensify/shift the environmental stresses: the new climate envelopes that result (e.g. northern shifts in suitable climate for warm condition species/changed precipitation regimes (IPCC, 2014)) will allow the ingress of pests and diseases that would have had difficulty in colonising under the 'pre-climate change' scenario (Anderson *et al.*, 2004; Rosenzweig *et al.*, 2001). Recent work has shown that plants and animals that are translocated from their traditional climate space to a different area (with suitable climate) can thrive. Examples of successful translocation include exotic trees in Iceland (Sæbø et al., 2003) and marbled white (*Melanargia galathea*) butterflies in the UK (Willis *et al.*, 2009). Such examples show that climate change presents opportunities for widening the range of tree species planted in streets, especially in northern European countries. But caution needs to be exercised in interpreting such results: part of the success of such translocations may be because pests/parasites/diseases of

**TABLE 7.3** The most frequent tree species (%) found in the streets of some towns and cities§

| Source reference: | Nagendra & Gopal (2010) | Sjöman et al. (2012c) | | | | | Most common (★) and occasionally (†) used trees in European Regions from Sæbo et al. (2005) | | | Kennedy & Southwood (1984) |
|---|---|---|---|---|---|---|---|---|---|---|
| Scientific name | Bangalore | Aarhus | Gothenberg | Malmö | Oslo | Turku | Northern Europe | Central Europe | Southern Europe | Phytophagous insects & mites on British trees |
| Acacia cyanophylla | | | | | | | | | † | |
| Acacia decurrens var. Dealbata | | | | | | | | | † | |
| Acer campestre | | | | | | | | ★ | | |
| Acer negundo | | | | | | | | | † | |
| Acer platanoides | | 10.3 | | | 11.6 | 9.1 | ★ | ★ | | |
| Acer pseudoplatanus | | | | | | | ★ | ★ | | 43 |
| Acer spp. | | | | | | | † | ★ | ★ | |
| Aesculus hippocastanum | | 4.0 | | 5.4 | 10.2 | | ★ | ★ | | 9 |
| Aesculus x carnea | | | | | | | | ★ | | |
| Albezia saman | **8.9** | | | | | | | | | |
| Albizzia julibrissin | | | | | | | | | † | |
| Alnus glutinosa | | | | | | | | | | 141 |
| Alnus x spaethii | | | | | | | | † | | |
| Bauhinia variegata | 6.3 | | | | | | | | | |
| Betula pendula | | | | | 5.5 | | ★ | | | |
| Betula pubescens | | | | | 4.7 | | ★ | | | |
| Betula spp. | | 4.1 | | | | | | | | 334‡ |
| Brachychiton acerifolius # | | | | | | | | | ★ | |
| Brachychiton diversifolius # | | | | | | | | | ★ | |
| Carpinus betulus | | | | 2.8 | | | | | | 51 |
| Cedrus atlantica | | | | | | | | | † | |
| Celtis australis | | | | | | | | † | ★ | |
| Ceratonia siliqua | | | | | | | | | ★ | |
| Cercis siliquastrum | | | | | | | | | † | |
| Citrus aurantium | | | | | | | | | † | |
| Corylus colurna | | | | | | | † | † | | |
| Crataegus spp. | | | | | | | † | † | † | 209‡ |
| Cupressus sempervirens | | | | | | | | | | |

| Species | | | | | | | |
|---|---|---|---|---|---|---|---|
| *Delonix regia* | 5.5 | | | | | | |
| *Fagus sylvatica* | | 3.9 | | | | | 98 |
| *Fraxinus angustifolia* | | | | 2.8 | | | |
| *Fraxinus excelsior* | | | | 3.3 | | | 68 |
| *Gleditsia triacanthos* | | | | | | † | ★ |
| *Koelreuteria paniculata* | | | | | | | ★ |
| *Laurus nobilis* | | | | | | † | † |
| *Malus spp.* | | | | | † | | |
| *Markhamia lutea* | 4.8 | | | | | | ★ |
| *Melia azedarach* | 4.4 | | | | | | ★ |
| *Millingtonia hortensis* | | | | | | | |
| *Morus alba* var. *pendula* | | | | | | | ★ |
| *Olea europaea* | 8.4 | | | | | | ★ |
| *Peltophorum pterocarpum* | | | | | | | ★ |
| *Phoenix canariensis* | | | | | | | ★ |
| *Phoenix dactylifera* | | | | | | | ★ |
| *Picea omorika* | | | | 2.7 | | | ★ |
| *Pinus halepensis* | | | | | | | ★ |
| *Pinus pinea* | | | | | | | † |
| *Pinus sylvestris* | | | | 4.9 | | | † | 172 |
| *Platanus orientalis* | | 4.2 | | | | † | |
| *Platanus x acerifolia/Platanus x hispanica* | | | 5.0 | | † | ★ | ★ |
| *Polyalthia longifolia* | 5.6 | | | | | | |
| *Pongamia pinnata* | 6.9 | | | | | | |
| *Populus alba* | | | | | | | ★ |
| *Populus canadensis* | | | | | | | ★ |
| *Populus nigra* ('Italica') | | | | 2.9 | | | (★)† |
| *Populus tremula* | | | | | ★ | | |
| *Populus trichocarpa* | | | | | † | | |
| *Prunus avium* | | | 5.3 | | † | | |
| *Prunus cerasifera* var *atropurpurea* | | | | | | | ★ |
| *Prunus padus* | | | | | † | | |
| *Prunus spp.* | 3.0 | | | | | | |
| *Quercus ilex* | | | | | | | † |

| Scientific name | Nagendra & Gopal (2010) Bangalore | Sjöman et al. (2012c) Aarhus | Gothenberg | Malmö | Oslo | Turku | Most common (★) and occasionally (†) used trees in European Regions from Sæbo et al. (2005) Northern Europe | Central Europe | Southern Europe | Kennedy & Southwood (1984) Pytophagous insects & mites on British trees |
|---|---|---|---|---|---|---|---|---|---|---|
| Quercus petraea | | | | | | | † | | | |
| Quercus robur | | 7.9 | | 4.0 | | | † | | | 423‡ |
| Quercus spp. | | | | | | | | ★ | | |
| Robinia pseudoacacia | | | | | | | † | ★ | ★ | 2 |
| Sophora japonica | | | | | | | | † | ★ | |
| Sorbus acuparia | | | | | | 9.3 | | | | 58 |
| Sorbus aria | | | | 2.7 | | | | | | |
| Sorbus latifolia | | 5.0 | | | | | | | | |
| Sorbus mougeotti | | **11.3** | | | | | | | | |
| Sorbus spp. | | | 3.7 | | | | ★ | † | | |
| Sorbus x intermedia | | 5.7 | 7.7 | 11.1 | | 2.9 | | | | |
| Spathodea campanulata | 8.4 | | | | | | | | | |
| Swietenia macrophylla | 5.7 | | | | | | | | | |
| Tilia cordata | | 8.0 | 7.0 | 3.3 | | | | ★ | | |
| Tilia platyphyllos# | | | | | | | | ★ | | |
| Tilia spp. | | | 13.9 | 2.9 | **45.3** | | † | | ★ | 57‡ |
| Tilia x europea (Tilia x vulgaris ('Pallida' in NE) Sæbo et al. (2005)) | | 7.3 | **35.5** | **16.5** | **16.7** | | ★ | ★ | | |
| Ulmus glabra | | | 5.8 | | 4.6 | 5.9 | | | | 124‡ |
| Ulmus minor | | | 2.8 | | | | | | | |
| Ulmus spp. | | | | | | | † | | | |
| Washingtonia filifera | | | | | | | | ★ | ★ | |
| Washingtonia robusta | | | | | | | | | ★ | |
| **Total % of ten (or fewer) species** | 64.9 | 67.8 | 80.3 | 59.0 | 87.7 | 70.6 | | | | |

Most dominant tree in bold; data on species richness of phytophagous insects and mites is for Britain only.

§Data is for the top ten (or fewer) species recorded. Note: species making up less than 2% of the population are not given; ‡ Kennedy & Southwood (1984)D for *Betula* the data is for two species (*B. pendula* and *B. pubescens*); for *Crataegus* the value is given for *C. monogyna* only, figure given for *Quercus robur* is also for *Q. petraea*, for *Tilia spp.* data is for two species, for *Ulmus glabra* the figure is for both *U. glabra* and *U. minor*; # these species were given as *Brachychiton acerifolium*, *B. diversifolium*, *B. diversifolium* and *Tilia platyphylla* in the source references.

TABLE 7.4 Tree selection criteria derived by Spellerberg and Green (2008) from a literature review

| Criterion | Considerations |
|---|---|
| Site conditions | Soil type, local climate, drought, frost and salt tolerance |
| Adjacent factors | Foundation materials used in construction of adjacent infrastructure (e.g. roads, avenues, boulevards) |
| Availability | Limited choice of species |
| Safety/Structural | Hazard reduction to people, traffic, underground infrastructure, foundations |
| Nuisance | Toxic plant parts/fruits; unpleasant smells; frequent bark/limb shedding |
| Maintenance | Minimisation, especially pruning and spraying |
| Design and aesthetics | Landscape design; shade/shelter; impact on property values |
| Historical | Maintaining continuity of plantings |
| Environmental | Ecosystem services (including biodiversity) |
| Socioeconomics | Societal value and land values |

the translocated species were not transferred at the same time. Also, little information is available on vitally important features of trees such as water requirements, a factor obviously important in relation to climate change planning and especially so in arid and semi-arid regions (McCarthy & Pataki, 2010). Sieghardt *et al.* (2005) give the water requirements of single trees in a temperate climate to be about 800 litres.$m^{-2}$ of crown/year and 600 litres if one of a row – they also note that trees surrounded by sealed surfaces probably get too little water in summer and too much in winter! Plants in pavements and plazas with sealed surfaces, whose roots have not reached the local water table, will be under greatest threat of water stress, exacerbated by compaction of soil pores (Sieghardt *et al.*, 2005). Comparisons of the native Californian sycamore *Platanus racemosa* and the Canary Island pine *Pinus canariensis* in Los Angeles and Orange County in the USA showed that the native species was less tolerant of drought conditions than the introduced Mediterranean species (McCarthy & Pataki, 2010).

Spellerberg and Green (2008) attempted to address the general question of tree selection criteria in a review paper and found information fragmented and dominated by lists of suitable/unsuitable species, but there was little information available on the scrutiny of the criteria underpinning the lists. Within New Zealand, they found that almost two-thirds of the responses from city councils they consulted revealed lists of suitable trees for planting, but no specific selection criteria. The criteria they developed following their review are outlined in Table 7.4.

Sjöman (2012) examined the selection of trees for 'urban paved environments' which obviously includes streets, but would also encompass plazas and other areas with sealed surface surroundings rather than urban parkland. Sjöman and Nielsen (2010) identified the information that practitioners wanted in relation to urban trees prior to their literature review: that the information on species and cultivars should:

- be derived from experience of plantings within the urban environment such that tree selection could be based on known performance;
- be regionally relevant, given that performance and suitability is likely to vary with the geographical location of plantings;
- be available on appropriate site conditions and environment.

In addition, that:

- reference plantings could be viewed so that the trees could be evaluated from personal 'impressions' rather than from tabulated information or images;
- a full catalogue of trees for the 'paved environment' should be available for each region.

As with the review by Spellerberg and Green (2008), Sjöman and Nielsen (2010), concentrating on Scandinavia, found the literature scattered and primarily at levels that were either too general (dendrology) or too specific (scientific literature) and sometimes inappropriate. For practical use the 'Goldilocks solution', neither too general nor too specific, was missing. Sæbø *et al.* (2005) summarise the findings of a pan-European study on selection criteria but specific comments on street tree selection are rather general: 'strong apical growth, strong branching angles, predictable growth rates, high overall aesthetic value and potential for a long lifespan'. In contrast to the recent trend for urban agriculture, they suggest the avoidance of species with large fruits; more importantly they stress the avoidance of species which produce known allergens (in pollen). In an earlier paper, Sæbø *et al.* (2003), in addition to the ability to withstand the factors in Table 7.1, stressed that the 'plasticity' of trees (their ability to grow in a wide range of conditions) was an important factor in selection of street trees, as was uniformity in growth form, wind resistance and low propensity to shed branches.

Although lists of tree species used in street environments can be quite substantial, it is clear from the studies above that a small number make up the majority of plantings in any given city. With new climate challenges on the way, and the need to diversify plantings to avoid wipe-out by pests and diseases, it is clearly important to identify new species and cultivars for urban situations (Sjöman *et al.*, 2012a). Given that the urban environment is inherently stressful for street trees, Sjöman *et al.* (2012b, 2010) carried out fieldwork in

**TABLE 7.5** Trees identified by Sjöman *et al.* (2012b, 2010) from fieldwork in Romania, Moldavia and China, as potentially valuable species for use in urban 'paved' environments

| *Source of tree* | |
| --- | --- |
| *Romania/Moldavia* | *China* |
| *Acer campestre*★ | *Ailanthus altissima*★ |
| *Acer tataricum*★ | *Carpinus turczaninowii* |
| *Carpinus betulus*★ | *Celtis bungeana* |
| *Carpinus orientalis* | *Fraxinus chinensis* |
| *Cornus mas*★ | *Koelreuteria paniculata*★ |
| *Crataegus monogyna*★ | *Morus mongolica* |
| *Fraxinus excelsior*★ | *Ostrya japonica* |
| *Quercus dalechampii* | *Quercus aliena* var. *acuteserrata* |
| *Quercus frainetto*★ | *Quercus baronii* |
| *Quercus pubescens* | *Quercus wutaishanica* |
| *Quercus robur*★ | *Sorbus folgneri* |
| *Sorbus torminalis*★ | *Syringa pekinensis* |
| *Tilia tomentosa*★ | *Ulmus glausescens* |
| | *Ulmus pumilia* |

★Species already used in northern European cities Sjöman *et al.* (2012a).

Romania, Moldavia and China, in areas whose environment could be equated with that of some northern European inner cities, to identify species that would be sufficiently drought tolerant in summer and be able to survive the winter. In China 14 species, and in Romania/ Moldavia 13 species, were identified as potentially useful (Table 7.5). The point of the exercise was not simply the identification of useful species but also the identification of an ecotype of the species, that subset of the genetic material of the species adapted to similar conditions as the inner-city environment, which was important. Little information, if any, is available on the performance in the city context of many of the trees that Sjöman *et al.* (2010, 2012b) found, and substantial evaluation work needs to be carried out on their actual, as opposed to potential, suitability and the specific growing conditions and optimum planting approaches required before they can be used in practice. 'Actual suitability' would include such issues as tolerance to air pollution; Jim and Chen (2008), for example, list 20 tree species commonly found in urban plantings in Guangzhou and rate their tolerance to $SO_2$, $NO_x$ and particulates.

## 7.6 Getting the most out of street trees

We have already seen that trees can have many different values in urban areas (Nowak & Heisler, 2010) and at the scale of the urban forest as a whole there are now decision support tools to assist in evaluating, planning and improving the ecosystem services that are delivered. Some support materials are traditionally paper-based guidance documents (e.g. Jaluzot, 2012) but some tools are much more dynamic and interactive. Perhaps the most well known of these tools is the 'i-Tree' software (USDAFS, 2014b) developed by the US Department of Agriculture Forestry Service, which was adapted for use in the UK in a study of Torbay (e.g. Treeconomics, 2011). The i-Tree software has also been adapted for use with street trees as 'i-Tree Streets' (USDAFS, 2014a). The software gives a financial value for the ecosystem services the street trees provide and can be seen as a cost/benefit tool balancing the costs of purchase, installation and management against the range of values delivered by trees to the human community. One of the free-standing components of i-Tree is the 'i-Tree Species' utility which contains data on 1,585 tree species with data outputs tailored for US states on the basis of whether the tree is appropriate for the climate zone and required ecosystem service (reduction of: air pollution, air temperature, UV exposure, wind, stream flow; improved carbon storage, building energy conservation; and knowledge of allergenic effects) (Nowak, 2008). Nowak (1999) pointed out that when drawing up the costs/benefits of trees of reducing the impact of fossil fuel emissions, it is worth remembering that on the debit side is the cost of maintenance. McPherson and Simpson (2002) showed that 50% of the annual costs of tree maintenance were for pruning. Pests and diseases can be an expensive component of maintenance, and McPherson and Kotow (2013) combined i-Tree Streets with the 'Pest Vulnerability Matrix' program and developed a 'Report card' to assist the public and local authority staff in understanding the results and adapting the species make-up to one which has more resilience.

The following sections describe some of the ecosystem services delivered by street trees.

## 7.7 Aesthetics and food

Ode and Fry (2002) point out that whilst the aesthetic experience associated with trees is primarily visual, it also incorporates the senses of smell and of hearing. Apart from the general aesthetic of greenery, street trees can add distinctiveness to a city or quarter, and generate economic benefits. Wolf (2005) summarised a series of her research projects, from large, medium and small US cities, that showed that visual quality ratings by potential shoppers increased with presence of trees in business districts, as did their perceptions of product value, product and shop quality, likelihood of visiting, length of stay and willingness to pay for parking. Economic benefits can also accrue through increased tourism and even, in some cases, result in an exportable product. One of the closest links between trees and the city aesthetic is perhaps in the city of Seville in Andalucía, Spain, where some 14,000 citrus trees perfume the streets in spring, shade the streets in summer, and gladden the eye (Padilla, 2011). The harvested fruits are exported, and in January and February enthusiastic cooks in the UK are sent into frenzied activity making Seville orange marmalade (Figure 7.4). Whilst the idea of urban agriculture is gaining ground (Brebbia *et al.*, 2002; Garnett, 2000), there may be legal problems that inhibit the planting of fruit- and nut-bearing trees in streets in some places. Despite movements towards public orchards in Seattle, apple *Malus* spp., cherries *Prunus* spp. and pears *Pyrus* spp. are on a planting blacklist as street trees (McLain *et al.*, 2012). The rationale for excluding such species – public safety (McLain *et al.*, 2012) – seems exceedingly flimsy and overly risk-averse. An interesting attempt to introduce fruit into urban streets is the concept of 'Guerrilla Grafters' in San Francisco (Faircompanies.com, 2012; GG, 2012) where fruit-bearing wood is being grafted onto ornamental species. Whilst this activity is not condoned by city authorities in San Francisco (HuffingtonPost, 2012), it is a technique that local authorities worldwide could easily adopt to rapidly introduce fruit into streetscapes by sidestepping the time taken for a fruit tree to grow to maturity.

Whilst trees are perceived as valuable to a city aesthetic (see review in Dandy, 2010), it is interesting that they also enhance the value of other attempts to 'green' urban areas. Todorova *et al.* (2004) in a questionnaire survey in Sapporo, Japan, using photomontages found that trees had a significant and positive effect on the perception of street plantings of flowers. Qin *et al.* (2013) compared responses to different areas of the Shanghai Botanical Garden (a lawn (grass green), a special tulip show (multicoloured), an 'arbor forest' (forest green), a bamboo grove (light green), and a Sakula show (pink)) and found that colour enhanced satisfaction with vegetated areas. Interesting age-specific differences were also found in the study, with younger and older people having higher satisfaction rates than 'adults'. Such studies suggest that careful attention to the planting designs of green spaces incorporating seasonal variations

FIGURE 7.4  Left: orange trees in Seville. © Alan Goodkin. Right: freshly made Seville orange marmalade in the author's kitchen (by the author's wife). © John Dover

in tree colour as well as incorporating colourful ground-level plantings may enhance their value to urban dwellers.

Of course not every tree is suitable for use in urban areas and reasons given include messy fruits and seed pods (Chakre, 2006), although a comment such as 'horse chestnuts (*Aesculus hippocastanum*) are totally unsuitable for urban areas', which a local authority staff member once opined to me, seems utterly ridiculous: yes they shed horse chestnuts over the streets in the autumn – so what? The pleasure such trees give to communities, and not least to children playing 'conkers' with the nuts, just has to outweigh the transient inconvenience of clearing them up or the irritation of motorists who park under conker-laden trees! Alternatively, plant them on wider verges (for those who are not familiar with the traditional game of conkers, see Arthur, 1981, or ACC, 2012). However, for a tree that probably needs careful siting, the female ginkgo, or maidenhair tree, (*Ginkgo biloba*) probably takes the biscuit – maturing after about 30–40 years, they produce seeds that 'smell like rotten eggs or vomit' and are slippery underfoot (Gray, 2009)! Nevertheless, the beauty of the tree is considered by some to outweigh the temporary olfactory inconvenience (Murdock, 2012). Chakre (2006) nominates the bastard poon tree *Sterculia foetida* as a species probably best not planted because of its smell – its generic name means 'manure' and its specific 'stinking' (Orwa *et al.*, 2009). There is evidence that poorly sited trees can be a hazard to drivers and this may be used as a reason for their removal; but tree planting has also been used as a traffic-calming measure (speed reduction) – a greater understanding of the implications of tree planting, both positive and negative, is probably required (see Dandy, 2010). The term 'traffic-calming' takes on another meaning when applied to driver frustration, as roadside greenery has been shown by Cackowski and Nasar (2003) to improve drivers' tolerance to frustration.

## 7.8 Trees and shade

### 7.8.1 Temperatures

Chaturvedi *et al.* (2013) give the major street trees along main roads in Nagpur, India, as *Dalbergia sissoo*, *Cassia siamea*, *Albizia lebbek*, *Azadirachta indica* and *Hardwickia binnata*, along link roads as *Delonix regia*, *C. siamea*, *Cassia fistula*, *A. lebbek*, *Acacia nilotica*, *A. indica*, *H. binnata* and *Pithocoellobium dulce* and state that they are planted primarily for shade. With the increasing trend of global warming and associated predicted increases in extreme events (O'Neill *et al.*, 2009) the importance of shade cast by street trees is likely to be increasingly important (Figure 7.5). Heatwaves increase mortality from heat stress and through the exacerbation of existing cardiovascular and pulmonary conditions (Hoffmann *et al.*, 2008; Ishigami *et al.*, 2008; Rey *et al.*, 2007).

Citrus trees are evergreen, so Seville's streets are green all year round; in many places, such as the UK, the majority of street trees are deciduous – this does have the advantage that streets, buildings and parks are shaded from the sun in the summer, but warmed during the winter. Shelter from the elements (rain, snow and cutting winter winds) is much reduced of course; evergreens would be needed for this purpose. Shading from the sun is an important feature and protects young and old alike from overheating and sunburn; street trees, as in parks and elsewhere, play their part in this (Nagendra & Gopal, 2010) and with suitable seating can create an important social focus (Figure 7.5). Peper *et al.* (2008) report that street trees shade 13.8% of Indianapolis's pavements. Poorly placed trees can have downsides,

FIGURE 7.5 Street trees can protect against the heat and sun and also provide a social focus for the community (Potes, northern Spain). © John Dover

however, such as impeding night-time heat loss and trapping cold air pockets (Stülpnagel *et al.*, 1990), but tall trees, with compact diameter canopies, can shade the walls of buildings in narrow streets without trapping pollutants (McPherson, 1994).

In July 2004, using a helicopter, Leuzinger *et al.* (2010) examined the temperature of tree crowns of ten species of street tree in Basel, Switzerland. Measurements were made from a height of about 400 m with a thermal-imaging camera. The streets registered mean surface temperatures of about 37°C and the average for tree crowns was about 27°C (air temperature of the order of 25°C). There was some variation in the crown temperatures of the tree species, with species with lower crown temperatures having a better prospect of reducing street temperatures. The trees identified as best for cooling were: silver lime *Tilia tomentosa*, the black locust, Scots pine, and small-leaved lime; and the worst: London plane, horse chestnut, honey locust *Gleditsia triacanthos* and Norway maple. In general, cooler crown (foliage) temperatures were correlated with smaller leaf size. This was not an invariable rule though, as silver lime which has large leaves also had a high transpiration rate whilst the honey locust had a high crown temperature, but also small leaves. Surroundings had an obvious effect on crown temperatures: trees located in parks had lower crown temperatures than those in streets. Leuzinger *et al.* (2010) caution that species that have lower crown temperatures at the present time may not behave in the same way under a future climate regime – factors such as heat and drought tolerance, stomatal response, leaf structure and canopy architecture are likely to be important in choosing trees for future, climate proofing, plantings.

Experimental work in Manchester, UK, reported by Armson *et al.* (2012) gave peak surface temperatures of trial plots covered in concrete with full sun to be 40°C (17°C above air temperature) with equivalent plots in tree shade at 28°C (4°C above air temp.). Similar plots surfaced with grass in sunlight were 23°C (1°C lower than air temp.) and 19°C in tree shade (4°C lower than air temp.). Differences were, of course, largest at higher temperatures. Globe thermometers measuring peak air temperatures at 1.1 m above ground level gave highest readings in full sun (when exposed to direct sunlight) of 32°C (above concrete) and 34°C (above grass) (air temp. 25°C), whilst air temperatures in the tree shade were only 2°C above air temperature at 27°C.

## 7.8.2 Ultraviolet

Melanoma cancers develop from overexposure to ultraviolet B (280–320 nm) radiation (De Fabo *et al.*, 2004), especially in genetically susceptible subjects, and increased doses of UV-B may result as a side-effect of cloud-free skies in heatwaves and through continued thinning of stratospheric ozone (Grant *et al.*, 2003b). Street trees are considered to reduce damaging ultraviolet radiation through shading (Sarajevs, 2011). Parisi and Kimlin (1999) found that tree shade was better at reducing UV-A than UV-B, unfortunately UV-B is the more damaging of the two wave ranges to human health (Heisler & Grant, 2000). But trees with other exposure reduction strategies (Stanton *et al.*, 2004) may help in lowering the incidence of erythemia (skin reddening), cataracts (clouding of the eye lens) and skin cancer. Grant *et al.* (2002) noted that tree shade reduces UV-B exposure by 40–60% compared with full sun and, of course, depends on sun angle (Parisi *et al.*, 2001). In the USA, Grant *et al.* (2003b) examined the transmission of UV radiation through leaves and also reflectance from the leaf surface for 20 common urban trees. They found that very little UV is transmitted through leaves with most UV-B being absorbed near the leaf surface in the cuticle. For most tree species, reflection of UV from the leaf surface is generally low: about 5–6% of the UV-A (320–400 nm) and UV-B fractions that reached the upper (adaxial) and lower (abaxial) leaf surfaces. Species with much higher reflectance than this general range were sugar maple *Acer saccharum* whose lower leaf surface reflected 13.8% of UV-B and 15.4% of UV-A, white oak *Quercus alba* (12.7% and 11.4%) and red maple *Acer rubrum* (12.9% and 13.0%). Higher reflectance was considered to be linked to leaves with filamentary or plate-like wax surfaces compared to those with smoother wax surfaces. Cloud scatters UV radiation such that exposure in shade does not come simply from direct sunlight from gaps in the tree canopy, but also from this 'diffuse' sunlight (Heisler *et al.*, 2003; Parisi *et al.*, 2000). Turnbull and Parisi (2006) also noted that subjects in shade will still be exposed to considerable levels of UV from scattered light and that reducing such radiation will require side shielding, perhaps using hedging (Yoshimura *et al.*, 2010). Also, as protection is required in the winter as well as summer, evergreen species are preferred to deciduous for year-round UV protection. Protection from UV is not, therefore, simply an issue of standing or sitting in direct shade, as substantial levels of UV can also reach a subject from the side. For example, Parisi and Kimlin (1999) calculated that the UV received by a person in a single hour sheltering under tree shade could still be more UV than the threshold limit for the Australian occupational exposure rate for eurythemia.

   Heisler and Grant (2000) point out that UV-B reflectance from high-albedo ground surfaces can be substantial (e.g. light concrete paving UV-B reflectance = 10–12% (Sliney, 1986)) and may cause eye problems. To avoid UV reflectance in urban areas grass, with its low UV reflectivity (2% (Chadyšienė & Girgždys, 2008)), is probably a useful ground cover to use in conjunction with trees. Of course if it snows, the grass is not much help; skiers readily appreciate that UV-B reflection from snow is high, as it can cause snow blindness (Atkinson, 1921), and reflectance can reach 88% depending on how fresh it is and time of day (Chadyšienė & Girgždys, 2008). UV-B exposure is greater at high altitudes (Gibson & McKenna, 2011). Perhaps fortunately the urban heat island effect tends to reduce snow in urban areas compared with the surrounding countryside.

The shadows that we see cast in the visual spectrum are not necessarily a good indication of the shade afforded in the UV-B range of the spectrum – which we cannot see (Heisler & Grant, 2000). However, it is clear that the general message is this: the more trees there are, and the denser the canopy, the deeper the shade and hence the greater the protection from direct and diffuse UV radiation.

## 7.9 Trees and gaseous pollutants

Street trees, as with all trees and other vegetation, remove carbon dioxide from the air and lock it up as biomass (see section 2.9.4). $CO_2$ is a natural, indeed essential, component of our atmosphere; the problem is that we produce too much of it by burning fossil fuels for energy, from agriculture, etc. The impact of $CO_2$ is global, and local action can help reduce it (just as urged in Agenda 21, the protocol to encourage local responses to global challenges which emerged from the Earth Summit in Rio de Janeiro in 1992 (e.g. NULBC, 2000)). For example, Indianapolis's street trees were reckoned to reduce $CO_2$ in the air by 14,146 tonnes.y$^{-1}$ (based on a 2005 inventory) (Peper et al., 2008). Street trees also remove gasses such as nitrogen dioxide and sulphur dioxide. Indianapolis's $NO_2$ load is reduced by 4,363 kg.y$^{-1}$ by street tree uptake; these, additionally, reduced the need for emission of a further 9,389 kg.y$^{-1}$ (Peper et al., 2008). Likewise, Indianapolis's street trees also remove about 2,147 kg of $SO_2$ a year and prevent the emission of a further 34,997 kg.y$^{-1}$ (see also section 2.9.4).

So, as with trees elsewhere, street trees help improve local air quality. But the choice of street trees in areas of high nitrogen oxide concentration is particularly important because low-level ozone is formed by a photochemical reaction between $NO_x$ and hydrocarbons. Hydrocarbons in the atmosphere come from a range of sources including un-burnt fuel in exhaust gasses, from industrial processes, and also from trees. Some trees produce more VOCs than others, so tree planting at busy roadsides needs to use species which will not promote ozone pollution; of course street trees can also reduce ozone pollution in the same way that they help reduce other gasses. This section explores this conundrum.

### 7.9.1 Trees as emitters of volatile organic compounds

Trees are known to naturally emit VOCs, and emissions can be increased by stress such as heavy metal pollution (Velikova et al., 2011). In 1981, US President Ronald Reagan famously tried to divert attention from pollution by humans by claiming 'Trees cause more pollution than automobiles do' (Radford, 2004; RationalWiki, 2012). Yes, trees do release volatile organic chemicals – quite a substantial amount; and globally vegetation does release far more VOCs than man (Ehhalt et al., 2001), but in developed countries with high-density populations, natural VOC emissions are dwarfed by those released by human activities (Logan, 1983). In the UK, in 1989, trees were estimated to release 72% of the 'non-methane' hydrocarbon (NMHC) emissions (151.4 ktC.y$^{-1}$) released from natural sources (Table 7.6). Of these NMHC emissions, 64% originated from coniferous species (mainly

TABLE 7.6 Sources of non-methane hydrocarbon (NMHC) sources in the UK for 1989

| Source of NMHC | $ktC.y^{-1}$ | % |
|---|---|---|
| Pine | 70.4 | 33.3 |
| Spruce | 64.5 | 30.5 |
| Deciduous | 16.5 | 7.8 |
| Agriculture | 59.8 | 28.3 |
| Total from natural sources | 211.2 | |

Source: Anastasi *et al.*, 1991.

monoterpenes) and 8% (mainly isoprene) from deciduous species (the balance, 28%, comes from agriculture) (Anastasi *et al.*, 1991). It was estimated that biogenic sources of VOCs (trees + agriculture) contributed 13% to the total UK non-methane VOC emissions. At the time, anthropogenic emissions were estimated at 1,675 kt VOC.y$^{-1}$ using 1980 data (Lubkert and Tilly, 1989, cited in Anastasi *et al.*, 1991). More recent information (MacCarthy *et al.*, 2012) suggests that there have been substantial reductions in NMHCs from anthropogenic sources since then. In the UK, there was a steady rise in NMHCs from 1970 to 1990 (Dore *et al.*, 2008) and then over the period 1990 to 2010 there was a 72% decline from 2,054.5 to 568 kt.y$^{-1}$ mainly as a result of changes in the transport sector and especially due to the increased use of catalytic converters and a switch to diesel from petrol as fuel (MacCarthy *et al.*, 2012). Natural and agricultural NMHC emissions also appear to have declined and were estimated by MacCarthy *et al.* (2012) as 97.5 kt.y$^{-1}$ in 1990 and 29.4 kt.y$^{-1}$ in 2010. However, the estimation of biogenic/natural emission rates is fraught with difficulty due to modelling assumptions, relevance of lab–based assessments, biomass estimates, high variability in samples, varietal differences, seasonal and temperature effects, etc. (Corchnoy *et al.*, 1992; Hakola *et al.*, 2001; Hewitt & Street, 1992). Whatever the current 'true' level of emissions from natural sources, future levels are unlikely to be static as climate change impacts bite; for example, changes in temperature, rainfall and plant community composition, and changes in tree stocks as a result of disease and alternate plantings for climate proofing (FC, 2012a; Hewitt & Street, 1992; Tubby, 2006).

## 7.9.2 Trees, nitrogen oxides, VOCs and ozone

The reason why Reagan's comments were contentious, and widely derided at the time, was because he was effectively claiming that trees were the main culprits behind the production of one of the components of photochemical smog: low–level ozone, along with other eye–irritating chemicals (e.g. peroxyacetyl nitrate and peroxybenzoyl nitrate) (Clarke, 1992). Plants can develop ecotypes that can tolerate elevated levels of ozone (Barnes *et al.*, 1999; Fuhrer & Bungener, 1999) and protection appears to be correlated with increased levels of foliar ascorbate (Zheng *et al.*, 2000). So whilst Reagan was correct that plants emit natural substances (including VOCs), the real problem is the $NO_x$ we produce. As with VOCs, a substantial quantity of nitrogen oxides are created naturally through a range of processes, but urban air has considerably elevated levels compared with background air levels – of the order of x10$^3$ or x10$^4$ (Logan, 1983) and is principally a result of human action. Natural sources of nitrogen oxides include: the result of lightning, biological processes in the soil, atmospheric oxidation of ammonia in the troposphere, biomass burning, and stratospheric transfer to the troposphere, whilst the primary source of anthropogenic $NO_x$ comes from the burning of fossil fuels in industry and transport (Logan, 1983). So, in the balance of things, are trees good or bad in relation to low level ozone? In a substantial report on the value of trees to Indianapolis, street trees were shown to remove ozone at the rate of 10,906 kg.y$^{-1}$ and to reduce anthropogenic emission of VOCs by 2,960 kg.y$^{-1}$ whilst, admittedly, emitting 21,500 kg.y$^{-1}$ (Peper *et al.*, 2008). Most of these Biogenic VOCs (BVOCs) were emitted by two species of tree: eastern cottonwood *Populus deltoides* (22%) and silver maple *Acer saccharinum* (17%) (Peper *et al.*, 2008).

Whether VOCs are of biogenic or anthropogenic origin, there is a great deal of sense in minimising the potential for ozone formation in urban areas with high $NO_x$ levels by planting trees with the lowest VOC emissions – especially, as Benjamin and Winer (1998) point out, planting of trees is being promoted as a way of mitigating climate change, and air-pollution issues such as particulates. A screening programme of the ozone-forming potential of 308 tree and shrubs was carried out in the Californian South Coast Air Basin by Benjamin and Winer (1998), taking into account the hourly variation in terpene and isoprene levels associated with elevated temperature and varying light levels, in order to provide information on the best and worst trees to grow. Data are complicated by such aspects as the VOC-emission level per unit of measurement; for example, weight of leaf when scaled up to the tree level. As Benjamin and Winer (1998) stress, a species might have a high emission weight per gram of leaf, but if total biomass is substantially less than another tree species with lower emission rates/g, the former might still be the better bet for planting in terms of reducing VOC levels. Also, trees and shrubs are not uniform in the number, nature or amount of volatiles emitted and some can be highly reactive – more so than some anthropogenic-derived chemicals. Some assumptions about emission rates, diurnality and seasonality of VOC emission, and relations with light and temperature which may be reasonable for southern California, may also not be uniformly applicable and may be quite different in other regions (Benjamin & Winer, 1998) such as northern Europe. Of the 308 species tested, 234 were trees; 'low' ozone trees were estimated to produce less than 1 g $O_3$ tree$^{-1}$.d$^{-1}$, 'medium' = 1–10 g $O_3$ tree$^{-1}$.d$^{-1}$ and 'high' more than 10 g $O_3$ tree$^{-1}$.d$^{-1}$. The results showed 87 trees were categorised as low, 72 medium and 75 species high ozone-forming potential – all are listed in rank order in Benjamin and Winer (1998) and they recommended low and medium-ranking species should be considered as appropriate for planting schemes. Sixty of the tree species were estimated to have no ozone-forming potential, whilst the most reactive was the palm oil tree *Elaeis guineensis* with an estimated ozone-producing potential of 406 g $O_3$ tree$^{-1}$.d$^{-1}$. VOC emissions were published by Corchnoy *et al.* (1992) for 12 of the species that Benjamin and Winer (1998) list; relative estimates of ozone-forming potential are not entirely comparable (Table 7.7).

Many, if not most, species tabulated by Benjamin and Winer (1998) are unlikely to be used in northern European areas, such as the UK, which have quite different temperature profiles to southern California, and of course emission profiles may differ with environment, ecotype and time (Hakola *et al.*, 2001). Having said that, it is interesting that the pedunculate oak *Quercus robur*, a classic tree of the English countryside with a high biodiversity value, is, unfortunately, one of the highest-scoring species for ozone formation listed by Benjamin and Winer (1998).

**TABLE 7.7** Comparison of hydrocarbon (VOC) emissions for 12 trees from Corchnoy et al. (1992) and ozone-forming potential for the same trees from Benjamin and Winer (1998) (note different units are used)

| Tree species | | Corchnoy et al. (1992) | Benjamin and Winer (1998) | | Ozone production ranking | | |
| Common name | Scientific name | VOC $\mu g^{-1}/h^{-1}$ | Isoprene and monoterpene $g/tree\text{-}1/d^{-1}$ | Ozone potential $g/tree^{-1}/d^{-1}$ | Low | Medium | High |
|---|---|---|---|---|---|---|---|
| Crape myrtle | Lagerstroemia indica | 0 | 0 | 0 | ★ | | |
| Camphor | Cinnamomum camphora | 0.03 | 0 | 0 | ★ | | |
| Aleppo pine | Pinus halepensis | 0.15 | 0.1 | 0 | ★ | | |
| Deodar cedar | Cedrus deodara | 0.29 | 0.4 | 2 | | ★ | |
| Italian stone pine | Pinus pinea | 0.42 | 0.1 | 0 | ★ | | |
| Monterey pine | Pinus radiata | 0.90 | 0.1 | 0 | ★ | | |
| Brazillian pepper | Schinus terebinthifolius | 1.3 | 0.9 | 3 | | ★ | |
| Canary Island pine | Pinus canariensis | 1.7 | 0.2 | 1 | ★ | | |
| Ginkgo (Maidenhair tree) | Ginko biloba | 3 | 0.2 | 1 | ★ | | |
| Californian pepper | Schinus molle | 3.7 | 0.2 | 1 | ★ | | |
| Liquidambar (Sweet gum) | Liquidambar styraciflua | 37 | 2.4 | 15 | | | ★ |
| Carrotwood | Cuopania anacardioides | 49 | 6.5 | 59 | | | ★ |

Hewitt and Street (1992) listed the eight most frequent deciduous species planted in the UK (by area) and showed sessile oak *Quercus petraea* and pedunculate oak to be isoprene emitters: whilst beech *Fagus sylvatica*, ash *Fraxinus excelsior*, hazel *Corylus avellana* and hawthorn *Crataegus monogyna* emitted neither isoprene nor terpenes. However, there was some doubt expressed by Hewitt and Street (1992) about the status of silver birch *Betula pendula* and subsequent work by Hakola *et al.* (2001) has shown that it and warty birch *Betula pubescens* (listed by Hewitt and Street (1992) as a non-emitter) are both mono-terpene emitters. The latter species also gives off linalool and sesquiterpenes; emissions from the trees varied during the season and with clone. Of the nine most common coniferous species, no data was available on the emission status of three (Japanese larch *Larix kaempferi*, hybrid larch *Larix x eurolepis* and European larch *Larix decidua*). However, five – Sitka spruce *Picea sitchensis*, Scots pine *Pinus sylvestris*, Norway spruce, Douglas fir, Corsican pine *Pinus nigra* var. *maritima* – were known to emit terpenes. Only Lodgepole pine *P. contorta* var. *latifolia* was shown not to emit terpenes. Isoprenes were only shown to be conclusively emitted by Sitka spruce. The lack of information on larch, at least in the UK, may now be irrelevant as Japanese larch has become infected with the pathogen *Phytophthora ramorum* (FC, 2012b; Shuttleworth *et al.*, 2012) and a steady decline in the species is expected, with the UK Forestry Commission having ceased planting and started felling on their holdings (FC, 2014).

Donovan *et al.* (2005) modelled the impact of VOC emissions of the 30 most common species of tree in the West Midlands metropolitan area of the UK (Birmingham) to develop an urban tree air quality score. The work included a survey of 32,000 trees, shrubs and hedges. The work was not restricted to street trees but the urban forest in its entirety. Of the 30 species studied, they considered alder *Alnus glutinosa*, field maple *Acer campestre*, hawthorn *C. monogyna*, laurel *Prunus laurocerasus*, Lawson cypress *Chamaecyparis lawsoniana*, Norway maple *Acer platanoides*, pine *P. nigra*, larch *L. decidua* and silver birch *B. pendula* to be the best species for improving air quality and oaks (*Quercus* spp.), willows (*Salix* spp.) and aspen *P. tremula* to be the worst. Donovan *et al.* (2005) took their VOC emission estimates from a periodically updated database of 112 references (Stewart *et al.*, undated-b) containing data on isoprene and monoterpene emissions from trees; the latest dated reference at the time of writing was 1999 and it is now probably out of date.

The development of NMHC ranking tables is clearly a useful tool for urban planners, and data is urgently needed on species from a wider range of environments and regions. Simpson and McPherson (2011) developed a BVOC index to help in the design of low-emission plantings. High VOC emission rates may appear to preclude the planting of such trees in urban areas but, as Benjamin and Winer (1998) say, trees have a range of attributes which may vary with intended function and make them more suited to one situation or planting location than another. It would be a shame if NMHC emissions become the sole determinant of planting suitability for urban areas; the ozone-forming potential of trees is but one consideration in the planning of plantings and is only relevant in areas of high $NO_x$ concentration.

As well as ensuring that trees with low BVOC scores are planted in urban areas with high $NO_x$, the best way to reduce the potential for ozone creation is to reduce both $NO_x$ levels and also VOCs at source. Sources of anthropogenic hydrocarbons in urban areas include commercial operations (e.g. industrial processes), but also unburned vehicle fuel in vehicle exhaust and evaporation from parked vehicles (Scott *et al.*, 1999). In Sacramento, motor vehicles emit 49% of reactive hydrocarbons, and 16% of emissions from vehicles (or 8.8 tonnes per day) are from when their motors are not running (Table 7.8). Such emissions

**TABLE 7.8** Sources of unburned emissions of hydrocarbons related to motor vehicles with internal combustion engines

| | |
| --- | --- |
| Diurnal loss | Daytime heating of fuel delivery systems |
| Resting loss | During constant or decreasing air temperature |
| Hot soak | During the hour after engine shut-down |
| Start loss | During the first few minutes of engine operation |

Source: Scott *et al.*, 1999.

are strongly influenced by air temperature, and in the case of 'start loss', how 'cold' the engine is.

For the majority of their 'lives', non-commercial vehicles are static and during the day parked at, or near, the owner's work or home. If parked in open car parks, they are subject to strongly varying temperature regimes. Scott *et al.* (1999) examined the potential of tree shading in reducing evaporative losses of fuel from cars using a shopping centre car park in Davis, California, which contained both shaded and unshaded parking bays. During the period of peak daytime temperatures (5–7 August 1997), shaded areas had air temperatures which were 1–2°C lower than unshaded areas. The maximum fuel tank temperatures for study vehicles were 41.6°C in unshaded bays and 2.1–3.7°C lower in shaded bays. Temperatures inside the cars reached 65°C for the unshaded vehicle and less than 50°C for the shaded. Average temperatures during the hottest part of the day, 1–4pm, in the shaded vehicles were 26.2°C cooler than the unshaded vehicle. Unshaded night-time temperatures were also warmer than shaded areas by 1°C. Modelling of the microclimate data showed that tree cover, at varying levels, could make modest contributions to lowering evaporative emissions from vehicles.

McPherson (2001) noted that about 10% of city land (in the USA) was dedicated to parking, and that some cities in the USA introduced requirements for 50% of the sealed surface of new car parks to be shaded within 15 years of development approval – in Sacramento the new rules were introduced in 1983. Setting rules is one thing, whether they work as anticipated is something else. McPherson (2001) surveyed 15 car parks in Sacramento 16 years after the introduction of the ordinance and found 13% of the sealed surface of the city land cover to be car parks; shade in these parks was estimated to be only 14% – and of that 44% was covered by built structures rather than tree cover. Modelling suggested that tree cover would increase to about 27% at the compliance point of 15-years post-approval rather than the required 50%; of the 15 car parks only two were expected to reach the 50% cover level (with one near miss of 49%). Essentially, anticipated benefits to Sacramento (in terms of aesthetics, reduced fuel evaporation, uptake of pollutants by trees, etc.) were not being delivered because of the failure to comply with the car park shading ordinance; McPherson (2001) estimated this deficit translated into US$1.4 to 2.5 million. Problems were identified in the shading ordinance itself, in training and resourcing of compliance staff, and in the advice, design, planting and management of car park trees.

## 7.10 Street trees and particulate pollution

### 7.10.1 Street trees and air quality

Vegetation can contribute to improving air quality (Litschke & Kuttler, 2008) and trees appear to be much more efficient in removing particulates from the air than grassland. Fowler *et al.* (2004) estimated that dry deposition of particulates to woodland was almost three times greater than for grassland (9 mm.s$^{-1}$ and 3.3 mm.s$^{-1}$ respectively). So, trees, along with other vegetation, capture airborne particulates that impact negatively on human health and also reduce the 'dustiness' of local streets (Beckett *et al.*, 1998; Smith, 1977). Decision support tools are now available to help planners include tree planting within air quality management plans (Bealey *et al.*, 2007; Donovan *et al.*, 2005), although some are rather crude and do not necessarily discriminate between species, capture efficiency, or whether in streets or urban parks. Choice of species may be important; for example, Lovasi *et al.* (2008) used data from 1999 on asthma in young children (4–5 years old) taken from school screening records and also hospital records of children of 14 and younger admitted for asthma in 1997 and compared them with the density of street trees from census counts in 1995 from the 42 hospital catchment areas in New York. Once confounding factors (population density and proximity to air pollution sources) had been corrected for, no link was found with hospitalisation data but there was a significant link between decreased incidence of asthma in younger children and increasing street-tree density. The findings indicated that an increase of 343 trees.km$^{-2}$ could potentially reduce asthma by 29%. Mitchell and Maher (2009) and especially Hofman *et al.* (2013) demonstrated that higher concentrations of particulates are captured at lower levels in tree canopies and this might be particularly relevant in terms of PM reduction at the pedestrian level – although this may be offset by reduced ventilation in narrow roads (see below). Lovasi *et al.* (2008) caution that there could be other confounding factors that might have affected their results, but the scale of potential health benefits indicates it could be a fruitful area of research. However, not all trees may be suitable for such plantings, especially if their pollen is linked to problems such as asthma and hay fever – as is the case for silver birch and a small number of other trees (Spellerberg *et al.*, 2006).

### 7.10.2 Capture efficiency

The majority of work on capture efficiency has concerned the leaves of trees, but bark has also been shown to be an effective part of trees for particulate capture (Becker *et al.*, 2000; Freer-Smith *et al.*, 2004). An early study of the particulate-capturing ability of London plane trees was carried out by Smith (1977) in New Haven in Connecticut; of particular interest was the following of deposition from bud burst to leaf fall, something more recent studies tend not to report. General findings showed greater particulate loads on the upper surface of leaves; particulate load increased with time, and sub-micron particles could be found adhering to leaf hairs (trichomes). In terms of leaf geography, more particulates were deposited in the 'mid-vein, centre portion' of leaves and especially just above the junction of the leaf blade with the petiole on the upper surface. Smith (1977) speculated that the observed variation in deposition on the leaf surface resulted from the effect of rainfall but also noted that trichomes, which were effective particulate collectors, were concentrated in the centre portions of both the upper and the lower leaf surfaces; trichomes also became damaged over time. In addition to natural inorganic and anthropogenic-originated particles, the leaf surfaces also removed pollen and fungal spores from the airstream.

Collectively, a settlement's street trees can have a substantial impact; for example, those of Indianapolis were estimated to capture 9,098 kg of particulates (PM10)/year and also prevented the emission of some 2,975 kg.year$^{-1}$ (Peper *et al.*, 2008). However, different tree species have different capture efficiencies. Beckett *et al.* (2000a, 2000b) demonstrated that of five species tested for particulate capture (whitebeam *Sorbus aria*, field maple *Acer campestre*, hybrid poplar *Populus deltoides-trichocarpa* 'Beaupré', Corsican pine *Pinus nigra* var. *maritima* and Leyland cypress *Cupressocyparis leylandii*), the evergreen Corsican pine was most efficient being, by far, the best at particulate capture (Figure 7.6). The performance of the other species was similar, with the exception of the poplar which appeared to be of less value than the others; however, the cypress would probably be a better choice than the deciduous species as it has foliage year-round. Beckett *et al.* (2000a) concluded that for pollution reduction, conifers would probably be 'the best choice'. Dzierzanowski *et al.* (2011) imply that this view may be overoptimistic and suggest that the long timescale of needle retention by conifers of, typically, 3+ years (Reich *et al.*, 1996) compared to deciduous species reduces the opportunity for removing adhered PMs through leaf fall. Particulates on conifer needles may be less likely to be remobilised by rain or wind compared with deciduous species, as Witherspoon and Taylor (1969) found in a comparative study of red oak *Quercus rubra* and small white pine *Pinus strobus* using quartz particles. They found that although oak was more efficient at removing particulates (fraction of particles initially retained, oak: 0.35; pine 0.24), it also lost them at a much faster rate and after only one hour had lost 90.5% of particles compared with pine's loss of 10%. The retention efficiency of the conifer may have been related to the thickness of the surface waxes (Kelsey *et al.*, 2009). Conifers tend to be much more sensitive to air pollution than deciduous species and Rhine (1924) investigated the 'clogging' of leaf stomata in conifers by a black substance and showed that it was actually a waxy deposit and not soot particles. He noted a potential relationship between the xerophytic (water retaining) nature of the plants and the susceptibility of conifers to damage by air pollution. Dzierzanowski *et al.* (2011) also point to the intolerance of conifers to air pollution, and especially, their lack of tolerance to de-icing salt, making them less appropriate for street tree plantings (see Sieghardt *et al.* (2005) for further information on salt impacts and recommendations for minimising damage).

Freer-Smith *et al.* (2004) compared the deposition velocities of particulates to five species widely used in European plantings (oak *Q. petraea*, alder *Alnus glutinosa*, ash *F. excelsior*, sycamore *Acer pseudoplatanus* and Douglas fir *P. menziesii*) and two from semi-arid areas, such as Egypt, (weeping fig *Ficus nitida* and Tasmanian blue gum *Eucalyptus globulus*). They found that deposition velocities varied considerably depending on windspeed (Figure 7.6) and that trees with smaller leaves and more complex architecture had greater capture efficiencies. However, Beckett *et al.* (2000a) and Freer-Smith *et al.* (2004) used salt (NaCl) as the model particulate source and it is possible that the behaviour of carbon-based particles in terms of capture by surface waxes differs. Later field studies (Freer-Smith *et al.*, 2005) compared the particulate capture by five tree species (*S. aria*, *A. campestre*, *P. deltoids-trichocarpa* 'Beaupré', *Pinus nigra* and *C. leylandii*) in both a rural and an urban situation and confirmed wind tunnel findings (Beckett *et al.*, 2000a; Freer-Smith *et al.*, 2004) that coniferous species were more efficient than deciduous species at particulate capture. Of particular importance was the finding that the fraction of particulates captured was highest in the ultra-fine, sub-micron range (i.e. less than 1 μm in diameter). These small particles travel further from the emission source and are particularly important because they are capable of getting deep into the lungs. These ultra-fine fine particles were predominantly soluble in water rather than being insoluble (e.g. carbon from diesel combustion) but much of the material was still predominantly from anthropogenic sources (e.g. fossil fuels: sulphates, nitrates and phosphates); natural sources of soluble materials included the sea for salt (the study site was coastal – Brighton, UK) and dust from soil for calcium; some of the phosphates, and ammonium, may have been from natural sources.

Dzierzanowski *et al.* (2011) compared the street trees field maple, ash, *Platanus x hispanica* London plane and *Tilia cordata* small-leaved lime for their ability to capture particulates in Warsaw, Poland. Trees were tested in 2007 and 2008 by washing off particulates from samples of leaf ranging from 300 to 400 m² and running the particulate-laden water through pre-dried filters designed to capture different size ranges of particulates (particles above 100 μm in size were removed in pre-treatment sieving). Samples were only taken at the end of the growing season (October) and from the side of the tree facing the road. Filters were dried and the mass of particles in the ranges 100–10 μm and 10–2.5 μm calculated. Interestingly, following water removal of particulates, the leaves were then subject to chloroform extraction to remove surface (epicuticular) leaf waxes and any particulates trapped in them. London plane was the least effective of the four species (capturing, on average, about 12 μg.cm⁻²) with small-leaved lime best (about 21 μg.cm⁻²), followed closely by field maple (just under 20 μg.cm⁻²) and with ash occupying an intermediate position (about 16 μg.cm²). The majority of the particulates found were in the larger size range. The different size ranges were retained by the leaves differently: the majority of the coarser fraction (100–10 μm) was removed by water and the remainder held by waxes (though the difference was minimal for small-leaved lime); the reverse was true for the smaller fraction (10–2.5 μm) for field maple, ash and London plane but not small-leaved lime. This means that the larger particulates can be more readily washed off leaves by rain (but also remobilised by wind) than the smaller fraction; for the latter,

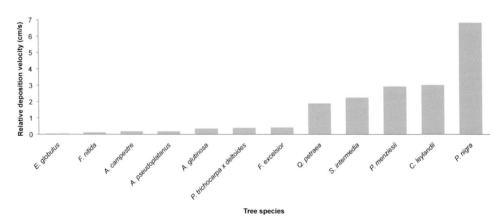

**FIGURE 7.6** Relative deposition velocities of salt (NaCl) particles to a range of deciduous and coniferous species in a wind tunnel (source Freer-Smith *et al.*, 2004). (Note: *P. nigra* = *Pinus nigra*)

incorporation in the soil/ground layer is delayed, especially for field maple and ash, until leaf fall. The study did not take into account the total leaf area presented by the different species as (say) trees of a given age or canopy dimension (e.g. volume) so the absolute differences in particulates captured per unit area of leaf may not be a good indicator of relative value *in situ*. However, as Sæbø *et al.* (2012) argue, there will be considerable variation between tree growth depending on geographical, climate and local growing conditions – making it difficult to directly compare studies carried out in different places. They suggest a standard comparison unit based on particulates/leaf area (e.g. $\mu g.cm^{-2}$) and leaf area index (Gower *et al.*, 1999); but, as tree morphology under different growth conditions and climates may also affect capture efficiency, some 'local' studies may be necessary to decide which species is most efficient for a given area.

A much more extensive test of the relative efficiency of particulate capture was carried out by Sæbø *et al.* (2012) in Norway and Poland on 22 tree species in experimental nursery plantings. Whilst the experimental trees were not grown on streets, they are species that the authors considered to be commonly used in street plantings. The study site in Stavanger, Norway, was near a busy motorway, but that in Poland was in a more rural location. Leaves were collected for analysis in the autumn just prior to leaf fall of the deciduous species. The procedure for removal of particulates was the same as for Dzierzanowski *et al.* (2011) except that an additional, very fine, fraction was collected giving the ranges 100–10 μm, 10–2.5 μm and 2.5–0.2 μm. In addition, data on waxiness, leaf dimensions and surface characteristics (hairiness and roughness) were recorded. The highest PM accumulators in Norway were silver birch and Scots pine (Table 7.9). The best species, in terms of the weight of very small particles (2.5–0.2 μm) captured, was silver birch and 83% of that fraction was held in the leaf waxes – effectively preventing remobilisation by wind or rain. In Poland, silver birch was the highest overall particulate accumulator, trapping 34 $\mu g.cm^{-2}$ (Scots pine was not part of the Polish study). The best species at holding the 2.5–0.2 μm fraction was, again, silver birch but yew *Taxus baccata*, hybrid alder *Alnus spaethii* and the Callery pear *Pyrus calleryana* were also strong accumulators (Table 7.9). Overall, the species with the highest wax levels on their surfaces tended to have the highest particulate accumulation; the dry weight/unit area of leaf and leaf surface hair density were also good predictors of the total weight of particulates captured in the 100–10 μm and 10–2.5 μm ranges, but not the 2.5–0.2 μm range. Sæbø *et al.* (2012) also found that particulate capture differed between years – possibly due to weather differences. Given that the smaller particulates are the most damaging to human health, plants which immobilise them via the surface waxes may be particularly valuable as street trees.

Studies based on end-of-year (autumn) particulate levels are likely to give a good feel for the cumulative load of particulates trapped in waxes, especially for species with very waxy surfaces; they may be less good at estimating the overall particulate capture capacity of different species, as particles that are not held by waxes can be remobilised. Remobilisation by rain means that particulates should largely be captured on (mainly) the soil surface or enter the drainage system, by wind they are returned to the air where they add to the pollution load or are re-impacted on other surfaces. Wind tunnel trials on 0.4 m high Norway spruce saplings, using 1.85 μm particles of silica doped with a dysprosium tracer as model particulates, indicate that resuspension by wind is relatively low and unlikely to be a hazard to humans or substantially add to particulate loads on other surfaces (Ould-Dada & Baghini, 2001). Over a season, the removal of surface particulates by rain may be substantial and, by

**TABLE 7.9** Approximate* mean total particulate accumulation performance in 22 tree species in Norway and Poland over a two-year period

| Species | Norway | | Poland | |
|---|---|---|---|---|
| | *2009* | *2010* | *2009* | *2010* |
| *Acer campestre* | | | 20 | 19 |
| *Acer platanoides* | 7 | 6 | 9 | 10 |
| *Acer pseudoplatanus* | 24 | 8 | 7 | 5 |
| *Aesculus hippocastanum* | 23 | 11 | | |
| *Alnus x spaethii* | | | 14 | 17 |
| **Betula pendula** | **25** | **25** | **38** | **21** |
| *Carpinus betulus* | 24 | 16 | | |
| *Fagus sylvatica* | 15 | 18 | | |
| *Fraxinus excelsior* | | | 13 | 4 |
| *Malus 'Van Eseltine'* | | | 10 | 18 |
| **Pinus sylvestris** | **34** | **72** | | |
| *Populus tremula* | 14 | 15 | 15 | 15 |
| *Prunus avium* | 12 | 7 | | |
| *Prunus padus* | 13 | 12 | | |
| *Pyrus calleryana* | | | 20 | 15 |
| *Quercus robur* | | | 17 | 19 |
| *Robinia pseudoacacia* | | | 9 | 3 |
| *Sorbus x intermedia* | | | 23 | 12 |
| *Taxus baccata* | | | 9 | 21 |
| *Tilia cordata* | 7 | 8 | 7 | 4 |
| *Tilia x europea* | | | 19 | 14 |
| *Ulmus glabra* | 29 | 20 | | |

*Data taken from graphs in Sæbø *et al.* (2012) and should be taken as indicative rather than exact values. The groups identified by cluster analysis as best accumulators are in bold and outlined (mean of two years: 24–55 µg.cm$^{-2}$), worst (darkest shading, 6–13 µg.cm$^{-2}$) and those in between (lightest shading, 14–23 µg.cm$^{-2}$). Data from Sæbø *et al.* (2012).

cleaning the leaf surface, increase its overall capture efficiency. Having said that, we simply do not know very much about the dynamics of particulate capture, and particulate-laden leaves may even be more efficient in capturing additional particles by providing aggregation niches – although the photosynthetic efficiency of the leaf would be likely to decrease with increasing particulate load and phytotoxic effects increase.

The geometry of streets and particulate capture has recently been examined by Hofman et al. (2013) who addressed concerns that in narrow streets tree crowns may inhibit dispersal of particulates by reducing air speed and increasing particulate concentrations – the so-called 'reduced ventilation hypothesis'. Their work confirmed the capture of particulates by street trees but also provided evidence to support the view that trees might indeed slow the dispersal of pollutants by dilution in the air. Their suggestions for narrow street canyons included using trees with narrow crowns, ensuring wide spacing between trees and also between trees and adjacent walls. This advice assumes that such roads are heavily used by traffic, but in pedestrianised areas or streets with very low traffic volumes other attributes of trees such as shade and aesthetics may be more important such that closer spacings and wider crowns are preferred.

## 7.11 Trees and heavy metals

Heavy metals are toxic to humans; lead, for example, is a neurotoxin (Maher et al., 2008). The ability of trees to act as sinks for toxic heavy metals has been known for some time (Greszta, 1982) and reflects the use of such elements as essential micro-nutrients by trees (Riddell-Black, 1994). Lu et al. (2006) showed that the heavy metals found in particulates on the leaves of street trees in Hangzhou, China, included cadmium, copper, zinc and lead along with iron, exceeded background levels and originated from industrial activity and vehicle exhaust. In Rome, Gratani et al. (2000) showed that leaves from evergreen holm oak Quercus ilex trees subject to varying levels of traffic density had accumulated particles containing lead, strontium, iron, copper, zinc and aluminium. Ugolini et al. (2013) working in Florence, Italy, also demonstrated uptake of lead, iron, manganese, chromium and barium by Q. ilex from traffic pollution, but additionally demonstrated that the retention of old leaves was particularly useful: increasing the leaf area available for collection. Younger leaves also tended to be higher in copper whilst older leaves had higher lead levels. Leaves of P. laurocerasus have been shown by Fumagalli et al. (2013), working in Varese, Italy, to collect particles containing platinum, palladium and rhodium from vehicle exhaust. The source of the metals was from catalytic converters, in particular those from diesel-fuelled engines. Leaves are not the only parts of a plant able to capture metals: Becker et al. (2000) showed that tree bark was also capable of capturing platinum from catalytic converters and uptake via the roots has also been documented (Riddell-Black, 1994). Despite the removal of lead from petrol in 1986 in the UK, Maher et al. (2008) reported the element still turning up in dust on roadside tree leaves in 1999 (along with iron, zinc, manganese and barium). Simon et al. (2011) tested for the presence of 19 elements on A. pseudoplatanus leaves in Vienna, Austria, and found six (aluminium, barium, iron, lead, phosphorus and selenium) in significantly higher concentration in urban Vienna compared with rural areas. Soils under trees have been shown to have substantially higher levels of lead than grassland soils (Fowler et al., 2004). Lead on leaves could be from remobilised soil particles from previous emissions and Maher et al. (2008) also suggested lead in fuel tanks, fuel hoses, piston rings, valve seats and spark plugs as potential sources. They also took samples of leaves at various heights above ground and noted that the highest concentrations were about 0.3 m above ground, where young children are more easily exposed. Concentrations were highest on the uphill sides of roads and lowest on downhill sides.

The study by Simon *et al.* (2011) also demonstrated that the leaf tissues of *A. pseudoplatanus* grown in urban Vienna contained higher concentrations of manganese and strontium than leaves of trees in the rural surroundings. Whist the uptake of heavy metals by trees is beneficial from a human point of view, by removing them from the general environment, trees do experience a cost. For example poplar (*P. deltoides*) from metal-contaminated sites may have reduced establishment and growth rates and potentially require more maintenance (Renninger *et al.*, 2013). Heavy dust from a cement factory has been shown to have an impact on walnut *Juglans regia* growth in India (Wani & Khan, 2010). Chlorophyll and ascorbic acid (vitamin C) levels have been shown to drop and rise, respectively, with increasing dust levels in a number of Indian species (sacred fig *Ficus religiosa*, Indian banyan *F. benghalensis*, mango *Mangifera indica*, Indian rosewood *D. sissoo*, guava *Psidium guajava* and male bamboo *Dendrocalamus strictus*) grown along roads in the city of Varanasi (Prajapati & Tripathi, 2008). The vitamin C increase was probably a response to the stress imposed by the physical or chemical properties of the dust (Prajapati & Tripathi, 2008). Gratani *et al.* (2000) showed that *Q. ilex* trees from areas of high and medium traffic levels in Rome had premature senescence with primarily one-year-old leaves and only 5% being two years old; leaves taken from trees growing in green spaces, which acted as controls, had a different age structure with 77% of leaves being one year old, 20% two years old and even some at three years old (3%). The younger leaf-age structure in trees from polluted areas was compensated for by having larger leaves and more shoots compared with those growing in green spaces. Velikova *et al.* (2011) studied the effect of nickel uptake by poplars *Populus nigra* and showed reduced photosynthetic activity especially in newly developing leaves (which had higher nickel concentrations than mature leaves). Nickel-related stress resulted in increased emissions of the volatiles linalool, isoprene 2-methyl-1,3-butadiene (a compound known to have a protective effect against a range of plant stresses such as heat, ozone and 'other reactive oxygen species' (Sharkey *et al.*, 2008)) and cis-ß-ocimene; isoprene only increased in young leaves, whilst ocimene was emitted in greater quantities from mature leaves. Not every tree species incorporates heavy metals into woody tissues: Riddell-Black (1994) note that some deciduous tree species shunt metals taken up from the soil to the leaves so they can be voided when leaves are shed, although most accumulate heavy metals in their roots – some such species retain substantial concentrations and are known as 'hyperaccumulators'. In known 'hot spots' for particulates rich in heavy metal contamination, the planting of trees that are particularly efficient at scavenging heavy metals may be worth considering. However, Riddell-Black (1994) considered that root uptake by *Salix* species was such that plantings were best done in large areas with relatively low levels of contamination – which clearly does not mesh with streetscape design. The willows in his work (*Salix viminalis* 'Bowles hybrid', *S. viminalis* 'SQ683', *S. viminalis x triandra* 'Q83' and *S. dasyclados*) were also of the type used in short-rotation coppice and their morphology does not make them suitable as street trees – but they might be worth considering on parts of large reclamation sites as part of an urban regeneration strategy.

## 7.12 Street trees and water

Street trees, along with other urban vegetation, can be considered to be components of sustainable urban drainage: intercepting rainfall, reducing peak loads, improving infiltration, acting as an evaporative surface, and actively removing water from the ground by transpiration (see section 2.8). Peper *et al.* (2008) quantified the value of Indianapolis's street trees in terms of stormwater reduction at 1,207,258,099 litres/year. In financial terms this was a saving of US$1,977,467 a year at the city level or US$16.83 per tree per year.

Livesley *et al.* (2014) compared the interception characteristics of two eucalypt species (*Eucalyptus nicholii* and *E. saligna*) in Melbourne, Australia, over a five-month period. The two species differ in canopy density with *E. nicholii* being more dense and with a rougher bark than *E. saligna*. The more dense canopy of the former species meant that it could capture more rain than the latter, especially the lighter intensity rain, such that it was capable of intercepting 44% of annual rainfall compared with 29% for *E. saligna*. Bark characteristics also impacted on flow down woody elements with the rough-barked *E. nicholii* transmitting less than *E. saligna*. In both cases the water transmitted to ground level via bark was relatively trivial, but whilst unimportant in a runoff limitation context, water delivery in this way is important for tree health. The work showed quite clearly that tree choice would have a direct impact on street trees' contribution to sustainable urban drainage and that trees with a denser canopy are superior in this context.

Day and Dickinson (2008) reported on work at Virginia Tech, Cornell University and the University of California at Davis which investigated how structural soils used for street trees could be used to store runoff which could then be both infiltrated and taken up by trees. Essentially the use of structural soils, designed to be load-bearing whilst providing large pores for root growth, was not restricted to the tree pits but extended underneath much larger areas – such as car parks. The concept integrated the multiple values of trees (interception, evaporation, delay of peak flows and uptake/removal via transpiration) and infiltration storage in structural soils. The much greater extent of the structural soils allows for much greater lateral tree root growth which, in turn, promotes the growth of larger canopies – a feature desirable where trees are used as SUDS components. Day and Dickinson (2008) suggest that the diameter of the root zone should be twice that of the desired canopy. The large pores in the structural soils also act as attenuation storage for stormwater. Provided that water in the structural soils is infiltrated/transpired off within two days, the trees do not suffer damage from root anoxia. The roots also create channels for water to disperse through (Day & Dickinson, 2008). Suspended pavement systems such as Silva Cells are also promoted as having similar values for stormwater storage (Deeproot, 2010).

## 7.13 Street trees and biodiversity

### 7.13.1 Native vs 'exotic' species

For biodiversity it has generally been considered best practice to use native trees and preferably those of local provenance in plantings. However, recent work by Quine and Humphrey (2010), working in UK forest ecosystems, suggest that this is an oversimplification and that exotic species may have biodiversity benefits – but that community structures and species composition of associated plants and animals may well be different. Given that the work of Quine and Humphrey (2010) reported species richness and not species composition, the view that native species and local ecotypes should be encouraged in plantings to support native species that feed on them is likely to continue.

Spellerberg and Green (2008) were surprised that biodiversity/planting native species was not a major criterion in urban tree selection in general. In New Zealand most plantings were primarily non-native for the location. Indeed, the use of natives was often actively resisted: probably due to the northern-hemisphere-oriented training of horticulturalists and arbori-culturalists, and the colonial heritage of New Zealand. They cite some exceptions to the non-native planting 'rule' that they found, including a deliberate policy in the development of Warrington in the UK, when it was given 'New Town' status, to ensure that the primary drive for plantings was to ensure they were naturalistic (Gustavsson, 1995). For New Zealand, they urge the reappraisal of selection criteria such that there is an emphasis on planting native species wherever possible to provide habitat for native fauna.

In the UK, Britt and Johnson (2008a) found only 23% of local authorities had a formal policy that promoted native species plantings over exotics although 'a small majority' did so in practice. Old trees are especially valuable as they will be habitat for saproxylic, or dead-wood-exploiting, species but because of concerns over safety they are unlikely to be retained in urban situations (Barkham, 2007; GLA, 2007).

### 7.13.2 Trees as components of biodiversity

Biodiversity can be considered to be an ecosystem service delivery system as well as an ecosystem service in its own right, and it is easy to forget that street trees are themselves elements of biodiversity. Whilst streetscapes can be very rich in tree species, Seattle, for example, having over 300 species (McLain et al., 2012), they tend to be dominated by a small number of common species and are usually less species rich than urban woodlands (Sjöman et al., 2012c).

In their study of trees in streets and parks of ten Nordic cities, Sjöman et al. (2012c) found that non-native species richness was higher in the streets of seven of the cities, but in Oslo and Tampere the street-scene was dominated by native species. In terms of tree numbers, however, natives dominated in the streets of all cities (Table 7.2). For Bangalore, non-natives dominated the streets (Nagendra & Gopal, 2010).

### 7.13.3 Trees as habitat

It is well known that different species of tree support different invertebrate assemblages (Kennedy & Southwood, 1984; Southwood, 1961; Southwood et al., 1982) (Table 7.3). Southwood (1961) correlated species richness of insects on trees (from inventory data) with length of time since a tree species had colonised Britain and its relative abundance, but noted that some species were exceptions to this generalisation with poor insect complements, probably because of either physical or chemical defences, and some had more species than expected. Southwood et al. (1982) made a complete inventory of the arthropods on six tree species in Britain and South Africa using a 'knock-down' technique using the insecticide pyrethrum. They observed that the arthropod fauna in all feeding guilds, except the herbi-vores, was similar whether the trees were native or non-native; with herbivores there were more species on native trees. Kennedy and Southwood (1984) in an updating and reanalysis confirmed that abundance and time since colonisation of a tree increased its herbivorous species richness (insects and mites) but also that evergreen or coniferous species had reduced species richness compared to deciduous species. Helden *et al.* (2012) compared native and non-native species in their ability to support insect assemblages, in this case from the Hemiptera (bugs), which Southwood *et al.* (1982) noted contained the dominant herbivorous species on native species of tree, and insectivorous birds (Paridae) in Bracknell, UK. Street trees were not used, as such, but samples came from 13 roundabouts, two areas on roadsides and two parks; in the latter only edge trees were sampled. The results were not clear cut, with native and non-native trees showing no significant differences in hemipteran richness when aggregated. However, at the individual tree species level more hemipteran species were present on native than non-native trees, and also the higher the proportion of native trees in an area the higher the hemipteran and bird abundance (Helden *et al.*, 2012). In this study, Helden *et al.* (2012) used beating as a sampling methodology, whereby tree branches were vigorously knocked with a stick and the dislodged insects caught in a sweep net with an opening of 0.27 m² held below the foliage. Whilst this technique is perfectly acceptable, differences in branch structure and density per unit area sampled may have resulted in higher data variability than, for example, Southwood *et al.* (1982) would have experienced with insecticide knock-down.

## 7.14 Societal value

Nagendra and Gopal (2010) in their list of services provided by trees mention not only that the shade cast is important for pedestrians and street vendors, but also that trees may provide the only accessible green space for many city dwellers, especially those in the middle-low income bracket. Larsen *et al.* (2009), in a study of schoolchildren aged 11–13 from 21 schools in London, Ontario, showed that 62% travelled to school, and 72% to home from school, through physical activity rather than via motorised transport. The majority of those walked (94.2% going; 93.4% returning) rather than cycling, using a scooter, skateboarding or roller-blading. As might be expected, the distance to school from home was the most important influence on the use of motorised vs non-motorised transport; males were also less likely to travel via vehicles than girls. Of environmental variables, only the land-use mix and the presence of street trees along the route increased the likelihood that children would walk or cycle (etc.) to school, although this was not apparent on the return journey. Todorova *et al.* (2004) found that street plantings of flowers were considered to improve the aesthetics of the streetscape and feelings of well-being, but particularly when they were planted in association with trees, in long rather than short plots, and in low, bright and formal plantings with blocks of single species. Interestingly some respondents to their photomontage/questionnaire survey indicated that having plantings between the walkway and road improved their feeling of safety.

## 7.15 Summary

Trees have an extremely wide range of documented values to humans, and especially relating to human health, potentially providing the following:

• Aesthetic value, e.g. greenery, blossom.

- Food – primarily in the form of fruit, but also potentially nuts.
- Shade – from direct sunlight, from ultraviolet radiation and providing cooler conditions.
- Opportunities for socialisation (arising from shading effects), but also from traditional activities such as playing 'conkers'.
- Improved air quality:
  - through directly removing gaseous and particulate pollutants from the air;
  - through reducing hydrocarbon emissions by shading fuel tanks and reducing evaporation;
  - by promoting walking over motorised transport;
  - by reducing the need for energy consumption from heating and air conditioning.
- Reduced flash flooding by intercepting rainfall.
- Habitat for wildlife.

However:

- Most urban areas do not have enough trees.
- Trees are often seen as a cost, not as providing multifunctional benefits.
- The urban environment is hostile to trees requiring careful planting preparation, species selection and maintenance.
- In areas with heavy pollution with nitrogen oxides, tree planting should not include species which emit high levels of VOCs.
- Some species' pollen can act as allergens.
- Future plantings need to be evaluated for their ability to withstand climate change.
- New pests and diseases are constantly emerging which threaten street trees – diversifying the species mix should provide more resilience to such attacks.
- Health and safety fears may be overplayed unnecessarily causing tree removal especially compared with the health and safety benefits.

# 8

# POLICY, REGULATION AND INCENTIVES

This chapter briefly considers the value of policy, incentives and regulations in encouraging the use of green infrastructure elements.

## 8.1 Public good vs Private good

Whilst we have seen that using a green infrastructure approach can save money (e.g. heating costs or public health) and even increase value (e.g. sales and rental value of buildings), there is a question of who makes those savings or gains those benefits (and who bears the costs). There are clearly two types of beneficiary: individuals and commercial organisations (Private), and Society (Public) (Perini & Rosasco, 2013). Many of the ecosystem services that are delivered through a green infrastructure approach (e.g. moderating local climate, reducing pollution, conserving pollinators) are not the kind of values that would motivate a business and can be classed as primarily Public benefits. Other services are clearly of direct Private benefit: such as a reduction in heating or air-conditioning costs, faster sales/take-up of commercial premises, and one can see how, provided the return was clear, Private interests would find them attractive.

Whilst we can divide benefits into a crude Public:Private dichotomy, the reality is more complicated: some benefits are shared. So, for example, Society benefits when Private organisations save money on energy, because less less $CO_2$ and heat is wasted: thus reducing the impact on the local and global climate. Likewise, Private organisations may benefit through improved recruitment or customers if the Public realm implements a planting scheme which improves (say) air quality or the aesthetics of a neighbourhood. The Public realm also has its own estate and both Society and Private organisations benefit through savings and reduced taxes if the Public administration buildings, for example, reduce energy costs and release less $CO_2$. The Public realm has a big role to play in using its resources to implement green infrastructure, and to act in a leadership role by encouragement and facilitation, and by creating demonstrators (e.g. public buildings with green roofs). However, the Public realm cannot make the scale of change required on its own.

The big question is thus: how to encourage the Private sector to carry out works which are largely for the Public benefit? Given that most businesses are not created to generate Public goods, but to generate owner and shareholder value, Society must create an environment whereby there is a reason for businesses to engage in such a way. One powerful motivator is Public image. Organisations often wish to show how they are conscious of the importance of Society and engage with its concerns. In more 'corporate speak' terms this is 'Corporate Social Responsibility' or 'CSR' (Fernandez-Feijoo et al., 2014). For large organisations this is an increasing area of interest and green infrastructure is now featuring as a component in CSR reports. In the UK, Barclays Bank installed a green roof on the top of the 32 floor, 160 m high, office complex at 1 Churchill Place in Canary Wharf, London. The value to Barclays had a large CSR component: biodiversity, staff engagement, brand and reputation – and it did a good job for them, they won the 2004 National Business Award for CSR and gained a lot of good publicity (Patel, 2006). The green roof also featured in their 2005 CSR report (Barclays, 2006). However, CSR is voluntary and it depends on the interests of the individual organisation – it may not be consistent either: with different areas attracting the organisation's attention from year to year. As a result, a different approach will probably be necessary where Society wishes to influence Private behaviour.

## 8.2 Creating the environment for implementation

As we have seen with sustainable urban drainage (section 2.8), definitions and policy are important. Voluntary inclusion of green infrastructure elements in high-profile landmark developments can be visually stunning and inspire other developers and architects, but it does not necessarily make it easier to ensure that such features are included in the more mundane, run-of-the-mill building projects. I lived in Cologne, Germany, for six months over the winter of 1985 to 1986 and was astounded at the extensive use of UPVC (PVCu) windows with double-glazing; back home in Southampton, England, it was something of a rarity with most houses being single-glazed. Now, in the UK, all new builds have to be double-glazed, and most replacements as well, on energy efficiency grounds (FENSA, undated). If double-glazing had been left to purely voluntary action, it is likely that many buildings now with double-glazing would have remained single-glazed.

The history of green roofs (section 6.1.3) shows the value of introducing standards for confidence in a technology (FLL, 2008), but especially the introduction of planning regulations and incentives for effective implementation. In Germany, a building code was introduced in the 1980s that required green roofs on blocks of flats in the city centre of Berlin and financial incentives, provided from 1983 to 1996, assisted in the installation of some 63,500 m² of extensive roofing. Unfortunately, the incentive programme was discontinued when Germany (and Berlin) was reunified, but green roofs are still compulsory for large developments (Köhler, 2006). Building regulations were introduced in Linz in 1985 for green roofs (LivingRoofs, 2014; Maurer, 2006); in 1989 the city started giving subsidies (up to 30% of costs), although these were reduced in 2005 to 5%. The results have been dramatic and almost certainly would not have been achieved without regulation and incentives (Figure 8.1). Examples of other areas with regulation/incentives include: Stuttgart (grass roofs on industrial buildings in 1989) (Johnston & Newton, 2004); Mannheim (as a SUDS component); and Vienna, Austria (financial incentives) (Peck et al., 1999). Since 2002 it has been compulsory to install green roofs on new builds with flat roofs in Basel, Switzerland.

This followed research into their potential biodiversity value which provided the evidence base for the development of the land-use regulation changes (Brenneisen, 2006; Kazmierczak & Carter, 2010). For roofs over 500 m² in extent, the growing media have to be composed of soil from the local area and vary in depth (Brenneisen, 2006). There is now a significant policy drive in favour of greening roofs in Basel with a triple-track approach: incentives from an energy saving fund (garnered from a 5% levy on energy bills), grants for research into benefits, and changes to the building regulations; there have even been 'best looking' green roof contests (Kazmierczak & Carter, 2010)!

In 2009 Toronto, Canada, became the first North American city to make green roofs compulsory (Toronto Municipal Code Chapter 492 Green Roof). The introduction was phased in with the requirement affecting new commercial and residential developments of at least 2,000 m² from 31 January 2010 and new industrial developments from 30 April 2012. The proportion of roof that had to be greened varied depending on the building's footprint: from 20% of the roof for 2,000–4,999 m², rising in 5,000 m² steps up to 60% for buildings of 20,000 m² or more (CoT, 1998–2014). In 2011 in the UK, Sheffield introduced supplementary planning guidance on green roofs (section 6.1.3), but this policy can be 'trumped' by others and it remains to be seen how effective it will be.

Incentive systems are important: Berlin's 'Courtyard Greening Programme' carried out between 1983 and 1996 gave incentives for installing green roofs, green façades and community gardens. Over the project lifetime, 185 courtyards were improved and in total

FIGURE 8.1 Impact of the introduction of green roofs in Linz, Austria. © Edmund Maurer, Municipality of Linz

54 ha of green roof and garden was created and 32.5 ha of walls greened. The total subsidy provided was €16.5 million and equated to just over €19 per m$^2$ (Johnston & Newton, 2004; Kopetzki, 2010). The cities of Groningen, Rotterdam, Amsterdam, The Hague, Berlin, Bonn, Munich, Stuttgart, Copenhagen and Linz all have green roof incentive schemes (Prokop et al., 2011). But a 'stick' approach, the use of fees, is also used to effect change. In Berlin there is now a tax on sealed surfaces because of the impact on runoff and this approach has resulted in reduced sealed surfaces with new builds and when areas are renovated (TCPA, 2004). A combination of tax and incentive has been taken up elsewhere in Germany as a 'Rain Water Tax' with tax incentives for use of permeable surfaces, although Prokop et al. (2011) consider that the incentives are often set too low. The example they gave was from Wuppertal in Germany: €3 annually for switching a parking area from sealed to permeable. They suggested encouraging rainwater harvesting might make more economic sense (reducing water flow to sewers and thus making it available for irrigation). The rainwater tax approach has been adopted outside of the EU. For example, in Maryland, USA, the Atlantic General Hospital in the town of Berlin had a yearly bill in excess of $5,000 for rainwater discharge from its 9.1 ha (24 acre) site (WBOC, 2013). Whilst homes in Berlin, Maryland, pay a flat rate of $50, organisations pay on an area basis (EFC, 2012) and bills, as the hospital have found, can be substantial. Reducing the amount of sealed surface in a landholding to reduce annual taxation is clearly a sensible approach for organisations. For householders, planning regulations may be more effective: the UK has introduced new laws to prevent the sealing of front gardens to provide car parking space (DCLG, 2008).

It is interesting that policy can also drive a change in attitudes, as long as the background work has been done. According to Seamans (2013), the relatively recent change of viewing vegetation as primarily an aesthetic issue to one of considering it an infrastructure component that delivers public goods (i.e. ecosystem services) in the USA has been delivered primarily by changes in policy backed up by a scientific evidence base that has been effectively disseminated. The green roof incentive and regulation history shows just how important regulation is in delivering a desired result (more green roofs), and also the security it brings to product developers in knowing that a market is stable and is worthy of investment. A self-regulating, laissez-faire approach to the introduction of green infrastructure will probably not deliver the enhancement of ecosystem services that are becoming increasingly urgent and important in towns and cities.

Long-term planning is important for most public infrastructure, but especially so for green infrastructure. The length of a development cycle can be very short, perhaps lasting as little as 30 years for some buildings from starting to build to eventual demolition, but trees can live much longer. Because of this Kelly (2012) recommends that developers should be encouraged to plant large trees. However, large trees cost far more than smaller trees and establishment can take much longer. Perhaps it is far better to have a long-term green infrastructure plan and pre-plant trees wherever possible so that some mature vegetation remains outside individual development parcels (whilst also encouraging developers to plant trees) no matter what the point in any given development cycle. Jaluzot (2012) stresses the need to integrate tree-related issues (standards, protection, care, planting, replacement, removal) within local plans, key policy documents and investment documents.

Given the importance of local control of the planning process, there is one really outstanding example of best practice. This comes from the Bo01 development in the Västra Hamnen or Western Harbour area in Malmö, Sweden. Two complementary approaches were used for this development: the introduction of a 'Green Space Factor' modelled on the 'Biotope Area Factor' used in Berlin and Hamburg (Kopetzki, 2010) and 'Green Points' (Kruuse, 2004, 2011). The Green Space Factor (GSF) is primarily a 'quantity' approach and the Green Points a 'quality' approach. Essentially the Green Space Factor is used to encourage developers to incorporate as much permeable, rather than sealed, surface in their developments as possible. The planning authority determines the overall Green Space Factor to be achieved. In the case of the Bo01 development, which is an urban centre district, this was to be 0.5 (i.e. half of a development's footprint should be permeable), whereas more outlying areas might be expected to achieve a GSF of 0.7. Different types of surface were given different weightings (GSF Bo01 column in Table 8.1). The total GSF was calculated on an area basis using the following formula:

$$\text{GSF} = \frac{((\text{m}^2 \text{ of area A x factor A}) + (\text{m}^2 \text{ of area B x factor B}) + (\text{m}^2 \text{ of area C x factor C}) \text{ etc.})}{\text{Total property area (m}^2)}$$

The Bo01 development was acclaimed for its environmental values but criticised for the expensive purchase and renting costs and so a rather looser version of the GSF was developed and used in the Flagghusen development, still within the Western Harbour area. Unfortunately the results were disappointing and the GSF was tightened up again for subsequent developments (see the GSF 2011 column in Table 8.1) (Kruuse, 2006, 2011).

The 'Green Points' system that was used in the Bo01 development was composed of 35 separate actions, and developers had to pick a minimum of ten. These actions delivered a range of environmental goods although the emphasis was on aesthetics, biodiversity and sustainable urban drainage (Table 8.2).

The UK's North West Regional Development Agency adapted the Malmö Green Space Factor and Green Points approach to ensure that any developments funded in its area under the European Regional Development Fund (ERDF) produced high-quality green infrastructure (Kruuse, 2011). The toolkit itself is based on an Excel spreadsheet and is freely downloadable (GINW, 2014). The spreadsheet has two sections: one for areas which have already been developed and one for those that have not (for other north-west related documents see also NENW (undated-a, b, c, e) and Barton & Jones (2009)). The spreadsheet includes suggestions for green infrastructure components (e.g. green roofs, permeable and semi-permeable surfaces, green walls, trees, etc.) which are linked to 11 economic values (Table 8.3).

But of course, implementing a green infrastructure approach costs money. However, as we have seen with green roofs, once regulations are enacted, such as in Berlin, Germany, a marketplace has something of a level playing field. Once a big enough market emerges, the costs of required actions are reduced by economies of scale in manufacture, architects learn how to design features in, and the increase is passed on to the customer or absorbed by the developer depending on the economic circumstances. Schemes such as BREEAM (Building Research Establishment Environmental Assessment Methodology) in the UK and LEED (Leadership in Energy & Environmental Design) in the USA are exceptionally useful (BRE, 2010–14; USGBC, 2014). Such schemes assess and rate the environmental performance of buildings and can play an important part in encouraging better-quality buildings and the inclusion of green infrastructure components that contribute to their scoring systems. For BREEAM, ratings are 'Unclassified', 'Pass', 'Good', 'Very Good', 'Excellent'

TABLE 8.1 The Green Space Factor as used in the Bo01 development in the Western Harbour district of Malmö, and as later modified

| Type of surface | *GSF Bo01 | *GSF 2011 |
|---|---|---|
| Vegetation: area where the plant roots have direct contact with deeper soil layers, and water can freely percolate to ground water level. | 1 | 1 |
| Vegetation (on beams): area where the plant roots do not have direct contact with deeper soil layers; e.g. on top of an underground car park. Soil depth less than 800 mm. | 0.6 | 0.7 |
| Vegetation (on beams): area where the plant roots do not have direct contact with deeper soil layers; e.g. on top of an underground car park. Soil depth more than 800 mm. | 0.8 | 0.9 |
| Green roofs, brown roofs, ecoroofs: calculated for the real area covered by plants, not the area of the roof as projected on the ground surface. | 0.8 | 0.6 |
| Open water in ponds, trenches and so on: the area should be under water for at least six months/year. | 1 | 1 |
| Non-permeable areas, including the house built on the plot. | 0 | 0 |
| Stone paved areas, with joints where water can infiltrate. | 0.2 | 0.2 |
| Semi permeable areas: sand, gravel, etc. | 0.4 | 0.4 |
| Green walls: climbing plants with or without support. The area of a wall that can be expected to be covered by vegetation within five years. Maximum calculated height: 10 metres. | 0.7 | 0.7 |
| Trees with a stem girth of more than 35 cm: calculated for the maximum area of 25 $m^2$ for each tree. | 0.4 | |
| Shrubs higher than three meters: calculated for the maximum area of 5 $m^2$ for each shrub. | 0.2 | |
| Collection and retention of stormwater. | | 0.2 |
| Draining of sealed surfaces to surrounding vegetation. | | 0.2 |
| Tree, stem girth 16–20 centimetres (20 square metres for each tree). | | 20 |
| Tree, stem girth 20–30 centimetres (15 square metres for each tree). | | 15 |
| Tree, stem girth more than 30 centimetres (10 square metres for each tree). | | 10 |
| Solitary bush higher than 3 metres (2 square metres for each bush). | | 2 |

Sources: Kruuse (2006, 2011).

*GSF Bo01: the Green Space Factor as originally used in the Bo01 development; GSF 2011: the modified version introduced after a more relaxed planning approach to that in Bo01 lead to poor environmental outcomes in the Flagghusen development.

**TABLE 8.2** The Malmö Green Points System with estimated primary ecosystem service(s) delivered (★★) and additional values (★)

| Point No. | Type of feature | Biodiversity | SUDS | Food | Aesthetics | Recycling |
|---|---|---|---|---|---|---|
| 1 | Each flat provided with a bird box | ★★ | | | ★ | |
| 2 | Habitat for specific insects provided in the courtyard (aquatic insects in a pond) | ★★ | ★ | | ★ | |
| 3 | Bat boxes in the courtyard | ★★ | | | | |
| 4 | All courtyard ground surfaces are permeable | | ★★ | | ★ | |
| 5 | Unpaved areas must have soil deep enough to grow vegetables | ★ | | ★★ | ★ | |
| 6 | The courtyard has a country garden divided into different sections | | | ★★ | ★★ | |
| 7 | All walls, where practicable, have climbing plants | ★★ | | | ★★ | |
| 8 | For every 5 m² of sealed surface in the courtyard there is 1 m² of pond | ★★ | ★ | | ★★ | |
| 9 | Courtyard vegetation must be rich in nectar and provide food for butterflies | ★★ | | | ★ | |
| 10 | No more than five of the same tree or shrub species to be planted | ★★ | ★ | | ★★ | |
| 11 | Courtyard vegetation is designed as a wetland | ★★ | ★ | | ★ | |
| 12 | Courtyard vegetation is designed as a dry biotope | ★ | | | ★★ | |
| 13 | Courtyard vegetation is designed to be semi-natural | ★★ | ★ | | ★ | |
| 14 | Stormwater runs for at least 10 m on the ground surface before entering piped drainage | | ★★ | | ★ | |
| 15 | The courtyard is fully vegetated, but not composed of mown lawns | ★ | ★ | ★ | ★ | |
| 16 | Rainwater is harvested from buildings and sealed surfaces and used for irrigation | ★★ | ★★ | ★ | ★ | ★★ |
| 17 | All plants grown have some household value | | | ★★ | | |
| 18 | Habitat for frogs is provided in the courtyard including hibernacula | ★★ | ★ | | ★ | |
| 19 | The courtyard has a minimum of 5 m² of conservatory per flat | | | ★ | ★ | |
| 20 | The courtyard provides bird food throughout the year | ★★ | | | ★★ | |
| 21 | For every 100 m² of courtyard there are at least two old varieties of fruit and berry | ★ | ★ | ★★ | ★ | |
| 22 | Building façades are fitted with swallow nest boxes | ★★ | | | ★ | |
| 23 | The whole courtyard is a fruit and vegetable garden | ★ | | ★★ | ★ | |
| 24 | Developers liaise with ecologists | ★★ | | | | |
| 25 | Greywater is captured and reused | | ★★ | | | ★★ |
| 26 | Household and garden waste is composted | | | ★★ | | ★★ |
| 27 | The courtyard is built only with recycled materials | | | | | ★★ |
| 28 | Each flat has at least 2 m² of growing space or flower boxes on its balcony | ★ | | ★★ | ★ | |
| 29 | At least half the courtyard is aquatic | ★★ | ★★ | | | |
| 30 | The garden has a specific colour and texture as a theme | | | | ★★ | |
| 31 | All courtyard trees and shrubs bear fruit | | | ★★ | | |
| 32 | The courtyard trees and shrubs are trimmed as a theme | | | | ★★ | |
| 33 | Part of the courtyard is left to natural succession (not deliberately planted or managed) | ★★ | | | | |
| 34 | Courtyards are planted up with at least 50 Swedish wild herbs | ★★ | | ★ | ★ | |
| 35 | All buildings have green roofs | ★★ | ★★ | | ★★ | |

Source: adapted from Kruuse (2011).

**TABLE 8.3** The 11 economic areas used to promote the use of green infrastructure in ERDF-funded developments in north-west England

Economic growth and investment
Land and property values
Labour productivity
Tourism
Products from the land
Health and well-being
Recreation and leisure
Quality of place
Land and biodiversity
Flood alleviation and management
Climate change adaptation and mitigation

Sources: GINW, 2014; NENW, undated-d.

and 'Outstanding'; for LEED 'pass' grades are 'Certified', 'Silver', 'Gold' and 'Platinum'. Clients may specify a particular rating they want from a development which gives a steer to the developer and architect. The range of parameters that are measured through BREEAM are management, health and well-being, energy efficiency, transport, water, materials, waste, land use and ecology, and pollution (BRE, 2010–2014) and many clearly link to ecosystem services delivered by green roofs, green walls, permeable pavements, etc. as well as linking in with other green infrastructure elements (e.g. cycleways). Different aspects of BREEAM are weighted differently, with energy efficiency and health and well-being having the highest impact on the scores (Table 8.4).

Green infrastructure, like any kind of infrastructure, has long-term maintenance and renewal costs. One of the big questions is how to protect the green infrastructure which is implemented by public bodies (e.g. street trees and permeable pavements) from the inevitable cycle of bursts of local/national and sometimes international (e.g. EU) capital spend and then low levels of recurrent expenditure which are vulnerable to budget cuts. For public-realm actions such as tree planting, some funding sources may be well signposted. In the UK, for example, the Forestry Commission publish a guide to funding for tree planting schemes and management (National Urban Forestry Unit, 2001), but this does not help with maintenance. Additionally, the competition for above-ground and, especially, below-ground space (Kelly, 2012) requires a multidisciplinary approach to the planning of all types of infrastructure as proposed in the best management practice guidelines for New York City (Brown *et al.*, 2005). The New York guidance document recognises the need to reduce conflicts, an approach sensible in itself but also with economic benefits. The same approach is needed for budgetary issues.

**TABLE 8.4** The nine parameters measured by the BREEAM scheme with their relative weightings (%) and features measured

| Parameter | Issue | Parameter | Issue |
|---|---|---|---|
| **Management (12%)** | | **Water (6%)** | |
| | Sustainable procurement | | Water consumption |
| | Responsible construction practices | | Water monitoring |
| | Construction site impacts | | Water leak detection and prevention |
| | Stakeholder participation | | Water efficient equipment |
| | Life cycle cost and service life planning | **Materials (12.5%)** | |
| **Health and well-being (15%)** | | | Life cycle impacts |
| | Visual comfort | | Hard landscaping and boundary protection |
| | Indoor air quality | | Responsible sourcing of materials |
| | Thermal comfort | | Insulation |
| | Water quality | | Designing for robustness |
| | Acoustic performance | **Waste (7.5%)** | |
| | Safety and security | | Construction waste management |
| **Energy (19%)** | | | Recycled aggregates |
| | Reduction of emissions | | Operational waste |
| | Energy monitoring | | Speculative floor and ceiling finishes |
| | External lighting | **Land use and ecology (10%)** | |
| | Low and zero carbon technologies | | Site selection |
| | Energy efficient cold storage | | Ecological value of site and protection of ecological features |
| | Energy efficient transportation systems | | Mitigating ecological impact |
| | Energy efficient laboratory systems | | Enhancing site ecology |
| | Energy efficient equipment | | Long term impact on biodiversity |
| | Drying space | **Pollution (10%)** | |
| **Transport (8%)** | | | Impact of refrigerants |
| | Public transport accessibility | | $NO_x$ emissions |
| | Proximity to amenities | | Surface water runoff |
| | Cyclist facilities | | Reduction of night-time light pollution |
| | Maximum car parking capacity | | Noise attenuation |
| | Travel plan | | |

Source: BRE, 2011.

## 8.3 Summary

- Policy, regulations, incentives and taxes all have a part to play in encouraging the take-up of green infrastructure and especially the greening of sealed surfaces.
- Solutions to cross-departmental boundaries in funding capital and recurrent costs in the public realm need to be identified and resolved to ensure best value is achieved and high quality is maintained.

# REFERENCES

Abdullah, A., Bahauddin, B., Mohamed, B., & Abbas, K. (2002) Sustainable city: blending nature with business. In *The Sustainable City II. Urban Regeneration and Sustainability* (ed. by C.A. Brebbia, J.F. Martin-Duque & L.C. Wadhwa), pp. 745–754. WIT Press, Southampton.

Abe, K. & Ozaki, Y. (1999) Evaluation of plant bed filter ditches for the removal of T-N and T-P from eutrophic pond water containing particulate N and P. *Soil Science and Plant Nutrition*, **45**, 737–744.

ABG (undated) Webwall ABG Ltd, Meltham.

ACC (2012) *All About Conkers.* Ashton Conker Club, Ashton.

Ackerfield, J. & Wen, J. (2003) Evolution of *Hedera* (The Ivy Genus, Araliaceae): insights from Chloroplast DNA Data. *International Journal of Plant Science*, **164**, 593–602.

Ackerman, K., Conard, M., Culligan, P., Plunz, R., Sutto, M.P., & Whittinghill, L. (2014) Sustainable Food systems for future cities: the potential of urban agriculture. *Economic and Social Review*, **45**, 189–206.

Adam, S.E.H., Shigeto, J., Sakamoto, A., Takahashi, M., & Morikawa, H. (2008) Atmospheric nitrogen dioxide at ambient levels stimulates growth and development of horticultural plants. *Botany*, **86**, 213–217.

Ahmed, M. & Durrani, P.K. (1970) The flora of the walls in Srinigar. *Botanische Jahrbucher*, **89**, 608–615.

Aitkenhead-Peterson, J.A., Dvorak, B.D., Voider, A., & Stanley, N.C. (2011) Chemistry of growth medium and leachate from green roof systems in south-central Texas. *Urban Ecosystems*, **14**, 17–33.

Akbari, H. (2002) Shade trees reduce building energy use and $CO_2$ emissions from power plants. *Environmental Pollution*, **116**, S119–S126.

Akbari, H., Davis, S., Doranso, S., Huang, J., & Winnett, eds. (1992) *Cooling Our Communities.* EPA, Washington, DC.

Akbari, H., Menon, S., & Rosenfeld, A. (2009) Global cooling: increasing world-wide urban albedos to offset $CO_2$. *Climatic Change*, **94**, 275–286.

Akbari, H., Pomerantz, M., & Taha, H. (2001) Cool surfaces and shade trees to reduce energy use and improve air quality in urban areas. *Solar Energy*, **70**, 295–310.

Akbari, H., Rose, L.S., & Taha, H. (2003) Analyzing the land cover of an urban environment using high-resolution orthophotos. *Landscape and Urban Planning*, **63**, 1–14.

ALA (undated) *Indoor Air Pollution: An Introduction for Health Professionals.* The American Lung Association, the American Medical Association, the U.S. Product Safety Commission & the U.S. EPA, New York.

Alaimo, K. (2001) Food insufficiency, family income, and health in US preschool and school-aged children. *American Journal of Public Health*, **91**, 781–786.

Alexandri, E. & Jones, P. (2008) Temperature decreases in an urban canyon due to green walls and green roofs in diverse climates. *Building and Environment*, **43**, 480–493.

Alsup, S., Ebbs, S., & Retzlaff, W. (2010) The exchangeability and leachability of metals from select green roof growth substrates. *Urban Ecosystems*, **13**, 91–111.

Alumasc (2004) Alumasc Green Roof Systems Zinco. Alumasc Exterior Building Products Ltd, St Helens, UK.

Alumasc (2010) ZInCo Green Roof Systems. Alumasc Exterior Building Products Ltd, St Helens, UK.

Ampim, P.A.Y., Sloan, J.J., Cabrera, R.I., Harp, D.A., & Jaber, F.H. (2010) Green roof growing substrates: types, ingredients, composition and properties. *Journal of Environmental Horticulture*, **28**, 244–252.

Anastasi, C., Hopkinson, L., & Simpson, V.J. (1991) Natural hydrocarbon emissions in the United Kingdom. *Atmospheric Environment Part a-General Topics*, **25**, 1403–1408.

Anderson, G.B., Krall, J.R., Peng, R.D., & Bell, M.L. (2012) Is the relation between ozone and mortality confounded by chemical components of particulate matter? Analysis of 7 components in 57 US communities. *American Journal of Epidemiology*, **176**, 726–732.

Anderson, H.R., Favarato, G., & Atkinson, R.W. (2013) Long-term exposure to air pollution and the incidence of asthma: meta-analysis of cohort studies (vol 6, pg 47, 2013). *Air Quality Atmosphere and Health*, **6**, 541–542.

Anderson, P.K., Cunningham, A.A., Patel, N.G., Morales, F.J., Epstein, P.R., & Daszak, P. (2004) Emerging infectious diseases of plants: pathogen pollution, climate change and agrotechnology drivers. *Trends in Ecology & Evolution*, **19**, 535–544.

Angold, P.G., Sadler, J.P., Hill, M.O., Pullin, A., Rushton, S., Austin, K., Small, E., Wood, B., Wadsworth, R., Sanderson, R., & Thompson, K. (2006) Biodiversity in urban habitat patches. *Science of the Total Environment*, **360**, 196–204.

Anonymous (1872) *Suggestions and Instructions in Reference to a Proposed Lunatic Asylum at St. Anne's Heath, Near Virginia Water Station, Surrey*. Surrey History Centre, Woking, Accession No. 2620/6/1.

Anonymous (1950) *Country Life Picture Book of Britain*, Second Series. Country Life Ltd, London.

Anonymous (2011) National Gallery goes green. *Water, Energy & Environment*, June/July, **56**.

Antupit, S., Gray, B., & Woods, S. (1996) Steps ahead: making streets that work in Seattle. *Landscape and Urban Planning*, **35**, 107–22.

Araujo, J.A. (2011) Particulate air pollution, systemic oxidative stress, inflammation, and atherosclerosis. *Air Quality, Atmosphere & Health*, **4**, 79–93.

Armson, D., Stringer, P., & Ennos, A.R. (2012) The effect of tree shade and grass on surface and globe temperatures in an urban area. *Urban Forestry & Urban Greening*, **11**, 245–255.

Arnfield, A.J. (2003) Two decades of urban climate research: A review of turbulence, exchanges of energy and water, and the urban heat island. *International Journal of Climatology*, **23**, 1–26.

Arnold, C.L. & Gibbons, C.J. (1996) Impervious surface coverage – The emergence of a key environmental indicator. *Journal of the American Planning Association*, **62**, 243–258.

Arthur, T. (1981) *All the Year Round. A compendium of games, customs and stories*. Puffin Books, Harmondsworth.

Asaeda, T. & Ca, V.T. (2000) Characteristics of permeable pavement during hot summer weather and impact on the thermal environment. *Building and Environment*, **35**, 365–375.

Atkinson, E.L. (1921) Snow-blindness: its causes, effects, changes, prevention and treatment. *The British Journal of Opthalmology*, **5**, 49–54.

Attrill, M., Bilton, D.T., Rowden, A.A., & Rundle, S.D. (1997) *Tidal Thames Foreshore Macroinvertebrate Survey*. Environment Agency, London.

Attrill, M., Bilton, D.T., Rowden, A.A., Rundle, S.D., & Thomas, R.M. (1999) The impact of encroachment and bankside development on the habitat complexity and supralittoral invertebrate communities of the Thames Estuary foreshore. *Aquatic Conservation: Marine and Freshwater Ecosystems*, **9**, 237–247.

Ayala, A., Brauer, M., Mauderly, J.L., & Samet, J.M. (2012) Air pollutants and sources associated with health effects. *Air Quality Atmosphere and Health*, **5**, 151–167.

Aydogan, A. & Montoya, L.D. (2011) Formaldehyde removal by common indoor plant species and various growing media. *Atmospheric Environment*, **45**, 2675–2682.

Bäckström, M. (2000) Ground Temperature in porous pavement during freezing and thawing. *Journal of Transportation Engineering-ASCE*, **126**, 375–381.

Badeja, P.J.E. (1986) The provision of verdure on the Credit Suisse administrative building 'Uetlihof', Zurich. *Anthos*, **1/86**, 11–16.

Bais, H.P., Weir, T.L., Perry, L.G., Gilroy, S., & Vivanco, J.M. (2006) The role of root exudates in rhizosphere interactions with plants and other organisms. *Annual Review of Plant Biology*, **57**, 233–266.

Bakker, M.G., Manter, D.K., Sheflin, A.M., Weir, T.L., & Vivanco, J.M. (2012) Harnessing the rhizosphere microbiome through plant breeding and agricultural management. *Plant and Soil*, **360**, 1–13.

Ball, D.J. (2009) *The public perception of risk*. National Tree Safety Group.

Bamfield, B. (2005) *Whole life costs and living roofs*. The Springboard Centre, Bridgewater. Report No. Sarnafil/001/006.2. The Solution Organisation, Cheshunt.

Bandy, L. (2003) *Non-infrastructural Solutions to San Francisco's Combined Sewer Overflow Problem*. PlantSF, San Francisco.

Barber, J.L., Kurt, P.B., Thomas, G.O., Kerstiens, G., & Jones, K.C. (2002) Investigation into the importance of the stomatal pathway in the exchange of PCBs between air and plants. *Environmental Science & Technology*, **36**, 4282–4287.

Barber, J.L., Thomas, G.O., Kerstiens, G., & Jones, K.C. (2003) Study of plant-air transfer of PCBs from an evergreen shrub: Implications for mechanisms and modeling. *Environmental Science & Technology*, **37**, 3838–3844.

Barber, P.A., Paap, T., Burgess, T.I., Dunstan, W., & Hardy, G.E.S.J. (2013) A diverse range of *Phytophthora* species are associated with dying urban trees. *Urban Forestry & Urban Greening*, **12**, 569–575.

Barclays (2006) *Barclays corporate responsibility report 2005: responsible banking*. Barclays, London.

Barkham, P. (2007) Chainsaw massacre. *The Guardian*, G2, pp. 4–8.

Barnes, J., Bender, J., Lyons, T., & Borland, A. (1999) Natural and man-made selection for air pollution resistance. *Journal of Experimental Botany*, **50**, 1423–1435.

Barnett, L. (2011) The shop growing food on its roof. *The Guardian*, 10 March, G2, p. 3.

Barr, C. & Petit, S. (2001) *Hedgerows of the World: Their Ecological Functions in Different Landscapes*. IALE (UK), Lymm.

Barr, C.J. & Gillespie, M.K. (2000) Estimating hedgerow length and pattern characteristics in Great Britain using Countryside Survey data. *Journal of Environmantal Management*, **60**, 23–32.

Bartens, J., Wiseman, P.E., & Smiley, E.J. (2010) Stability of landscape trees in engineered and conventional urban soil mixes. *Urban Forestry & Urban Greening*, **9**, 333–338.

Bartens, J., Day, S.D., Dove, J.E., Harris, J.R., & Wynn, T. (2008) Tree root penetration into compacted soils increases infiltration. In S.D. Day & S.B. Dickinson (eds) *Managing Stormwater for Urban Sustainability using Trees and Structural Soils*. Virginia Polytechnic Institute and State University, Blacksburg, VA, pp. 33–37.

Bartfelder, F. & Köhler, M. (1987) Experimentelle Untersuchungen an Fassadenbegrünungen. *Diss. TU Berlin. 612 S.*

Barthel, S., Colding, J., Elmqvist, T., & Folke, C. (2005) History and local management of a biodiversity-rich, urban cultural landscape. *Ecology & Society*, **10(2)**, 10. www.ecologyandsociety.org/vol10/iss2/art10/.

Barton, M. & Jones, N. (2009) *A guide to planning green infrastructure at the sub-regional level DRAFT*. Natural Economy Northwest.

Bass, B., Krayenhoff, S., Martilli, A., & Stull, R. (undated) *Mitigating the urban heat island with green roof infrastructure*. Clean Air Partnership, Toronto. www.cleanairpartnership.org/cooltoronto/pdf/finalpaper_bass.pdf.

Bass, B., Liu, K.K.Y., & Baskaran, B.A. (2003) *Evaluating Rooftop and Vertical Gardens as an Adaptation Strategy for Urban Areas*. National Research Council Canada, Ottawa.

Bassuk, N., Curtis, D.F., Marranca, B.Z., & Neal, B. (2009) *Recommended Urban Trees: Site Assessment and Tree Selection for Stress Tolerance*. Urban Horticulture Institute, New York.

Bastian, O. & Schreiber, K.F. (1999) *Analyse und Bewertung der Landschaft.* Spektrum, Heidelberg.

Bates, A.J., Sadler, J.P., & Mackay, R. (2013) Vegetation development over four years on two green roofs in the UK. *Urban Forestry & Urban Greening,* **12,** 98–108.

Bauder (2010) *Technical Data Sheet: Bauder Xero Flor XF301 Sedum Blanket.* Bauder Ltd, Ipswich.

Bauder (2012a) *Technical Data Sheet Bauder Xero Flor XF118 UK Native Species Wildflower Blanket.* Bauder Ltd, Ipswich.

Bauder (2012b) *Vegetation for Extensive and Biodiverse Green Roofs.* Bauder, Ipswich.

Baudry, J. & Merriam, H.G. (1988) Connectivity and connectedness: functional versus structural patterns in landscapes. In *Connectivity in Landscape Ecology Proceedings of the 2nd International Seminar of the 'International Association for Landscape Ecology',* pp. 23–28. Münster.

Baumann, N. (2006) Ground-nesting birds on green roofs in Switzerland: Preliminary observations. *Urban Habitats,* **4,** 37–50.

Baxter, L.K., Franklin, M., Ozkaynak, H., Schultz, B.D., & Neas, L.M. (2013) The use of improved exposure factors in the interpretation of fine particulate matter epidemiological results. *Air Quality Atmosphere and Health,* **6,** 195–204.

BDA (2001) *Use of Traditional Lime Mortars in Modern Brickwork.* Brick Development Association, Windsor.

Bealey, W.J., McDonald, A.G., Nernitz, E., Donovan, R., Dragosits, U., Duffy, T.R., & Fowler, D. (2007) Estimating the reduction of urban PM10 concentrations by trees within an environmental information system for planners. *Journal of Environmental Management,* **85,** 44–58.

Becker, J.S., Bellis, D., McLeod, C.W., Dombovari, J., & Becker, J.S. (2000) Determination of trace elements including platinum in tree bark by ICP mass spectrometry. *Fresenius' Journal of Analytical Chemistry,* **368,** 4901–4905.

Beckett, K.P., Freer-Smith, P.H., & Taylor, G. (1998) Urban woodlands: their role in reducing the effects of particulate pollution. *Environmental Pollution,* **99,** 347–360.

Beckett, K.P., Freer-Smith, P.H., & Taylor, G. (2000a) Effective tree species for local air-quality management. *Journal of Arboriculture,* **26,** 12–19.

Beckett, K.P., Freer-Smith, P.H., & Taylor, G. (2000b) Particulate pollution capture by urban trees: effect of species and windspeed. *Global Change Biology,* **6,** 995–1003.

Begon, M., Harper, J.L., & Townsend, C.R. (1996) *Ecology.* Blackwell, Oxford.

Bekker, R.M., Verweij, G.L., Bakker, J.P., & Fresco, L.F.M. (2000) Soil seed bank dynamics in hayfield succession. *Journal of Ecology,* **88,** 594–607.

Bell, J.F., Wilson, J.S., & Liu, G.C. (2008) Neighborhood greenness and 2-year changes in body mass index of children and youth. *American Journal of Preventive Medicine,* **35,** 547–553.

Bell, J.N.B., Ayazloo, M., & Wilson, G.B. (1982) Selection for sulphur dioxide tolerance in grass populations in polluted areas. In *Urban Ecology* (ed. by R. Bornkamm, J.A. Lee & M.R.D. Seaward), pp. 171–180. Blackwell Scientific, Oxford.

Bell, J.N.B., Honour, S.L., & Power, S.A. (2011) Effects of vehicle exhaust emissions on urban wild plant species. *Environmental Pollution,* **159,** 1984–1990.

Benedict, M.A. & McMahon, E.T. (2006) *Green Infrastructure. Linking landscapes and Communities.* Island Press, Washington DC.

Beniston, M., Stephenson, D.B., Christensen, O.B., Ferro, A.T., Frei, C., Goyette, S., Halsnaes, K., Holt, T., Jylha, K., Koffi, B., Palutikof, J., Scholl, R., Semmler, T., & Woth, K. (2007) Future extreme events in European climate: an exploration of regional climate model projections. *Climatic Change,* **81,** 71–95.

Benjamin, M.T. & Winer, A.M. (1998) Estimating the ozone-forming potential of urban trees and shrubs. *Atmospheric Environment,* **32,** 53–68.

Benvenuti, S. & Bacci, D. (2010) Initial agronomic performances of Mediterranean xerophytes in simulated dry green roofs. *Urban Ecosystems,* **13,** 349–363.

Berlin.de (undated) *Potsdammer Platz.* Berlin.de, Berlin, www.berlin.de/orte/sehenswuerdigkeiten/potsdamer-platz/index.en.php.

Berman, J.D., Fann, N., Hollingsworth, J.W., Pinkerton, K.E., Rom, W.N., Szema, A.M., Breysse,

P.N., White, R.H., & Curriero, F.C. (2012) Health benefits from large-scale ozone reduction in the United States. *Environmental Health Perspectives*, **120**, 1404–1410.

Bertelsen, R.J., Carlsen, K.C.L., Calafat, A.M., Hoppin, J.A., Haland, G., Mowinckel, P., Carlsen, K.H., & Lovik, M. (2013) Urinary biomarkers for phthalates associated with asthma in Norwegian children. *Environmental Health Perspectives*, **121**, 251–256.

Bertin, C., Yang, X.H., & Weston, L.A. (2003) The role of root exudates and allelochemicals in the rhizosphere. *Plant and Soil*, **256**, 67–83.

Berto, R. (2005) Exposure to restorative environments helps restore attentional capacity. *Journal of Environmental Psychology*, **25**, 249–259.

Biggs, J. (2003) Christmas curiosity or medical marvel?: A seasonal review of mistletoe. *Biologist*, **50**, 249–254.

Bignal, K., Ashmore, M., & Power, S. (2004) *The ecological effects of diffuse air pollution from road transport*. English Nature Research Report 580. English Nature, Peterborough.

Bird, W. (2004) *Natural Fit*. RSPB, Sandy.

Bitter, S.D. & Bowers, J.K. (1994) Bioretention as a water quality Best Management Practice. *Watershed Protection Techniques*, **1**, 114–116.

Blakeman, J.P. & Fokkema, N.J. (1982) Potential for biological control of plant diseases on the phylloplane. *Annual Review of Phytopathology*, **20**, 167–192.

Blanc, P. (2008) *The Vertical Garden: In Nature and the City*. W.W. Norton & Co., London.

Blanc, P. (undated) *The Vertical Garden, from nature to Cities. A Botanical and Artistic approach by Patric Blanc*. self published: www.verticalgardenpatrickblanc.com/#/en/resources.

Boivin, M.-A., ed. (1992) *Geld vom staat fur grune dacher*. Peck & Associates.

Bolund, P. & Hunhammar, S. (1999) Ecosystem services in urban areas. *Ecological Economics*, **29**, 293–301.

Borland, A.M., Zambrano, V.A.B., Ceusters, J., & Shorrock, K. (2011) The photosynthetic plasticity of crassulacean acid metabolism: an evolutionary innovation for sustainable productivity in a changing world. *New Phytologist*, **191**, 619–633.

Bornkamm, R., Lee, J.A., & Seaward, M.R.D., eds. (1982) *Urban Ecology*. Blackwell Scientific, Oxford.

Boving, T.B., Stolt, M.H., Augenstern, J., & Brosnan, B. (2008) Potential for localized groundwater contamination in a porous pavement parking lot setting in Rhode Island. *Environmental Geology*, **55**, 571–582.

Brand, A.B. & Snodgrass, J.W. (2010) Value of artificial habitats for amphibian reproduction in altered landscapes. *Conservation Biology*, **24**, 295–301.

Brand, C., Goodman, A., & Ogilvie, D. (2014) Evaluating the impacts of new walking and cycling infrastructure on carbon dioxide emissions from motorized travel: A controlled longitudinal study. *Applied Energy*, **128**, 284–295.

Brattebo, B.O. & Booth, D.B. (2003) Long-term stormwater quantity and quality performance of permeable pavement systems. *Water Research*, **37**, 4369–4376.

Brazel, A., Selover, N., Vose, R., & Heisler, G. (2000) The tale of two climates – Baltimore and Phoenix urban LTER sites. *Climate Research*, **15**, 123–135.

BRE (2010–2014) BREEAM: *The world's leading design and assessment method for sustainable buildings*. Building Research Establishment, Watford, www.breeam.org/page.jsp?id=369.

BRE (2011) BREEAM *New construction. Non-domestic buildings. Technical manual SD5073–2.0.2011*. BRE, Watford.

Brebbia, C.A., Martin-Duque, J.F., & Wadhwa, L.C. (2002) *The Sustainable City II. Urban Regeneration and Sustainability*. WIT Press, Southampton.

Brengman, M., Willems, K., & Joye, Y. (2012) The impact of in-store greenery on customers. *Psychology & Marketing*, **29**, 807–821.

Brenneisen, S. (2006) Space for urban wildlife: designing green roofs as habitats in Switzerland. *Urban Habitats*, **4**, 27–36.

Brest, C.L. (1987) Seasonal albedo of an urban/rural landscape from satellite observations. *Journal of Climate and Applied Meteorology*, **26**, 1169–1187.

Briggs, D.J., de Hoogh, K., Morris, C., & Gulliver, J. (2008) Effects of travel mode on exposures to particulate air pollution. *Environment International*, **34**, 12–22.

Bright, P.W. (1998) Behaviour of specialist species in habitat corridors: arboreal dormice avoid corridor gaps. *Animal Behaviour*, **56**, 1485–1490.

Bringslimark, T., Hartig, T., & Patil, G.G. (2007) Psychological benefits of indoor plants in workplaces: putting experimental results into context. *HortScience*, **42**, 581–587.

Bringslimark, T., Hartig, T., & Patil, G.G. (2009) The psychological benefits of indoor plants: A critical review of the experimental literature. *Journal of Environmental Psychology*, **29**, 422–433.

Britt, C. & Johnson, M., eds. (2008a) *Trees in Towns II. A new survey of urban trees in England and their condition and management. Executive summary*, pp. 34. DCLG, London.

Britt, C. & Johnson, M. (2008b) *Trees in Towns II. A New Survey of Urban Trees in England and Their Condition and Management*. Research for Amenity Trees Series No. 9. DCLG, London.

Brown, H., Caputo, S.A.J., Carnahan, K., & Nielsen, S. (2005) *High Performance Infrastructure Guidelines*. New York City Department of Design and Construction, New York.

Brown, S.K. (2002) Volatile organic pollutants in new and established buildings in Melbourne, Australia. *Indoor Air*, **12**, 55–63.

Brunkeef, B. & Holgate, S.T. (2002) Air pollution and health. *The Lancet*, **360**, 1233–1242.

Bruse, M., Thönnessen, M., & Radke, U. (1999) Practical and theoretical investigation of the influence of façade greening on the distribution of heavy metals in urban streets. www.envi-met.com/documents/papers/façade1999.pdf.

BSDUD (2009) *Elements of Sustainability: Ecological Building in Berlin*. Senate Department for Urban Development, Berlin.

BSDUD (undated) *Institute of Physics in Berlin-Adlershof.* Urban Ecological Model Projects Berlin Senate Department for Urban Development, Berlin.

BTO (2009) *The Biodiversity in Glasgow Project*. British Trust for Ornithology, Thetford, www.bto.org/survey/special/glasgow_biodiversity/index.htm.

BTO (2014) *The Breeding Bird Survey*. British Trust for Ornithology, Thetford, www.bto.org/volunteer-surveys/bbs.

Buchta, E., Hirsch, K.W., & Buchta, C. (1984) *Laermmindernde Wirkung von Bewuchs in Strassenschluchten*. Institut fuer Laermschutz Buchta.

Bućko, M.S., Magiera, T., Johanson, B., Petrovský, E., & Pesonen, L.J. (2011) Identification of magnetic particulates in road dust accumulated on roadside snow using magnetic, geochemical and micro-morphological analyses. *Environmental Pollution*, **159**, 1266–1276.

Burchett, M., Torpy, F., Brennan, J., & Craig, A. (2010) *Greening the Great Indoors for Human health and wellbeing*. Horticulture Australia, Sydney

Burholt, V. & Windle, G. (2006) Keeping warm? Self-reported housing and home energy efficiency factors impacting on older people heating homes in North Wales. *Energy Policy*, **34**, 1198–1208.

Burkart, K., Nehls, I., Win, T., & Endlicher, W. (2013) The carcinogenic risk and variability of particulate-bound polycyclic aromatic hydrocarbons with consideration of meteorological conditions. *Air Quality Atmosphere and Health*, **6**, 27–38.

Burnett, R.T., Cakmak, S., & Brook, J.R. (1998) The effect of the urban ambient air pollution mix on daily mortality rates in 11 Canadian cities. *Canadian Journal of Public Health-Revue Canadienne De Sante Publique*, **89**, 152–156.

Busby, K. (1960) Brickwork. In *The Complete Handyman* (ed. by C.H. Hayward), pp. 36–59. Evans Brothers Ltd, London.

Butkovich, K., Graves, J., McKay, J., & Slopack, M. (2008) *An Investigation into the Feasibility of Biowall Technology*. George Brown College Applied Research & Innovation, Toronto, Canada.

Butler, C., Butler, E., & Orians, C.M. (2012) Native plant enthusiasm reaches new heights: Perceptions, evidence, and the future of green roofs. *Urban Forestry & Urban Greening*, **11**, 1–10.

Buyantuyev, A. & Wu, J.G. (2010) Urban heat islands and landscape heterogeneity: linking spatiotemporal variations in surface temperatures to land-cover and socioeconomic patterns. *Landscape Ecology*, **25**, 17–33.

CABE (2005) *Does Money Grow on Trees?* CABE, London.

Cackowski, J.M. & Nasar, J.L. (2003) The restorative effects of roadside vegetation implications for driver anger. *Environment & Behaviour*, **35**, 736–751.

Campbell, G. (2006) *Adnams' Eco Centre*. BBC Suffolk, Ipswich. Available at: www.bbc.co.uk/suffolk/content/articles/2006/10/26/adnams_distribution_centre_feature.shtml

Cañero, R.F., Urrestarau, L.P., & Salas, A.F. (2012) Assessment of the cooling potential of an indoor living wall using different substrates in a warm climate. *Indoor and Built Environment*, **21**, 642–650.

Carter, T. & Butler, C. (2008) Ecological impacts of replacing traditional roofs with green roofs in two urban areas. *Cities and the Environment*, **1**, Article 9. http://digitalcommons.lmu.edu/cgi/viewcontent.cgi?article=1020&context=cate.

Carter, T. & Keeler, A. (2008) Life-cycle cost-benefit analysis of extensive vegetated roof systems. *Journal of Environmental Management*, **87**, 350–363.

Carter, T.L. & Rasmussen, T.C. (2006) Hydrologic behaviour of vegetated roofs. *Journal of the American Water Resources Association* **42**, 1261–1274.

CDT (undated) *The Chicago Green Alley Handbook: An Action Guide to Create a Greener, Environmentally Sustainable Chicago*. Chicago Department of Transportation, Chicago.

Cekstere, G. & Osvalde, A. (2013) A study of chemical characteristics of soil in relation to street trees status in Riga (Latvia). *Urban Forestry & Urban Greening*, **12**, 69–78.

Černikovský, L., Krejčí, B., Targa, J., Kurfürst, P., & Volná, V. (2013) *Air Pollution by Ozone Across Europe During Summer 2012. Overview of exceedances of EC ozone threshold values for April-September 2012*. European Environment Agency, Copenhagen.

Chadyšienė, R. & Girgždys, A. (2008) Ultraviolet radiation albedo of natural surfaces. *Journal of Environmental Engineering and Landscape Management*, **16**, 83–88.

Chakre, O.J. (2006) Choice of eco-friendly trees in urban environment to mitigate airborne particulate pollution. *Journal of Human Ecology*, **20**, 135–138.

Chaloulakou, A., Kassomenos, P., Spyrellis, N., Demokritou, P., & Koutrakis, P. (2003) Measurements of PM10 and PM2.5 particle concentrations in Athens, Greece. *Atmospheric Environment*, **37**, 649–660.

Chang, C.Y. & Chen, P.-K. (2005) Human response to window views and indoor plants in the workplace. *Hortscience*, **40**, 1354–1359.

Changon, S.A. (1992) Inadvertent weather modification in urban areas: lessons for global climate change. *Bulletin of the American Meteorological Society*, **73**, 619–752.

Chapman, L. (2007) Transport and climate change: a review. *Journal of Transport Geography*, **15**, 354–367.

Chaturvedi, A., Kamble, R., Patil, N.G., & Chaturvedi, A. (2013) City–forest relationship in Nagpur: One of the greenest cities of India. *Urban Forestry & Urban Greening*, **12**, 79–87.

Chaudhuri, N. (1998) Child health, poverty and the environment: the Canadian context. *Canadian Journal of Public Health*, **89 Supplement 1**, 26–30.

Chauhan, A. (2010a) Photosynthetic pigment changes in some selected trees induced by automobile exhaust in Dehradun, Uttarakhand. *New York Science Journal*, **3**, 45–51.

Chauhan, A. (2010b) Tree as bio-indicator of automobile pollution in Dehradun City: a case study. *New York Science Journal*, **3**, 88–95.

Chen, J.C. & Schwartz, J. (2009) Neurobehavioral effects of ambient air pollution on cognitive performance in US adults. *Neurotoxicology*, **30**, 231–239.

Chen, Y. & Wong, N.H. (2006) Thermal benefits of city parks. *Energy and Buildings*, **38**, 105–120.

Cheng, C.Y., Cheung, K.K.S., & Chu, L.M. (2010) Thermal performance of a vegetated cladding system on façade walls. *Building and Environment*, **45**, 1779–1787.

Chiappini, L., Dagnelie, R., Sassine, M., Fuvel, F., Fable, S., Tran-Thi, T.-H., & George, C. (2011) Multi-tool formaldehyde measurement in simulated and real atmospheres for indoor air survey and concentration change monitoring. *Air Quality, Atmosphere & Health*, **4**, 211–220.

Chilton, C. (2011) Are diesel particulate filters more trouble than help? In *Car*, www.carmagazine.

co.uk/Community/Car-Magazines-Blogs/Chris-Chilton-Blog/Are-diesel-particulate-filters-more-trouble-than-theyre-worth/.

Chiquet, C., Dover, J.D., & Mitchell, P.M. (2012) Are hedgerows green wall analogues? In *Hedgerow Futures* (ed. by J.W. Dover), pp. 45–53. Hedgelink, Leeds.

Chiquet, C., Dover, J.W., & Mitchell, P. (2013) Birds and the urban environment: the value of green walls. *Urban Ecosystems*, **16**, 453–462.

Cho, H. & Choi, J.M. (2014) The quantitative evaluation of design parameter's effects on a ground source heat pump system. *Renewable Energy*, **65**, 2–6.

Chun, S.-C., Yoo, M.H., Moon, Y.S., Shin, M.H., Son, K.-C., Chung, I.-M., & Kays, S.J. (2010) Effect of bacterial population from rhizosphere of various foliage plants on removal of indoor volatile organic compounds. *Korean Journal of Horticultural Science & Technology*, **28**, 476–483.

Cimprich, B. (1993) Development of an intervention to restore attention in cancer patients. *Cancer Nursing*, **16**, 83–92.

Cimprich, B. (2003) An environmental intervention to restore attention in women with newly diagnosed breast cancer. *Cancer Nursing*, **26**, 284–292.

CityCo (2011) *Manchester's First Urban Orchard.* CityCo, Manchester, http://cityco.com/project/manchesters-first-urban-orchard/.

Clarke, A.G. (1992) The Atmosphere. In *Understanding our Environment: An introduction to environmental chemistry and pollution* (ed. by R.M. Harrison), pp. 5–51. Royal Society of Chemistry, Cambridge.

CLT (2013) *Hedgerows/Hedge Survey.* Cheshire Landscape Trust, Runcorn, http://cheshirelandscapetrust.org.uk/hedgerows/hedge-survey/.

Cockrall-King, J. (2012) *Food and the City: Urban Agriculture and the New Food Revolution.* Prometheus Books, New York.

Colla, S.R., Willis, E., & Packer, L. (2009) Can green roofs provide habitat for urban bees (Hymenoptera:Apidae)? *Cities and the Environment*, **2**, 12.

Comba, L., Corbet, S.A., Barron, A., Bird, A., Collinge, S., Miyazaki, N., & Powell, M. (1999a) Garden flowers: Insect visits and the floral reward of horticulturally modified variants. *Annals of Botany*, **83**, 73–86.

Comba, L., Corbet, S.A., Hunt, L., & Warren, B. (1999b) Flowers, nectar and insect visits: Evaluating British plant species for pollinator-friendly gardens. *Annals of Botany*, **83**, 369–383.

Connors, J.P., Galletti, C.S., & Chow, W.T.L. (2013) Landscape configuration and urban heat island effects: assessing the relationship between landscape characteristics and land surface temperature in Phoenix, Arizona. *Landscape Ecology*, **28**, 271–283.

Cook, D.I. & Van Haverbeke, D.F. (1974) *Tree-Covered Land-Forms for Noise Control.* Research Bulletin 263 University of Nebraska, Lincoln, USA.

Cook, D.I. & Van Haverbeke, D.F. (1977) Suburban noise control with plant materials and solid barriers. *USDA Forest Service General Technical Report NE*, **25**, 234–241.

Coombes, E., Jones, A.P., & Hillsdon, M. (2010) The relationship of physical activity and overweight to objectively measured green space accessibility and use. *Social Science & Medicine*, **70**, 816–822.

CoP (2014) *Clean River Rewards Program: frequently asked questions.* The City of Portland, Portland, Oregon.

Corchnoy, S.B., Arey, J., & Atkinson, R. (1992) Hydrocarbon emissions from 12 urban shade trees of the Los-Angeles, California, air basin. *Atmospheric Environment Part B-Urban Atmosphere*, **26**, 339–348.

Cornejo, J.J., Munoz, F.G., Ma, C.Y., & Stewart, A.J. (1999) Studies on the decontamination of air by plants. *Ecotoxicology*, **8**, 311–320.

Cosgrave, B. (2009) America's dream store has arrived. *Daily Telegraph*, 11 November, p. 27.

Costa, P. & James, R.W. (1995) Constructive use of vegetation in office buildings. In *Plants for People Symposium*, p. 22. The Hague, Holland.

CoT (1998–2014) *Toronto Green Roof Bylaw.* City of Toronto, Toronto. www1.toronto.ca/wps/portal/contentonly?vgnextoid=83520621f3161410VgnVCM10000071d60f89RCRD&vgnextchannel=3a7a036318061410VgnVCM10000071d60f89RCRD#know.

Coulthard, T.J. & Frostick, L.E. (2010) The Hull floods of 2007: implications for the governance and management of urban drainage systems. *Journal of Flood Risk Management*, **3**, 223–231.

Countryman, A. (2009) *An Edible Urban Forest: An Element of the Sustainability Equation*. BSES, University of Indiana, Bloomington.

Coupe, S.J., Smith, H.G., Newman, A.P., & Puehmeir, T. (2003) Biodegradation and microbial diversity within permeable pavements. *European Journal of Protistology*, **39**, 495–498.

Crisp, T.M., Clegg, E.D., Cooper, R.L., Wood, W.P., Anderson, D.G., Baetcke, K.P., Hoffmann, J.L., Morrow, M.S., Rodier, D.J., Schaeffer, E., Touart, L.W., Zeeman, M.G., & Patel, Y.M. (1998) Environmental endocrine disruption: an effects assessment and analysis. *Environmental Health Perspectives*, **106 (Supplement 1)**, 11–56.

Cronin, S. (2012) *Waitrose opens its lowest carbon store yet*. Waitrose, Bracknell.

Cronshey, R. (1986) *Urban hydrology for small watersheds*. US Department of Agriculture, Natural Resources Conservation Service, Conservation Engineering Division.

CROP (2009) *Chicago Rarities Orchard Project*. CROP, Chicago, www.chicagorarities.org/

CROP (2012) *Chicago Rarities Orchard Project: Logan Square Plaza*. CROP, Chicago, http://chicagorarities. org/crop_logan_general_pages.pdf

Croucher, K., Myers, L., & Bretherton, J. (2007) *The Links Between Greenspace and Health: a Critical Literature Review*. Greenspace Scotland, Stirling.

Cruikshank, M. (2001) Core subject. *Times Education Supplement*, 5 October. www.tes.co.uk/article. aspx?storycode=353672.

Cunningham, M.A., Menking, K.M., Gillikin, D.P., Smith, K.C., Freimuth, C.P., Belli, S.L., Pregnall, A.M., Schlessman, M.A., & Batur, P. (2010) Influence of open space on water quality in an urban stream. *Physical Geography*, **31**, 336–356.

Currie, B.A. & Bass, B. (2008) Estimates of air pollution mitigation with green plants and green roofs using the Ufore model. *Urban Ecosystems*, **11**, 409–422.

Dandy, N. (2010) *Climate Change and Street Trees Project. Social Research Report. The social and cultural values, and governance, of street trees*. Forest Research, Farnham.

Dandy, N., Marzano, M., Moseley, M., Stewart, A.J., & Lawrence, A. (2012) Exploring the role of street trees in the improvement and expansion of green networks. In *Trees, People and the Built Environment* (ed. by M. Johnston & G. Percival), pp. 73–83. Forestry Commmission, Edinburgh.

Darius, F. & Drepper, J. (1984) Rasendächer in West-Berlin. *Das Gartenamt*, **33**, 309–315.

Darlington, A., Chan, M., Malloch, D., Pilger, C., & Dixon, M.A. (2000) The biofiltration of indoor air: Implications for air quality. *Indoor Air-International Journal of Indoor Air Quality and Climate*, **10**, 39–46.

Davidson, A.W. & Barnes, J.D. (1998) Effects of ozone on wild plants. *New Phytology*, **139**, 135–151.

Dawson, D. (1994) *Are habitat corridors conduits for animals and plants in a fragmented landscape? A review of the scientific evidence*. English Nature, Peterborough.

Day, S.D. & Dickinson, S.B., eds. (2008) *Managing Stormwater for Urban Sustainability using Trees and Structural Soils*. Virginia Polytechnic Institute and State University, Blacksburg, VA.

DCLG (2008) *Guidance on the permeable surfacing of front gardens*. Department for Communities and Local Government, London.

DCMS (2002) *Game Plan: a Strategy for Delivering the Government's Sport and Physical Activity Objectives*. Department for Culture, Media and Sport/Cabinet Office Strategy Unit, London.

De Fabo, E.C., Noonan, F.P., Fears, T., & Merlino, G. (2004) Ultraviolet B but not Ultraviolet A radiation initiates melanoma. *Cancer Research*, **64**, 6372–6376.

de Groot, W.T. & van den Born, R.J.G. (2003) Visions of nature and landscape type prefences: an exploration in The Netherlands. *Landscape and Urban Planning*, **63**, 127–138.

de Lange, P. (2011) *Urban Agriculture in Amsterdam: Understanding the Recent Trend in Food Production Activities Within the Limits of a Developed Nation's Capital*. Masters in Environment & Resource Management, VU University, Amsterdam.

de Vries, S., Verheij, R.A., Groenewegen, P.P., & Spreeuwenberg, P. (2003) Natural environments – healthy environments? An exploratory analysis of the relationship between greenspace and health. *Environment and Planning A*, **35**, 1717–1731.

Deaton, A. (2002) Policy implications of the gradient of health and wealth. *Health Affairs*, **21**, 13–30.

Declet-Barreto, J., Brazel, A., Martin, C., Chow, W.L., & Harlan, S. (2013) Creating the park cool island in an inner-city neighborhood: heat mitigation strategy for Phoenix, AZ. *Urban Ecosystems*, **16**, 617–635.

Deeproot (2010) FAQs: *Silva Cells & On-Site Stormwater Management*. Deeproot, San Francisco. www. deeproot.com/blog/blog-entries/faqs-silva-cells-on-site-stormwater-management.

Deeproot (2011) *How much soil to grow a big tree?* Deep Root Partners, San Francisco.

Deeproot (2012) *Silva Cell*. DeepRoot, London. www.deeproot.com/products/silva-cell/silva-cell-overview.html.

Defra (2007a) *The Air Quality Strategy for England, Scotland, Wales and Northern Ireland* (Volume 1). Defra, London.

Defra (2007b) *An economic analysis to inform the air quality strategy: updated third report of the Interdepartmental Group on Costs and Benefits*. Defra, London.

Defra (2010) *Flood and Water Management Act 2010*. HMSO, London.

Defra (2011a) *GB Non-native Species Secretariat Website*. Defra, London.

Defra (2011b) *The Natural Choice: securing the value of nature*. HMSO, London.

Defra (2012) *Air Pollution in the UK 2011*. Defra, London.

Defra (2013a) *The Heatwave Plan for England*. Defra, London.

Defra (2013b) *The National Adaptation Programme: making the Country Resilient to Climate Change*. Defra, London.

Dendy, T. (1987) The value of corridors (and design features of the same) and small patches of habitat. In *Nature Conservation: The Role of Remnants of Native Vegetation* (ed. by D. Saunders, G. Arnold, A. Burbidge & A. Hopkins), pp. 357–358. Surrey Beatty & Sons Pty Ltd, Chipping Norton, Australia.

Dennis, R.L.H., Dapporto, L., & Dover, J.W. (2014) Ten years of the resource-based habitat paradigm: the biotope-habitat issue and implications for conserving butterfly diversity. *Journal of Insect Biodiversity*, **2**, 1–32.

Deshmukh, D.K., Deb, M.K., & Mkoma, S.L. (2013) Size distribution and seasonal variation of size-segregated particulate matter in the ambient air of Raipur city, India. *Air Quality Atmosphere and Health*, **6**, 259–276.

Despommier, D. (2011) *The Vertical Farm: Feeding the World in the 21st Century*. Picador, New York.

Dhainaut, J.F., Claessens, Y.E., Ginsburg, C., & Riou, B. (2004) Unprecedented heat-related deaths during the 2003 heat wave in Paris: consequences on emergency departments. *Critical Care*, **8**, 1–2.

Diamanti-Kandarakis, E., Bourguignon, J.-P., Guidice, L.C., Hauser, R., Prins, G.S., Soto, A.M., Zoeller, R.T., & Gore, A.C. (2009) Endocrine-Disrupting Chemicals: An Endocrine Society Scientific Statement. *Endocrine Reviews*, **30**, 293–342.

Diamond, J.M. (1975) The island dilemma: lessons of modern biogeographic studies for the design of nature reserves. *Biological Conservation*, **7**, 129–145.

Dimitroulopoulou, C. (2012) Ventilation in European dwellings: A review. *Building and Environment*, **47**, 109–125.

Ding, L., Van Renterghem, T., Botteldooren, D., Horoshenkov, K., & Khan, A. (2013) Sound absorption of porous substrates covered by foliage: Experimental results and numerical predictions. *Journal of the Acoustical Society of America*, **134**, 4599–4609.

Dingle, P., Tapsell, P., & Hu, S. (2000) Reducing formaldehyde exposure in office environments using plants. *Bulletin of Environmental Contamination and Toxicology*, **64**, 302–308.

Dochinger, L.S. (1980) Interception of airborn particles by tree plantings. *Journal of Environmental Quality*, **9**, 265–268.

Doernach, R. (1979) On the use of 'biotectural systems'. *Garten+Landschaft*, **89**, 452–457.

Doernach, R. (1986) Nature as home Biolympiad of the plant houses. *Anthos*, **1/86**, 17–21.

Donovan, G.H. & Butry, D.T. (2009) The value of shade: Estimating the effect of urban trees on summertime electricity use. *Energy and Buildings*, **41**, 662–668.

Donovan, G.H., Butry, D.T., Michael, Y.L., Prestemon, J.P., Liebhold, A.M., Gatziolis, D., & Mao,

M.Y. (2013) The relationship between trees and human health evidence from the spread of the emerald ash borer. *American Journal of Preventive Medicine*, **44**, 139–145.

Donovan, R.G., Stewart, H.E., Owen, S.M., MacKenzie, A.R., & Hewitt, C.N. (2005) Development and application of an urban tree air quality score for photochemical pollution episodes using the Birmingham, United Kingdom, area as a case study. *Environmental Science & Technology*, **39**, 6730–6738.

Dore, C.J., Murrells, T.P., Passant, N.R., Hobson, M.M., Thistlethwaite, G., Wagner, A., Li, Y., Bush, T., King, K.R., Norris, J., Coleman, P.J., Walker, C., Stewart, R.A., Tsagatakis, I., Conolly, C., Brophy, N.C.J., & Hann, M.R. (2008) *UK Emissions of Air Pollutants 1970 to 2006*. AEA Technology, Didcot.

Dover, J.W. (1992) *Kirklees UDP: green corridors, final report*. The Environmental Advisory Unit Ltd, Liverpool.

Dover, J.W. (2000) Human, environmental and wildlife aspects of corridors with specific reference to UK planning practice. *Landscape Research*, **25**, 333–344.

Dover, J.W. (2006a) *Embedding Green Infrastructure in Housing Market Renewal. Biodiversity and Green Infrastructure Document Review*. IESR, Staffordshire University, Stoke-on-Trent.

Dover, J.W. (2006b) *Embedding Green Infrastructure in Housing Market Renewal: Linking Biodiverse Green Infrastructure Drivers with Actions: Final Report*. IESR, Staffordshire University, Stoke-on-Trent.

Dover, J.W., ed. (2012) *Hedgerow Futures*. Hedgelink, Leeds.

Dover, J.W. & Fry, G.L.A. (2001) Experimental simulation of some visual and physical components of a hedge and the effects on butterfly behaviour in an agricultural landscape. *Entomologia Experimentalis et Applicata*, **100**, 221–233.

Dramstad, W.E., Olson, J.D., & Forman, R.T.T. (1996) *Landscape Ecology Principles in Landscape Architecture and Land-use Planning,* 1st edn. Harvard University Graduate School of Design, Harvard.

Dravigne, A., Waliczek, T.M., Lineberger, R.D., & Zajicek, J.M. (2008) The effect of live plants and window views of green spaces on employee perceptions of job satisfaction. *HortScience*, **43**, 183–187.

Dunnett, N., Gedge, D., Little, J., & Snodgrass, E.C. (2011) *Small Green Roofs: Low-tech Options for Greener Living*. Timber Press, Portland, Oregon.

Dunnett, N.P. & Kingsbury, N. (2004) *Planting Green Roofs and Living Walls*. Timber Press, Portland, Oregon.

Dunning, J., Danielson, B., & Pulliam, H. (1992) Ecological processes that affect populations in complex habitats. *Oikos*, **65**, 169–175.

Dvorak, B. & Volder, A. (2010) Green roof vegetation for North American ecoregions: A literature review. *Landscape and Urban Planning*, **96**, 197–213.

Dvorak, B. & Volder, A. (2013) Rooftop temperature reduction from unirrigated modular green roofs in south-central Texas. *Urban Forestry & Urban Greening*, **12**, 28–35.

Dzierzanowski, K., Popek, R., Gawrońska, H., Saebø, A., & Gawroński, S.W. (2011) Deposition of particulate matter of different size fractions on leaf surfaces and in waxes of urban forest species. *International Journal of Phytoremediation*, **13**, 1037–1046.

EA (2008) Harvesting rainwater for domestic uses: an information guide. Environment Agency, Bristol.

EASAC (2009) *Ecosystem Services and Biodiversity in Europe*. The Royal Society for the European Academies Science Advisory Council, London.

Easterling, D.R., Evans, J.L., Groisman, P.Y., Karl, T.R., Kunkel, K.E., & Ambenje, P. (2000) Observed variability and trends in extreme climate events: A brief review. *Bulletin of the American Meteorological Society*, **81**, 417–425.

Edmondson, J.L., Davies, Z.G., Gaston, K.J., & Leake, J.R. (2014) Urban cultivation in allotments maintains soil qualities adversely affected by conventional agriculture. *Journal of Applied Ecology*, **51**, 880–889.

EFC (2012) *Financing Feasibility Study for Stormwater Management in Berlin, Maryland*. Maryland Department of Natural Resources/The Town Creek Foundation, Berlin.

EFF (2006) *Allotments Regeneration Initiative 2002–2006*. The Esmée Fairbairn Foundation, London.

Ehhalt, D., Prather, M., Dentener, F., Derwent, R., Dlugokencky, E., Holland, E., Isaksen, I., Katima,

J., Kirchhoff, V., Matson, P., Midgley, P., Wang, M., Berntsen, T., Bey, I., Brasseur, G., Buja, L., Collins, W.J., Daniel, J., DeMore, W.B., Derek, N., Dickerson, R., Etheridge, D., Feichter, J., Fraser, P., Friedl, R., Fuglestvedt, J., Gauss, M., Grenfell, L., Grübler, A., Harris, N., Hauglustaine, D., Horowitz, L., Jackman, C., Jacob, D., Jaeglé, L., Jain, A., Kanakidou, M., Karlsdottir, S., Ko, M., Kurylo, M., Lawrence, M., Logan, J.A., Manning, M., Mauzerall, D., McConnell, J., Mickley, L., Montzka, S., Müller, J.F., Olivier, J., Pickering, K., Pitari, G., Roelofs, G.J., Rogers, H., Rognerud, B., Smith, S., Solomon, S., Staehelin, J., Steele, P., Stevenson, D., Sundet, J., Thompson, A., Weele, M.v., Kuhlmann, R.v., Wang, Y., Weisenstein, D., Wigley, T., Wild, O., Wuebbles, D., & Yantosca, R. (2001) *Atmospheric Chemistry and Greenhouse Gases*. In *Climate Change 2001: The Scientific Basis. Contribution of Working Group I to the Third Assessment Report of the Intergovernmental Panel on Climate Change* (ed. by J.T. Houghton, Y. Ding, D.J. Griggs, M. Noguer, P.J.v.d. Linden, X. Dai, K. Maskell & C.A. Johnson). Cambridge University Press, Cambridge.

Ehrlich, P.R. & Raven, P.H. (1967) Butterflies and Plants. *Scientific American*, **216**, 104–113.

Endress, A.G. & Thomson, W.W. (1976) Ultrastructural and cytochemical studies on the developing adhesive disc of Boston Ivy tendrils. *Protoplasma*, **88**, 315–331.

Enwright, C. (2013) Green roofs are growing: standards support sustainability. *Standardization News*, May/June, pp. 32–35, West Conshohocken, PA.

EPA (1993) *Guidance Specifying Management Measures for Sources of Nonpoint Source Pollution in Coastal Waters*. US Environmental Protection Agency, Office of Water, Washington, DC.

EPA (2000) *Vegetated Roof Cover Philadelphia, Pennsylvania*. EPA-841-B-00–005D US EPA, Washington, DC.

EPA (2008) *Managing Wet Weather with Green Infrastructure*. Action Strategy 2008 EPA.

EPA (2009) *Buildings and their Impact on the Environment: A Statistical Summary*. EPA, Washington DC.

Espinosa, A.J.F., Rodriguez, M.T., de la Rosa, F.J.B., & Sanchez, J.C.J. (2001) Size distribution of metals in urban aerosols in Seville (Spain). *Atmospheric Environment*, **35**, 2595–2601.

EU (2000) Directive 2000/60/EC of the European Parliament and of the Council of 23 October 2000 establishing a framework for Community action in the field of water policy. *Official Journal of the European Communities*, **L327**, 1–72.

EU (2008) European Union Directive 2008/50/EC on ambient air quality and cleaner air for Europe *Official Journal of the European Union*, **L 151/1**, 1–44.

Eumorfopoulo, E.A. & Kontoleon, K.J. (2009) Experimental approach to the contribution of plant-covered walls to the thermal behaviour of building envelopes. *Building and Environment*, **44**, 1024–1038.

Faircompanies.com (2012) *Guerrilla grafters: splicing fruit onto a city's trees*. www.youtube.com/watch?v=0osO1_FR_24.

Falchi, F., Cinzano, P., Elvidge, C.D., Keith, D.M., & Haim, A. (2011) Limiting the impact of light pollution on human health, environment and stellar visibility. *Journal of Environmental Management*, **92**, 2714–2722.

Fang, C.F. & Ling, D.L. (2003) Investigation of the noise reduction provided by tree belts. *Landscape and Urban Planning*, **63**, 187–195.

Fang, C.F. & Ling, D.L. (2005) Guidance for noise reduction provided by tree belts. *Landscape and Urban Planning*, **71**, 29–34.

Fann, N. & Risley, D. (2013) The public health context for PM2.5 and ozone air quality trends. *Air Quality Atmosphere and Health*, **6**, 1–11.

Farag, M.A., Zhang, H., & Ryn, C.-M. (2013) Dynamic chemical communication between plants and bacteria through airborne signals: induced resistance by bacterial volatiles. *Journal of Chemical Ecology*, **39**, 1007–1018.

Fassman, E.A. & Blackbourn, S. (2010) Urban runoff mitigation by a permeable pavement system over impermeable soils. *Journal of Hydrologic Engineering*, **15**, 475–485.

FC (2012a) *Ash Dieback Disease*. Forestry Commission, Bristol.

FC (2012b) *Phytophthora ramorum* in larch trees – update. Forestry Commission, www.forestry.gov.uk/forestry/INFD-8EJKP4.

FC (2014) *Phytophthora ramorum*. Forestry Commission, www.forestry.gov.uk/pramorum.

FENSA (undated) *What are the new regulations?* FENSA Ltd London, www.fensa.co.uk/faqs.aspx#.

FERA (2013) *Plant Pests and Diseases*. FERA, York.

Fernandez-Canero, R. & Gonzalez-Redondo, P. (2010) Green roofs as a habitat for birds: a review. *Journal of Animal and Veterinary Advances*, **9**, 2041–2052.

Fernandez-Feijoo, B., Romero, S., & Ruiz, S. (2014) Commitment to Corporate social responsibility measured through global reporting initiative reporting: factors affecting the behaviour of companies. *Journal of Cleaner Production*, **81**, 244–254.

Fino, D. (2007) Diesel emission control: Catalytic filters for particulate removal. *Science and Technology of Advanced Materials*, **8**, 93–100.

Fino, D., Fino, P., Saracco, G., & Specchia, V. (2003) Innovative means for the catalytic regeneration of particulate traps for diesel exhaust cleaning. *Chemical Engineering Science*, **58**, 951–958.

Fintikakis, N., Gaitani, N., Santamouris, M., Assimakopoulos, M., Assimakopoulos, D.N., Fintikaki, M., Albanis, G., Papadimitriou, K., Chryssochoides, E., Katopodi, K., & Doumas, P. (2011) Bioclimatic design of open public spaces in the historic centre of Tirana, Albania. *Sustainable Cities and Society*, **1**, 54–62.

Fjeld, T. (2000) The effect of indoor planting on health and discomfort among workers and schoolchildren. *HortTechnology*, **10**, 46–52.

Fjeld, T., Veiersted, B., Sandvik, L., Riise, G., & Levy, F. (1998) The effect of indoor foliage plants on health and discomfort symptoms among office workers. *Indoor & Built Environment*, **7**, 204–209.

Flannigan, J. (2005) An evaluation of resident's attitudes to street trees in southwest England. *Arboricultural Journal*, **28**, 219–241.

Fleury, A. (2002) Agriculture as an urban infrastructure: a new social contract. In *The Sustainable City II. Urban Regeneration and Sustainability* (ed. by C.A. Brebbia, J.F. Martin-Duque & L.C. Wadhwa), pp. 935–944. WIT Press, Southampton.

FLL (2008) *Guidelines for the planning, execution and upkeep of green roof sites*. Forschüngsgesellschaft Landschaftsentwicklung Landschaftbau e.V., Bonn.

Fontana, C.S., Burger, M.I., & Magnusson, W.E. (2011) Bird diversity in a subtropical South-American City: effects of noise levels, arborisation and human population density. *Urban Ecosystems*, **14**, 341–360.

Forman, R.T.T. (1995) Some general principles of landscape and regional ecology. *Landscape Ecology*, **10**, 133–142.

Forrest, M. & Konijnendijk, C.C. (2005) A history of urban forests and trees in Europe. In *Urban Forests and Trees: A Reference Book* (ed. by C.C. Konijnendijk, K. Nilsson, T.B. Randrup & J. Schipperijn), pp. 23–48. Springer-Verlag, Berlin.

Fowler, D. (2002) Pollutant deposition and uptake by vegetation. In *Air Pollution and Plant Life* (ed. by J.N.B. Bell & M. Treshow), pp. 43–67. Wiley, Chichester.

Fowler, D., Skiba, U., Nemitz, E., Choubedar, F., Branford, D., Donovan, R., & Rowland, P. (2004) Measuring aerosol and heavy metal deposition on urban woodland and grass using inventories of 210PB and metal concentrations in soil. *Water, Air and Soil Pollution*, **4**, 483–499.

Francis, R.A. (2010) Wall ecology: A frontier for urban biodiversity and ecological engineering. *Progress in Physical Geography*, **35**, 43–63.

Francis, R.A. & Hoggart, S.P.G. (2008) Waste not, want not: the need to utilize existing artificial structures for habitat improvement along urban rivers. *Restoration Ecology*, **16**, 373–381.

Francis, R.A. & Hoggart, S.P.G. (2009) Urban river wall habitat and vegetation: observations from the River Thames through central London. *Urban Ecosystems*, **12**, 465–485.

Francis, R.A. & Lorimer, J. (2011) Urban reconciliation ecology: The potential of living roofs and walls. *Journal of Environmental Management*, **92**, 1429–1437.

Franco, A., Fernández-Cañero, R., Pérez-Urrestarazu, L., & Valera, D.L. (2012) Wind tunnel analysis of artificial substrates used in active living walls for indoor environment conditioning in Mediterranean buildings. *Building and Environment*, **51**, 370–378.

Freer-Smith, P.H., Beckett, K.P., & Taylor, G. (2005) Deposition velocities to *Sorbus aria, Acer campestre*,

*Populus deltoides* × *trichocarpa* 'Beaupré', *Pinus nigra* and × *Cambridge University Pressressocyparis leylandii* for coarse, fine and ultra-fine particles in the urban environment. *Environmental Pollution*, **133**, 157–167.

Freer-Smith, P.H., El-Khatib, A.A., & Taylor, G. (2004) Capture of particulate pollution by trees: a comparison of species typical of semi-arid areas (*Ficus nitida* and *Eucalyptus globulus*) with European and north American species. *Water, Air and Soil Pollution*, **155**, 173–187.

Freire, C., Ramos, R., Puertas, R., Lopez-Espinosa, M.J., Julvez, J., Aguilera, I., Cruz, F., Fernandez, M.F., Sunyer, J., & Olea, N. (2010) Association of traffic-related air pollution with cognitive development in children. *Journal of Epidemiology and Community Health*, **64**, 223–228.

Fries, G.F. & Morrow, G.S. (1981) Chlorobiphenyl movement from soil to soybean plants. *J. Agric. Food Chem.*, **29**, 33–40.

Fuhrer, J. & Bungener, P. (1999) Effects of air pollutants on plants. *Analusis*, **27**, 355–360.

Fuller, R.A., Irvine, K.N., Devine-Wright, P., Warren, P.H., & Gaston, K.J. (2007) Psychological benefits of greenspace increase with biodiversity. *Biology Letters*, **3**, 390–394.

Fulong, C.R.P. & Espino, M.P.B. (2013) Decabromodiphenyl ether in indoor dust from different microenvironments in a university in the Philippines. *Chemosphere*, **90**, 42–48.

Fumagalli, A., Faggion, B., Ronchini, M., Terzaghi, G., Lanfranchi, M., Chirico, N., & Cherchi, L. (2013) Platinum, palladium, and rhodium deposition to the *Prunus laurus cerasus* leaf surface as an indicator of the vehicular traffic pollution in the city of Varese area. *Environmental Science and Pollution Research*, **17**, 665–673.

Game, M. (1980) Best shape for nature reserves. *Nature*, **287**, 630–631.

Gange, A.C. & Smith, A.K. (2005) Arbuscular mycorrhizal fungi influence visitation rates of pollinating insects. *Ecological Entomology*, **30**, 600–606.

Garnett, T. (2000) Urban agriculture in London: rethinking our food economy. In *Growing Cities, Growing Food: Urban Agriculture on the Policy Agenda. A Reader on Urban Agriculture* (ed. by N. Bakker, M. Dubbeling, S. Gündel, U. Sabel-Koschella & H.d. Zeeuw), pp. 477–500. Deutsche Stiftung für Internationale Entwicklung (DSE), Zentralstelle für Ernahrung und Landwirtschaft, Feldafing.

Gatrell, J.D. & Jensen, R.R. (2002) Growth through greening: developing and assessing alternative economic development programmes. *Applied Geography*, **22**, 331–350.

Gedge, D., Grant, G., Kadas, G., & Dinham, C. (undated) *Creating Green roofs for Invertebrates: A Best Practice Guide*. Buglife, London.

Gedzelman, S.D., Austin, S., Cermak, R., Stefano, N., Partridge, S., Quesenberry, S., & Robinson, D.A. (2003) Mesoscale aspects of the Urban Heat Island around New York City. *Theoretical and Applied Climatology*, **75**, 29–42.

Gehring, U., Wijga, A.H., Brauer, M., Fischer, P., de Jongste, J.C., Kerkhof, M., Oldenwening, M., Smit, H.A., & Brunekreef, B. (2010) Traffic-related air pollution and the development of asthma and allergies during the first 8 years of life. *American Journal of Respiratory and Critical Care Medicine*, **181**, 596–603.

Gerlach-Spriggs, N., Kaufmann, R.E., & Warner, S.B.J. (1998) *Restorative Gardens: The Healing Landscape*. Yale University Press, New Haven, CT.

Getter, K.L. & Rowe, D.B. (2008) Media depth influences Sedum green roof establishment. *Urban Ecosystems*, **11**, 361–372.

GG (2012) *Guerrilla Grafters*. http://guerrillagrafters.org/.

Giannopoulou, K., Livada, I., Santamouris, M., Saliari, M., Assimakopoulos, M., & Caouris, Y.G. (2011) On the characteristics of the summer urban heat island in Athens, Greece. *Sustainable Cities and Society*, **1**, 16–28.

Gibson, A. & McKenna, M. (2011) The effects of high altitude on the visual system. *Journal of the Royal Army Medical Corps*, **157**, 49–52.

Gibson, C.W.D. (1998) *Brownfield: Red Data. The values artificial habitats have for uncommon invertebrates*. English Nature, Peterborough.

Gilbert, O. (1996) *Rooted in Stone*. English Nature, Peterborough.

GINW (2014) *Green infrastructure toolkit green infrastructure northwest*. www.ginw.co.uk/resources/gi_toolkit.xls.

GLA (2005) *Connecting Londoners with trees and woodlands: a tree and woodland framework for London.* Mayor of London/Greater London Authority, London.

GLA (2007) *Chainsaw massacre. A review of London's street trees.* Greater London Authority, London.

GLA (2012) *Boris marks delivery of his trees and parks promises by planting 10,000th street tree.* Greater London Authority, London.

Godoi, R.H.M., Godoi, A.F.L., de Quadros, L.C., Polezer, G., Silva, T.O.B., Yamamoto, C.I., van Grieken, R., & Potgieter-Vermaak, S. (2013) Risk assessment and spatial chemical variability of PM collected at selected bus stations. *Air Quality Atmosphere and Health*, **6**, 725–735.

Golden, J.S. (2004) The built environment induced urban heat island effect in rapidly urbanizing arid regions – a sustainable urban engineering complexity. *Journal of Integrative Environmental Sciences*, **1**, 321–349.

Golden, R. (2011) Identifying an indoor air exposure limit for formaldehyde considering both irritation and cancer hazards. *Critical Reviews in Toxicology*, **41**, 672–721.

Golding, G. & Christensen, E. (2000) Volatile organic compounds: relevance and measurement in Australia. In *Indoor Air Quality: A Report on Health Impacts and Management Options*, pp. 85–98. Department of Health and Age Care, Commonwealth of Australia, Canberra.

Gombert, S., Asta, J., & Seaward, M.R.D. (2003) Correlation between the nitrogen concentration of two epiphytic lichens and the traffic density in an urban area. *Environmental Pollution*, **123**, 281–290.

Gong, N. (2007) *Green Roofs and Bumblebees: An Observation of Bumblebees on Green Roofs.* The Green Roof Centre, Sheffield.

Goode, D. (2006) *Green infrastructure.* Report to the Royal Commission on Environmental Pollution, London.

Goodman, A., Sahlqvist, S., & Ogilvie, D. (2014) New walking and cycling routes and increased physical activity: one- and 2-year findings from the UK iConnect Study. *American Journal of Public Health*, **104**, E38-E46.

Goodrich, B.A. & Jacobi, W.R. (2012) Foliar damage, ion content, and mortality rate of five common roadside tree species treated with soil applications of magnesium chloride. *Water, Air and Soil Pollution*, **223**, 847–862.

Gore, T., Ozdemiroglu, E., Eadson, W., Gianferrara, E., & Phang, Z. (2013) *Green Infrastructure's Contribution to Economic Growth: A Review.* Defra/Natural England, London.

Gower, S.T., Kucharik, C.J., & Norman, J.M. (1999) Direct and indirect estimation of leaf area index, $f_{APAR}$, and net primary production of terrestrial ecosystems. *Remote Sensing of Environment*, **70**, 29–51.

Grahn, P. & Stigsdottir, U.A. (2003) Landscape planning and stress. *Urban Forestry & Urban Greening*, **2**, 1–18.

Grant, G. (2006a) Extensive green roofs in London. *Urban Habitats*, **4**, 51–65.

Grant, G. (2006b) *Green Roofs and Façades.* IHS BRE Press, Watford.

Grant, G. (2010) *Westfield Green Living wall.* Shepherd's Bush, London. http://livingroofs.org/20101018240/the-grant-column/grantgreenwall.html.

Grant, G., Engleback, L., Nicholson, B., Gedge, D., Frith, M., & Harvey, P. (2003a) *Green roofs: their existing status and potential for conserving biodiversity in urban areas.* English Nature Research Report 498, English Nature, Peterborough.

Grant, R.H., Heisler, G.M., & Gao, W. (2002) Estimation of pedestrian level UV exposure under trees. *Photochemistry and Photobiology*, **75**, 369–376.

Grant, R.H., Heisler, G.M., Gao, W., & Jenks, M. (2003b) Ultraviolet leaf reflectance of common urban trees and the prediction of reflectance from leaf surface characteristics. *Agricultural and Forest Meteorology*, **120**, 127–139.

Gratani, L., Crescente, M.F., & Petruzzi, M. (2000) Relationship between leaf life-span and photosynthetic activity of *Quercus ilex* in polluted urban areas (Rome). *Environmental Pollution*, **110**, 19–28.

Graves, H.M., Watkins, R., Westbury, P., & Littlefair, P.J. (2001) *Cooling Buildings in London.* CRC Ltd., London.

Gray, L. (2009) Smelly ginkgo trees have got to go say American planners. *Daily Telegraph*, October, p. 17.

GRC (2014) *Sheffield bus shelter*. The Green Roof Centre, Sheffield. www.greenroofs.com/projects/pview.php?id=760.

GreenRoofs (2013) *Rhypark*. @green roofs.com, www.greenroofs.com/projects/pview.php?id=141.

Greszta, J. (1982) Accumulation of heavy metals by certain tree species. In *Urban Ecology* (ed. by R. Bornkamm, J.A. Lee & M.R.D. Seaward), pp. 161–165. Blackwell Scientific, Oxford.

Grinde, B. & Patil, G.G. (2009) Biophilia: does visual contact with nature impact on health and well-being? *International Journal of Environmental Research and Public Health*, **6**, 2332–2343.

Grisey, E., Laffray, X., Contoz, O., Cavalli, E., Mudry, J., & Aleya, L. (2011) The Bioaccumulation Performance of Reeds and Cattails in a Constructed Treatment Wetland for Removal of Heavy Metals in Landfill Leachate Treatment (Etueffont, France). *Water, Air & Soil Pollution*, **223**, 1723–1741.

Groenewegen, P.P., den Berg, A.E., de Vries, S., & Verheij, R.A. (2006) Vitamin G: effects of green space on health, well-being, and social safety. *BMC Public Health*, **6**.

Grossi, C.M. & Brimblecombe, P. (2007) Effect of long-term changes in air pollution and climate on the decay and blackening of European stone buildings. In *Building Stone Decay: From Diagnosis to Conservation* (ed. by R. Prikryl and B.J. Smith), pp. 117–130. Special Publications, London.

Groundwork (2011) *The GRO Green Roof Code*. Groundwork Sheffield, Sheffield.

Groundwork (2014) *Green Roof Guide*. Groundwork Sheffield, Sheffield.

GSS (2008) *Health Impact Assessment of Greenspace: A Guide*. Green Space Scotland, Stirling.

Guieysse, B., Hort, C., Platel, V., Munoz, R., Ondarts, M., & Revah, S. (2008) Biological treatment of indoor air for VOC removal: Potential and challenges. *Biotechnology Advances*, **26**, 398–410.

Gulliver, J. & Briggs, D.J. (2007) Journey-time exposure to particulate air pollution. *Atmospheric Environment*, **41**, 7195–7207.

Guo, Z.W., Xiao, X.M., & Li, D.M. (2000) An assessment of ecosystem services: Water flow regulation and hydroelectric power production. *Ecological Applications*, **10**, 925–936.

Gustavsson, R. (1995) A structural approach to woodland plantations – steps towards a new approach when restoring or creating new urban and rural landscapes. In *Landscape Ecology: Theory and Application* (ed. by G.H. Griffiths), pp. 113–120. IALE(UK), University of Reading.

Haeger-Eugensson, M. & Holmer, B. (1999) Advection caused by the urban heat island circulation as a regulating factor on the nocturnal urban heat island. *International Journal of Climatology*, **19**, 975–988.

Hagedoorn, J. & Zucchi, H. (1989) Untersuchungen zur Besiedlung von Kletterpflanzen durch Insekten (Insecta) und Spinnen (Araneae) an Hauswänden. *Landschaft + Stadt*, **21**, 41–55.

Hagler, G.S.W., Lin, M.Y., Khlystov, A., Baldauf, R.W., Isakov, V., Faircloth, J., & Jackson, L.E. (2012) Field investigation of roadside vegetative and structural barrier impact on near-road ultrafine particle concentrations under a variety of wind conditions. *Science of the Total Environment*, **419**, 7–15.

Hakola, H., Laurila, T., Lindfors, V., Hellen, H., Gaman, A., & Rinne, J. (2001) Variation of the VOC emission rates of birch species during the growing season. *Boreal Environment Research*, **6**, 237–249.

Hall, A., Richards, L., Dalbey, M., Desautels, L., Nelson, K., Susman, M., Thomas, J., Wilson, C., Dils, R., Goo, R., Hair, L., Weitman, D., Molloy, J., Malm, S., Estornell, P., & Voigt, G. (2009) *Water Quality Scorecard: Incorporating Green Infrastructure Practices at the Municipal, Neighborhood, and Site Scales*. United States Environmental Protection Agency, Washington DC.

Handley, J.F. & Gill, S.E. (2009) Woodlands helping society to adapt. In *Combating Climate Change – a Role for UK Forests. An assessment of the potential of the UK's trees and woodlands to mitigate and adapt to climate change* (ed. by D.J. Read, P.H. Freer-Smith, J.I.L. Morison, N. Hanley, C.C. West & P. Snowdon), pp. 180–194. The Stationary Office, Edinburgh.

Handley, J.F., Pauleit, S., Slinn, P., Barber, A., Baker, M., Jones, C., & Lindley, S. (2003) *Accessible Natural Green Space Standards in Towns and Cities: a review and toolkit for their implementation*. English Nature Research Report 526, English Nature, Peterborough.

Hänninen, O., Hoek, G., Mallone, S., Chellini, E., Katsouyanni, K., Gariazzo, C., Cattani, G., Marconi, A., Molnár, P., Bellander, T., & Jantunen, M. (2011) Seasonal patterns of outdoor PM infiltration into indoor environments: review and meta-analysis of available studies from different climatological zones in Europe. *Air Quality, Atmosphere & Health*, **4**, 221–233.

Hansen, G. (2010) *Arbor, Trellis, or Pergola—What's in Your Garden? A Mini-Dictionary of Garden Structures and Plant Forms*. University of Florida, Gainesville.

Harrison, M.J. (2005) Signalling in the arbuscular mycorrhizal symbiosis. *Annual Review of Microbiology*, **59**, 19–42.

Harrison, R.M. (1992) *Understanding our Environment: An Introduction to Environmental Chemistry and Pollution*. Royal Society of Chemistry, Cambridge.

Haughton, G. (1999) Environmental justice and the sustainable city. *Journal of Planning Education and Research*, **18**, 233–243.

HederaScreens (2009) *Naturally grown ivy screens for indoor and outdoor use*. Hedera Screens Moira. www. hederascreens.co.uk.

Heisler, G.M. (1986a) Effects of individual trees on the solar radiation climate of small buildings. *Urban Ecology*, **9**, 337–359.

Heisler, G.M. (1986b) Energy saving with trees. *Journal of Arboriculture*, **12**, 113–125.

Heisler, G.M. (1990) Mean wind speed below building height in residential neighborhoods with different tree densities. *ASHRAE Transactions*, **96**, 1389–1396.

Heisler, G.M. & Grant, R.H. (2000) *Ultraviolet Radiation, Human Health and the Urban Forest*. USDA Forest Service, Newtown Square, PA.

Heisler, G.M., Grant, R.H., & Gao, W. (2003) Individual- and scattered-tree influences on ultraviolet irradiance. *Agricultural and Forest Meteorology*, **120**, 113–126.

Helden, A.J., Stamp, G.C., & Leather, S.R. (2012) Urban biodiversity: comparison of insect assemblages on native and non-native trees. *Urban Ecosystems*, **15**, 611–624.

Helm, T. (2014) The science is clear, but Britain is sleepwalking to a climate crisis. *The Observer*, February, pp. 6–7.

Helm, T. & Doward, J. (2014) Climate change is an issue of national security, warns Ed Miliband. *The Observer*, February, pp. 1, 7.

Helmschlager, E. & Kirisits, T. (2008) First report of the ash dieback pathogen *Chalara fraxinea* on *Fraxinus excelsior* in Austria. *Plant Pathology*, **57**, 1177.

Henschel, J.R., Ward, D., & Lubin, Y. (1992) The importance of thermal factors for nest-site selection, web construction and behavior of *Stegodyphus lineatus* (Araneae, Eresidae) in the Negev Desert. *Journal of Thermal Biology*, **17**, 97–106.

Henwood, K. (2001) *Exploring Linkages Between the environment and Health: Is there a role for environmental and Countryside Agencies in Promoting Benefits to Health?* Forestry Commission, Bristol.

Herrera-Montes, M.I. & Aide, T.M. (2011) Impacts of traffic noise on anuran and bird communities. *Urban Ecosystems*, **14**, 415–427.

Herrington, L.P. (1974) Trees and acoustics in urban areas. *Journal of Forestry*, **72**, 462–465.

Hesselink, J.K., Duijn, B.v., Bergen, S.v., Hooff, M.v., & Cornelissen, E. (undated) *Plants enhance productivity in case of creative work: Results of a laboratory experiment in The Netherlands*. Available at: www.landscapeontario.com/attach/1301596822.Plants_Enhance_Productivity.pdf.

Hewitt, A.J. (2001) *Drift Filtration by Natural and Artificial Collectors: a Literature Review*. Stewart Agricultural Research Services, Inc., Macon, MO.

Hewitt, C.N. & Street, R.A. (1992) A qualitative assessment of the emission of nonmethane hydrocarbon compounds from the biosphere to the atmosphere in the UK – present knowledge and uncertainties. *Atmospheric Environment Part a-General Topics*, **26**, 3069–3077.

Hill, D. (2012) London's population up by 12% in 10 years, pp. Blog. *The Guardian*. www.guardian.co.uk/uk/davehillblog/2012/jul/16/london-population-rises-12.

Hill, M.O., Carey, P.D., Eversham, B.C., Arnold, H.R., Preston, C.D., Telfer, M.G., Brown, N.J., Veitch, N., Welch, R.C., Elmes, G.W., & Buse, A. (1993) *The role of corridors, stepping stones, and islands for species conservation in a changing climate*, rep. no. 75. English Nature, Peterborough.

Hirons, A.D. & Percival, G.C. (2012) Fundamentals of tree establishment: a review. In *Trees, People and the Built Environment* (ed. by M. Johnston & G. Percival), pp. 51–62. Forestry Commission, Edinburgh.

Hitchings, R. (2010) Urban greenspace from the inside out: An argument for the approach and a study with city workers. *Geoforum*, **41**, 855–864.

Hoffmann, B., Hertel, S., Boes, T., Weiland, D., & Jockel, K.H. (2008) Increased cause-specific mortality associated with 2003 heat wave in Essen, Germany. *Journal of Toxicology and Environmental Health-Part a-Current Issues*, **71**, 759–765.

Hofman, J., Stokkaer, I., Snauwaert, L., & Samson, R. (2013) Spatial distribution assessment of particulate matter in an urban street canyon using biomagnetic leaf monitoring of tree crown deposited particles. *Environmental Pollution*, **183**, 123–132.

Hoggart, S.P.G., Francis, R.A., & Chadwick, M.A. (2012) Macroinvertebrate richness on flood defence walls of the tidal River Thames. *Urban Ecosystems*, **15**, 327–346.

Honour, S.L., Bell, J.N.B., Ashenden, T.W., Cape, J.N., & Power, S.A. (2009) Responses of herbaceous plants to urban air pollution: Effects on growth, phenology and leaf surface characteristics. *Environmental Pollution*, **157**, 1279–1286.

Horbert, M., Dlume, H.P., Elvers, H., & Sukopp, H. (1982) Ecological contributions to urban planning. In *Urban Ecology* (ed. by R. Bornkamm, J.A. Lee & M.R.D. Seaward), pp. 255–276. Blackwell Scientific, Oxford.

Hosker, R.P. & Lindberg, S.E. (1982) Review – Atmospheric deposition and plant assimilation of gases and particles. *Atmospheric Environment*, **16**, 889–910.

Howard, L. (1833) *The Climate of London, Deduced from Meteorological Observations, Made in the Metropolis and at Various Places Around It*. Harvey & Darton; J & A. Arch; Longman & Co; Hatchard & Son; S. Highley; R. Hunter. New edition by IAUC, London, http://urban-climate.com/wp3/wp-content/uploads/2011/04/LukeHoward_Climate-of-London-V1.pdf.

Huang, Y.J., Akbari, H., Taha, H., & Rosenfeld, A.H. (1987) The potential of vegetation in reducing summer cooling loads in residential buildings. *Journal of Climate and Applied Meteorology*, **26**, 1103–1116.

HuffingtonPost (2012) *Guerilla grafters secretly splicing San Francisco trees with fruit-bearing branches. Huffington Post*, www.huffingtonpost.com/2012/01/04/guerilla-grafters-san-francisco_n_1184790.html.

Humphreys, D.K., Goodman, A., & Ogilvie, D. (2013) Associations between active commuting and physical and mental wellbeing. *Preventive Medicine*, **57**, 135–139.

Hunova, I., Maly, M., Rezacova, J., & Branis, M. (2013) Association between ambient ozone and health outcomes in Prague. *International Archives of Occupational and Environmental Health*, **86**, 89–97.

Hutchinson, D., Abrams, P., Retzlaff, R., & Liptan, T. (2003) Stormwater monitoring two ecoroofs in Portland, Oregon, USA. *Greening Rooftops for Sustainable Communities*, Chicago, 29–30 May, pp. 372–389.

Iaquinta, D.L. & Drescher, A.W. (2002) Food security in cities – a new challenge to development. In *The Sustainable City II. Urban Regeneration and Sustainability* (ed. by C.A. Brebbia, J.F. Martin-Duque & L.C. Wadhwa), pp. 983–994. WIT Press, Southampton.

Ichinose, T., Shimodozono, K., & Hanaki, K. (1999) Impact of anthropogenic heat on urban climate in Tokyo. *Atmospheric Environment*, **33**, 3897–3939.

Ip, K., Lam, M., & Miller, A. (2010) Shading performance of a vertical deciduous climbing plant canopy. *Building and Environment*, **45**, 81–88.

IPCC (2007) Summary for policymakers. In *Contribution of Working Group I to the Fourth Assessment Report of the Intergovernmental Panel on Climate Change, 2007* (ed. by S. Solomon, D. Qin, M. Manning, Z. Chen, M. Marquis, K.B. Averyt, M. Tignor & H.L. Miller). Cambridge University Press, Cambridge.

IPCC (2014) Summary for policymakers. In *Climate Change 2014: Impacts, Adaptation, and Vulnerability. Part A: Global and Sectoral Aspects. Contribution of Working Group II to the Fifth Assessment Report of the Intergovernmental Panel on Climate Change* (ed. by C.B. Field, V.R. Barros, D.J. Dokken, K.J. Mach, M.D. Mastrandrea, T.E. Bilir, M. Chatterjee, K.L. Ebi, Y.O. Estrada, R.C. Genova, B. Girma,

E.S. Kissel, A.N. Levy, S. MacCracken, P.R. Mastrandrea & L.L. White), pp. 1–32. Cambridge University Press, Cambridge.

Irga, P.J., Torpy, F.R., & Burchett, M.D. (2013) Can hydroculture be used to enhance the performance of indoor plants for the removal of air pollutants. *Atmospheric Environment*, **77**, 267–271.

Ishigami, A., Hajat, S., Kovats, R.S., Bisanti, L., Rognoni, M., Russo, A., & Paldy, A. (2008) An ecological time-series study of heat-related mortality in three European cities. *Environmental Health*, **7**, 5.

ISI (2014) State of the Earth Innovation! Invisible Structures Inc, Golden, Colorado, www. invisiblestructures.com/grasspave2.html.

Jackson, J.I. & Boutle, R. (2008) Ecological functions within a sustainable urban drainage system. In *11th International Conference on Urban Drainage*, Edinburgh, Scotland, UK, pp. 1–10.

Jacobs, J.H., Clark, S.J., Denholm, I., Goulson, D., Stoate, C., & Osborne, J.L. (2010) Pollinator effectiveness and fruit set in common ivy, *Hedera helix* (Araliaceae). *Arthropod-Plant Interactions*, **4**, 19–28.

Jacobson, M. (2002) *Atmospheric Pollution: history, science, and regulation*. Cambridge University Press, Cambridge.

Jaluzot, A. (2012) Trees in the townscape: a guide for decision makers trees and design action group. www.tdag.org.uk.

Jansson, M., Fors, H., Lindgren, T., & Wiström, B. (2013) Perceived personal safety in relation to urban woodland vegetation – A review. *Urban Forestry & Urban Greening*, **12**, 127–133.

Jekyll, G. (1982) *Wall, Water and Woodland Gardens*. Reprint of 1933 Edition with 16 additional colour plates. Antique Collectors' Club, Woodbridge.

Jenkins, P.L., Phillips, T.J., Mulberg, E.J., & Hui, S.P. (1992) Activity patterns of Californians – use of and proximity to indoor pollutant sources. *Atmospheric Environment Part a-General Topics*, **26**, 2141–2148.

Jim, C.Y. (2004) Spatial differentiation and landscape-ecological assessment of heritage trees in urban Guangzhou (China). *Landscape and Urban Planning*, **69**, 51–68.

Jim, C.Y. & Chen, W.Y. (2008) Assessing the ecosystem service of air pollutant removal by urban trees in Guangzhou (China). *Journal of Environmental Management*, **88**, 665–76.

Jim, C.Y. & Peng, L.H. (2012) Weather effect on thermal and energy performance of an extensive tropical green roof. *Urban Forestry & Urban Greening*, **11**, 73–85.

JNCC (2010) *UK Biodiversity Action Plan Priority Habitat Descriptions: Open Mosaic Habitats on Previously Developed Land*. JNCC/Defra, Peterborough. http://jncc.defra.gov.uk/PDF/UKBAP_PriorityHabitatDesc-Rev2011.pdf.

Johnson, D.P. & Wilson, J.S. (2009) The socio-spatial dynamics of extreme urban heat events: The case of heat-related deaths in Philadelphia. *Applied Geography*, **29**, 419–434.

Johnston, J. & Newton, J. (2004) *Building Green. A guide to using plants on roofs, walls and pavements*. Greater London Authority, London.

Jonas, R., Horbert, M., & Pflug, W. (1985) Die Filterwirkung von Wäldern gegenüber staubbelasteter Luft. *Forstwissenschaftliches Centralblatt*, **104**, 289–299.

Jones, R.A. (2002) *Tecticolous invertebrates. The invertebrate fauna of ecoroofs in urban London*. English Nature, Peterborough.

Joshi, N. & Bora, M. (2011) Impact of air quality on physiological attributes of certain plants. *Report and Opinion*, **3**, 42–47.

Jovan, S. (2008) *Lichen Bioindication of Biodiversity, Air Quality, and Climate: Baseline Results From Monitoring in Washington, Oregon, and California*. Gen.Tech. Rep. PNW-GTR-737 USDA Forest Service, Pacific Northwest Research Station, Portland, OR.

Joye, Y., Willems, K., Brengman, M., & Wolf, K. (2010) The effects of urban retail greenery on consumer experience: Reviewing the evidence from a restorative perspective. *Urban Forestry & Urban Greening*, **9**, 57–64.

Kadas, G. (2002) Study of invertebrates on green roofs – how roof design can maximise biodiversity in an urban environment. Unpublished Msc Thesis, University College London, London.

Kadas, G. (2006) Rare invertebrates colonizing green roofs in London. *Urban Habitats*, **4**, 66–86.

Kadas, G. (2010) *Green Roofs and Biodiversity: Can Green Roofs Provide Habitat for Invertebrates in an Urban Environment*. LAP Lambert Academic Publishing, Saarbrüken.

Kadiyala, A. & Kumar, A. (2012) An examination of the sensitivity of sulfur dioxide, nitric oxide, and nitrogen dioxide concentrations to the important factors affecting air quality inside a public transportation bus. *Atmosphere*, **3**, 266–287.

Kadiyala, A. & Kumar, A. (2013) Quantification of in-vehicle gaseous contaminants of carbon dioxide and carbon monoxide under varying climatic conditions. *Air Quality Atmosphere and Health*, **6**, 215–224.

Kalbitz, K., Popp, P., Geyer, W., & Hanschmann, G. (1997) ß-HCH mobilisation in polluted wetland soils as influenced by dissolved organic matter. *The Science of the Total Environment*, **204**, 37–48.

Kanechi, M., Fujiwara, S., Shintani, N., Suzuki, T., & Uno, Y. (2014) Performance of herbaceous *Evolvulus pilosus* on urban green roof in relation to substrate and irrigation. *Urban Forestry & Urban Greening*, **13**, 184–191.

Kaplan, R. & Kaplan, S. (1989) *The Experience of Nature: A Psychological Perspective*. Cambridge University Press, Cambridge.

Katsouyanni, K., Touloumi, G., Spix, C., Schwartz, J., Balducci, F., Medina, S., Rossi, G., Wojtyniak, B., Sunyer, J., Bacharova, L., Schouten, J.P., Ponka, A., & Anderson, H.R. (1997) Short term effects of ambient sulphur dioxide and particulate matter on mortality in 12 European cities: Results from time series data from the APHEA project. *British Medical Journal*, **314**, 1658–1663.

Katz, R.W. & Brown, B.G. (1992) Extreme events in a changing climate: variability is more important than averages. *Climatic Change*, **21**, 289–302.

Kaur, S., Nieuwenhuijsen, M.J., & Colvile, R.N. (2005) Pedestrian exposure to air pollution along a major road in Central London, UK. *Atmospheric Environment*, **39**, 7307–7320.

Kaur, S., Nieuwenhuijsen, M.J., & Colvile, R.N. (2007) Fine particulate matter and carbon monoxide exposure concentrations in urban street transport microenvironments. *Atmospheric Environment*, **41**, 4781–4810.

Kazemi, F., Beecham, S., & Gibbs, J. (2011) Streetscape biodiversity and the role of bioretention swales in an Australian urban environment. *Landscape and Urban Planning*, **101**, 139–148.

Kazmierczak, A. & Carter, J. (2010) *Basel, Switzerland: Building Regulations for Green Roofs*. University of Manchester, Manchester. www.grabs-eu.org/membersArea/files/basel.pdf.

Kelly, M. (2012) Urban trees and the green infrastructure agenda. In *Trees, People and the Built Environment. Proceedings of the Urban Trees Research Conference 13–14 April 2011* (ed. by M. Johnson & G. Percival), pp. 166–180. Forestry Commission, Edinburgh.

Kelsey, R.G., Forsman, E.D., & Swingle, J.K. (2009) Terpenoid resin distribution in conifer needles with implications for red tree vole, *Arborimus longicaudus*, foraging. *The Canadian Field-naturalist*, **123**, 12–18.

Kennedy, C.E.J. & Southwood, T.R.E. (1984) The number of species of insects associated with British trees – a re-analysis. *Journal of Animal Ecology*, **53**, 455–478.

Kheirbek, I., Wheeler, K., Walters, S., Kass, D., & Matte, T. (2013) PM2.5 and ozone health impacts and disparities in New York City: sensitivity to spatial and temporal resolution. *Air Quality Atmosphere and Health*, **6**, 473–486.

Kilbourne, E.M., Choi, K., Jones, T.S., & Thacker, S.B. (1982) Risk-factors for heatstroke – a case-control study. *Journal of the American Medical Association*, **247**, 3332–3336.

Kim, K.J., Il Jeong, M., Lee, D.W., Song, J.S., Kim, H.D., Yoo, E.H., Jeong, S.J., Han, S.W., Kays, S.J., Lim, Y.W., & Kim, H.H. (2010) Variation in formaldehyde removal efficiency among indoor plant species. *Hortscience*, **45**, 1489–1495.

Kim, K.J., Kil, M.J., Song, J.S., Yoo, E.H., Son, K.-C., & Kays, S.J. (2008) Efficiency of volatile formaldehyde removal by indoor plants: contribution of aerial plant parts versus the root zone. *Journal of the American Society for Horticultural Science*, **133**, 521–526.

Kim, K.J., Yoo, E.H., & Kays, S.J. (2012) Decay kinetics of toluene phytoremediation stimulation. *Hortscience*, **47**, 1195–1198.

Knowles, L., MacLean, P., Rosato, M., Stanley, C., Volpe, S., Yousif, D., & Wismer, S. (2003) *Living Wall: A Feasibility Study for the SLC University of Waterloo*. Waterloo, Ontario.

Köhler, M. & Poll, P.H. (2010) Long-term performance of selected old Berlin greenroofs in comparison to younger extensive greenroofs in Berlin. *Ecological Engineering*, **36**, 722–729.

Köhler, M. (2006) Long-term vegetation research on two extensive green roofs in Berlin. *Urban Habitats*, **4**, 3–26.

Köhler, M. (2008) Green façades—a view back and some visions. *Urban Ecosystems*, **11**, 423–436.

Köhler, M., Barth, G., Brandwein, T., Gast, D., Joger, H.G., Seitz, U., & Vowinkel, K. (1993) *Fassaden-und Dachbegrünung Ulmer*. Stuttgart.

Köhler, M. & Poll, P.H. (2010) Long-term performance of selected old Berlin green roofs in comparison to younger extensive green roofs in Berlin. *Ecological Engineering*, **36**, 722–729.

Kolb, W. & Schwarz, T. (1986) New habitats on the roof – the possibilities for the provision of extensive verdure. *Anthos*, **1/86**, 4–10.

Kong, F.H., Yin, H.W., & Nakagoshi, N. (2007) Using GIS and landscape metrics in the hedonic price modeling of the amenity value of urban green space: A case study in Jinan City, China. *Landscape and Urban Planning*, **79**, 240–252.

Kontoleon, K.J. & Eumorfopoulou, E.A. (2010) The effect of the orientation and proportion of a plant-covered wall layer on the thermal performance of a building zone. *Building and Environment*, **45**, 1287–1303.

Kopetzki, S. (2010) Berlin: The Biotope Area Factor. In *Adaptation to Climate Change Using Blue and Green Infrastructure: A Database of Case Studies* (ed. by A. Kazmierczak & J. Carter), pp. 9. University of Manchester, Manchester. www.grabs-eu.org/membersarea/files/berlin.pdf.

Kotsiris, G., Nektarios, P.A., Ntoulas, N., & Kargas, G. (2013) An adaptive approach to intensive green roofs in the Mediterranean climatic region. *Urban Forestry & Urban Greening*, **12**, 380–392.

Kovats, S. (ed.) *Health Effects of Climate Change in the UK: An Update of the Department of Health Report 2001/2002*. Department of Health, London.

Kruuse, A. (2004) Sweden's Green Space Factor. In *Greening Rooftops for Sustainable Communities*, 2–4 June, Portland, Oregon.

Kruuse, A. (2006) Green space factor as used in Western Harbour, Malmö, Sweden. Unpublished document sent to JWD, p. 2.

Kruuse, A. (2011) *GRaBS expert paper 6: the green space factor and the green points system*. Town and Country Planning Association, Malmö.

Ksiazek, K., Fant, J., & Skogen, K. (2012) An assessment of pollen limitation on Chicago green roofs. *Landscape and Urban Planning*, **107**, 401–408.

Kuchelmeister, G. (2000) Trees for the urban millennium: urban forestry update. *Unasylva*, **51**, 49–55.

Kumar, T.B., Rahul, Kumar, M.A., & Chandrajit, B. (2011) Biofiltration of volatile organic compounds (VOCs) – an overview. *Research Journal of Chemical Sciences*, **1**, 83–92.

Kuo, F.E. (2003) The role of arboriculture in a healthy social ecology. *Journal of Arboriculture*, **29**, 148–155.

Kuo, F.E. & Taylor, A.F. (2004) A potential natural treatment for attention-deficit/hyperactivity disorder: Evidence from a national study. *American Journal of Public Health*, **94**, 1580–1586.

Kusaka, H. (2008) Recent progress on urban climate study in Japan. *Geographical Review Japan*, **81**, 361–374.

Kweon, B.-S., Sullivan, W.C., & Wiley, A.R. (1998) Green common spaces and the social integration of inner-city older adults. *Environment and Behaviour*, **30**, 832–858.

LaBadie, K.T. & Scully, M. (2012) *Welcome to urban chickens*. Urban Chickens, Iowa City. www.urbanchickens.org.

Larsen, K., Gilliland, J., Hess, P., Tucker, P., Irwin, J., & He, M.Z. (2009) The influence of the physical environment and sociodemographic characteristics on children's mode of travel to and from school. *American Journal of Public Health*, **99**, 520–526.

Larson, D.W., Matthes, U., & Kelly, P.E. (2000) *Cliff Ecology*. Cambridge University Press, Cambridge.

Larson, D.W., Matthes, U., Kelly, P.E., Lundholm, J.T., & Gerrath, J.A. (2004) *The Urban Cliff Revolution: new findings on the origins and evolution of human habitats*. Fitzhenry & Whiteside, Markham, Ontario.

Leather, S. (2010) Do shifting baselines in natural history knowledge threaten the environment? *Environmentalist*, **30**, 1–2.

Lee-Smith, D. (2010) Cities feeding people: an update on urban agriculture in equatorial Africa. *Environment & Urbanization*, **22**, 483–499.

Lee, Z., Viles, H., & Wood, C., eds. (undated) *Soft Capping Historic Walls*. English Heritage, London.

Lees, S. & Evans, P. (2003) *Biodiversity's contribution to the quality of life.* English Nature Research Report 510. English Nature, Peterborough.

Leonard, R.E. & Parr, S.B. (1970) Trees as a sound barrier. *Journal of Forestry* **68**, 282–283.

Lercher, P., Evans, G.W., & Widmann, U. (2013) The ecological context of soundscapes for children's blood pressure. *Journal of the Acoustical Society of America*, **134**, 773–781.

Leuzinger, S., Vogt, R., & Körner, C. (2010) Tree surface temperature in an urban environment. *Agricultural and Forest Meteorology*, **150**, 56–62.

Levins, R. (1970) Extinction. In *Some Mathematical Questions in Biology. Lectures on Mathematics in Life Sciences* (ed. by M. Gerstenhaber), Vol. II, pp. 77–107. American Mathematical Society, Providence.

Lewis, T. (1965) The effects of an artificial windbreak on the aerial distribution of flying insects. *Annals of Applied Biology*, **55**, 503–512.

Lewis, T. & Stephenson, J.W. (1966) The permeability of artificial windbreaks and the distribution of flying insects in the leeward sheltered zone. *Annals of applied Biology*, **58**, 355–365.

Li, G.Q., Huang, H.C., Miao, H.J., Erickson, R.S., Jiang, D.H., & Xiao, Y.N. (2006) Biological control of sclerotinia diseases of rapeseed by aerial applications of the mycoparasite *Coniothyrium minitans*. *European Journal of Plant Pathology*, **114**, 345–355.

Li, H.N., Chau, C.K., & Tang, S.K. (2010) Can surrounding greenery reduce noise annoyance at home? *Science of the Total Environment*, **408**, 4376–4384.

Lindberg, F. & Grimmond, C.S.B. (2011) Nature of vegetation and building morphology characteristics across a city: Influence on shadow patterns and mean radiant temperatures in London. *Urban Ecosystems*, **14**, 617–634.

Lindberg, N., Bengtsson, J., & Persson, T. (2003) Effects of experimental irrigation and drought on the composition and diversity of soil fauna in a coniferous stand. *Journal of Applied Ecology*, **40**, 192–192.

Liptan, T. (2006) Integrating water, soil, and plants with the urban fabric. In *Green Roof Conference 2006.* www.green-roof.group.shef.ac.uk/pdf/TomLipton.pdf, University of Sheffield, UK.

Liptan, T. & Strecker, E. (2003) EcoRoofs (greenroofs) – a more sustainable infrastructure. In *National Conference on Urban Storm Water: Enhancing Programs at the Local Level*, Chicago, pp. 198–214.

Lisci, M., Monte, M., & Pacini, E. (2003) Lichens and higher plants on stone: a review. *International Biodeterioration & Biodegradation*, **51**, 1–17.

Lisci, M. & Pacini, E. (1993) Plants growing on the walls of Italian towns 1. Sites and distribution. *Phyton (Horn, Austria)*, **33**, 15–26.

Litschke, T. & Kuttler, W. (2008) On the reduction of urban particle concentration by vegetation – a review *Meteorologische Zeitschrift*, **17**, 229–240.

Liu, K. & Baskaran, B. (2003) Thermal performance of green roofs through field evaluation. In *Proceedings of the First North American Green Roof Infrastructure Conference, Awards and Trade Show*, Chicago, IL, May 29–30, pp. 1–10. Institute for Research in Construction, National Research Council of Canada, Ottawa.

Liu, Y.J., Mu, Y.-J., Zhu, Y.-G., Ding, H., & Arens, N.C. (2007) Which ornamental plant species effectively remove benzene from indoor air? *Atmospheric Environment*, **41**, 650–654.

Livesley, S.J., Baudinette, B., & Glover, D. (2014) Rainfall interception and stem flow by eucalypt street trees – The impacts of canopy density and bark type. *Urban Forestry & Urban Greening*, **13**, 192–197.

LivingRoofs (2014) *One of the first green roof policies in the World – Linz, Austria.* Living Roofs, www.livingroofs.org/linz-green-roof-policy.

Logan, J.A. (1983) Nitrogen oxides in the trophosphere: global and regional budgets. *Journal of Geophysical Research*, **88**, 10785–10807.

Loh, S. (2008) Living walls – a way to green the built environment. In *BEDP Environment Design Guide*, pp. 1–7. Institute for Built Environment Design Professions Ltd.

Lohr, V.I. & Pearson-Mims, C.H. (1996) Particulate matter accumulation on horizontal surfaces in interiors: influence of foliage plants. *Atmospheric Environment*, **30**, 2565–2568.

Lohr, V.I., Pearson-Mims, C.H., & Goodwin, G.K. (1996) Interior plants may improve worker productivity and reduce stress in a windowless environment. *Journal of Environmental Horticulture*, **14**, 97–100.

Longcore, T. & Rich, C. (2004) Ecological light pollution. *Frontiers of Ecology and Environment*, **2**, 191–198.

Lovasi, G.S., Quinn, J.W., Neckerman, K.M., Perzanowski, M.S., & Rundle, A. (2008) Children living in areas with more street trees have lower prevalence of asthma. *Journal of Epidemiology and Community Health*, **62**, 647–649.

Lovett, G.M. (1994) Atmospheric deposition of nutrients and pollutants in North-America – an ecological perspective. *Ecological Applications*, **4**, 629–650.

LS (undated) *Green Routes – a walking and cycling network for Southwark*. Living Streets. www.livingstreets. org.uk/sites/default/files/content/library/conferencepresentations/GreenRoutesbySouthwarkLivin gStreetsGroup.pdf, London.

Lu, S.G., Zheng, Y.W., & Bai, S.Q. (2006) A HRTEM/EDX approach to identification of the source of dust particles on urban tree leaves. *Atmospheric Environment*, **42**, 6431–6441.

Lubkert, B. & Tilly, d.S. (1989) The OECD-MAP emission inventory for $SO_x$, $NO_x$ and VOC in Western Europe. *Atmospheric Environment*, **23**, 3–15.

Luck, G.W., Davidson, P., Boxall, D., & Smallbone, L. (2011) Relations between Urban Bird and Plant Communities and Human Well-Being and Connection to Nature. *Conservation Biology*, **25**, 816–826.

Luley, C.J. & Bond, J. (2002) *A plan to integrate management of urban trees into air quality planning: a report to North East State Foresters Association*. Davey Resource Group, Naples, NY.

Lundholm, J., MacIvor, J.S., MacDougall, Z., & Ranalli, M. (2010) Plant species and functional group combinations affect green roof ecosystem functions. *Plos One*, **5**.

Lundholm, J.T. (2006) Green roofs and façades: a habitat template approach. *Urban Habitats*, **4**, 87–101.

Lyytimäki, J. (2013) Nature's nocturnal services: Light pollution as a non-recognised challenge for ecosystem services research and management. *Ecosystem Services*, **3**, e44–e48.

Maas, J., Verheij, R.A., Groenewegen, P.P., de Vries, S., & Spreeuwenberg, P. (2006) Green space, urbanity, and health: how strong is the relation? *Journal of Epidemiology and Health*, **60**, 587–592.

Maas, J., Verheij, R.A., Spreeuwenberg, P., & Groenewegen, P.P. (2008) Physical activity as a possible mechanism behind the relationship between green space and health: a multilevel analysis. *BMC Public Health*, **8**, 206.

MacCarthy, J., Thistlethwaite, G., Salisbury, S., Pang, Y., & Misselbrook, T. (2012) *Air quality pollutant inventories for England, Scotland, Wales and Northern Ireland: 1990–2010*. A report of the National Atmospheric Emissions Inventory AEA Technology, Didcot.

MacIvor, J.S. & Lundholm, J. (2011) Performance evaluation of native plants suited to extensive green roof conditions in a maritime climate. *Ecological Engineering*, **37**, 407–417.

Maco, S.E. & McPherson, E.G. (2003) A practical approach to assessing structure, function and value of street tree populations in small communities. *Journal of Arboriculture*, **29**, 84–97.

Madaleno, I.M. (2002) Urban agriculture, an environmentally sustainable land use – case studies from Europe, Latin America and Africa. In C.A. Brebbia, J.F. Martin-Duque & L.C. Wadhwa (eds), *The Sustainable City II: Urban Regeneration and Sustainability*. WIT Press, Southampton.

Madre, F., Vergnes, A., Machon, N., & Clergeau, P. (2013) A comparison of 3 types of green roof as habitats for arthropods. *Ecological Engineering*, **57**, 109–117.

Magill, J.D., Midden, K., Groninger, J., & Therrell, M. (2011) *A History and Definition of Green Roof Technology with Recommendations for Future Research*. Southern Illinois University Carbondale, Carbondale.

Maher, B.A., Moore, C., & Matzka, J. (2008) Spatial variation in vehicle-derived metal pollution identified by magnetic and elemental analysis of roadside tree leaves. *Atmospheric Environment*, **42**, 364–373.

Manes, F., Incerti, G., Salvatori, E., Vitale, M., Ricotta, C., & Costanza, R. (2012) Urban ecosystem services: tree diversity and stability of tropospheric ozone removal. *Ecological Applications*, **22**, 349–360.

Mansfield, C., Pattanayak, S.K., McDow, W., McDonald, R., & Halpin, P. (2005) Shades of green: Measuring the value of urban forests in the housing market. *Journal of Forest Economics*, **11**, 177–199.

Martens, R., Bass, B., & Alcazar, S.S. (2008) Roof–envelope ratio impact on green roof energy performance. *Urban Ecosystems*, **11**, 399–408.

Martin, A., Goryakin, Y., & Suhrcke, M. (in press) Does active commuting improve psychological wellbeing? Longitudinal evidence from eighteen waves of the British Household Panel Survey. *Preventive Medicine*, http://dx.doi.org/10.1016/j.ypmed.2014.08.023.

Matteson, K.C. & Langellotto, G.A. (2011) Small scale additions of native plants fail to increase beneficial insect richness in urban gardens. *Insect Conservation and Diversity*, **4**, 89–98.

Maurer, E. (2006) *Green Roofs in Linz Municipality of Linz*. Linz, Austria.

McBride, M.B., Shayler, H.A., Spliethoff, H.M., Mitchell, R.G., Marquez-Bravo, L.G., Ferenz, G.S., Russell-Anelli, J.M., Casey, L., & Bachman, S. (2014) Concentrations of lead, cadmium and barium in urban garden-grown vegetables: The impact of soil variables. *Environmental Pollution*, **194**, 254–261.

McCarthy, H.R. & Pataki, D.E. (2010) Drivers of variability in water use of native and non-native urban trees in the greater Los Angeles area. *Urban Ecosystems*, **13**, 393–414.

McCarthy, K. & Lathrop, R.G. (2011) Stormwater basins of the New Jersey coastal plain: subsidies or sinks for frogs and toads? *Urban Ecosystems*, **14**, 395–413.

McDonald, A.G., Bealey, W.J., Fowler, D., Dragosits, U., Skiba, U., Smith, R.I., Donovan, R.G., Brett, H.E., Hewitt, C.N., & Nemitz, E. (2007) Quantifying the effect of urban tree planting on concentrations and depositions of PM10 in two UK conurbations. *Atmospheric Environment*, **41**, 8455–8467.

McLain, R., Poe, M., Hurley, P.T., Lecompte-Mastenbrook, J., & Enery, M.R. (2012) Producing edible landscapes in Seattle's urban forest. *Urban Forestry & Urban Greening*, **11**, 187–194.

McMichael, A.J., Woodruff, R.E., & Hale, S. (2006) Climate change and human health: present and future risks. *Lancet*, **367**, 859–869.

McPherson, E.G. (1988) Functions of buffer plantings in urban environments. *Agriculture, Ecosystems and Environment*, **22/23**, 281–291.

McPherson, E.G. (1994) Cooling urban heat islands with sustainable landscapes. In *Sustainable cities: preserving and restoring biodiversity* (ed. by R. Platt & R.A. Rowntree), pp. 151–171. University of Massachusetts Press, Amherst.

McPherson, E.G. (2001) Sacramento's parking lot shading ordinance: environmental and economic costs of compliance. *Landscape and Urban Planning*, **57**, 105–123.

McPherson, E.G. & Kotow, L. (2013) A municipal forest report card: Results for California, USA. *Urban Forestry & Urban Greening*, **12**, 134–43.

McPherson, E.G. & Rowntree, R.A. (1989) Using structural measures to compare twenty-two U.S. street tree populations. *Landscape Journal*, **8**, 13–23.

McPherson, E.G. & Rowntree, R.A. (1993) Energy conservation potential of urban tree planting. *Journal of Arboriculture* **19**, 321–331.

McPherson, E.G. & Simpson, J.R. (2002) A comparison of municipal forest benefits and costs in Modesto and Santa Monica, California, USA. *Urban Forestry and Urban Greening*, **1**, 61–74.

McPherson, E.G. & Simpson, J.R. (2003) Potential energy savings in buildings by an urban tree planting programme in California. *Urban Forestry and Urban Greening*, **2**, 73–86.

McPherson, E.G., Simpson, J.R., Peper, P.J., Gardner, S.L., Vargas, K.E., Maco, S.E., & Xiao, Q. (2006) *Piedmont Community Tree Guide: Benefits, Costs and Strategic Planting*. USDA Forest Service, Davis, CA.

McVeigh, T. & Layton, J. (2012) *More forest sites infected as ash disease takes hold. The Observer*. www.theguardian.com/environment/2012/oct/28/more-forest-sites-ash-disease.

Melichar, J. & Kaprova, K. (2013) Revealing preferences of Prague's homebuyers toward greenery amenities: The empirical evidence of distance-size effect. *Landscape and Urban Planning*, **109**, 56–66.

Mell, I.C. (2007) Green Infrastructure Planning: What are the Costs for Health and Well-Being? *The International Journal of Environmental, Cultural, Economic and Social Sustainability*, **3**, 117–124.

Melzer, B., Seidel, R., Steinbrecher, T., & Speck, T. (2011) Structure, attachment properties, and ecological importance of the attachment system of English ivy (*Hedera helix*). *Journal of Experimental Botany*, **63**, 191–201.

Mentens, J., Raes, D., & Hermy, M. (2006) Green roofs as a tool for solving the rainwater runoff problem in the urbanized 21st century? *Landscape and Urban Planning*, **77**, 217–226.

Middelie, G.-J., ed. (2009) *Groene gevels (Green Façades)*. University of Applied Sciences Van Hall Larenstein, Velp.

Milligan, C., Gatrell, A., & Bingley, A. (2004) 'Cultivating health': therapeutic landscapes and older people in northern England. *Social Science and Medicine*, **58**, 1781–1793.

Minke, G. & Witter, G. (1982) *Häuser mit grünen pelz. Ein handbuch zue hausbegrünung*. Verlag Dieter Fricke GmbH, Frankfurt.

Mir, M.A. (2011) *Green façades and building structures*. MSc, Delft University of Technology, Delft.

Mitchell, R. & Maher, B.A. (2009) Evaluation and application of biomagnetic monitoring of traffic-derived particulate pollution. *Atmospheric Environment*, **43**, 2095–2103.

Mitchell, R. & Popham, F. (2007) Greenspace, urbanity and health: relationships in England. *Journal of Epidemiology and Community Health*, **61**, 681–683.

Mitchell, R. & Popham, F. (2008) Effect of exposure to natural environment on health inequalities: an observational population study. *The Lancet*, **372**, 1655–1660.

Mitchell, R.G., Spliethoff, H.M., Ribaudo, L.N., Lopp, D.M., Shayler, H.A., Marquez-Bravo, L.G., Lambert, V.T., Ferenz, G.S., Russell-Anelli, J.M., Stone, E.B., & McBride, M.B. (2014) Lead (Pb) and other metals in New York City community garden soils: Factors influencing contaminant distributions. *Environmental Pollution*, **187**, 162–169.

MITTMH (2009) *Reglamento de Instalaciones Termicas de los Edificios (RITE)*. Ministry of Industry, Tourism & Trade & Ministry of Housing, Madrid.

Mobilane (undated-a) *Case Study: Monaco Living Wall*. Mobilane, Kidsgrove.

Mobilane (undated-b) *Live Panel Living Wall*. Mobilane (UK) Ltd, Kidsgrove.

Moll, G. (1989) The state of our urban forest. *American Forests*, **November 1**, 61–64.

Moran, A., Hunt, B., & Jennings, G. (2004) A North Carolina field study to evaluate greenroof runoff quantity, runoff quality, and plant growth. In *Proceedings of the 2nd North American Green Roof Conference: Greening rooftops for sustainable communities*, pp. 446–460. The Cardinal Group, Toronto.

Morikawa, H., Higaki, A., Nohno, M., Takahasi, M., Kamada, M., Nakata, M., Toyohara, G., Okamura, Y., Matsui, K., Kitani, S., Fujita, K., Irifune, K., & Goshima, N. (1998) More than a 600-fold variation in nitrogen dioxide assimilation among 217 plant taxa. *Plant Cell and Environment*, **21**, 180–190.

Morikawa, H., Takahashi, M., Hakata, M., & Sakamoto, A. (2003) Screening and genetic manipulation of plants for decontamination of pollutants from the environments. *Biotechnology Advances*, **22**, 9–15.

Morris, N. (2003) *Health, Well-being and Open Space. OPENspace: the research centre for inclusive access to outdoor environments*. Edinburgh College of Art and Heriot-Watt University, Edinburgh.

MSU (2008) *Acanthopanax sieboldianus*: fiveleaf aralia. Michigan State University, East Lansing.

Muir, D. (1998) PM10 particulates in relation to other atmospheric pollutants. *Environmental Monitoring and Assessment*, **52**, 29–42.

Müller, N. (2007) *Bestandsentwicklung von Brutvögeln ausgewählter Kleingartenanlagen Leipzigs*. Diploma thesis. Hochschule Anhalt (FH), Bernburg, Germany.

Murdock, V. (2012) *Ginkgo trees making a stink in New York City Parks*. CBS New York. http://newyork.cbslocal.com/2012/10/11/ginkgo-trees-making-a-stink-in-new-york-city-parks/.

Murphy, J.M., Sexton, D.M.H., Jenkins, G.J., Boorman, P.M., Booth, B.B.B., Brown, C.C., Clark, R.T., Collins, M., Harris, G.R., Kendon, E.J., Betts, R.A., Brown, S.J., Howard, T.P., Humphrey, K.A., McCarthy, M.P., McDonald, R.E., Stephens, A., Wallace, C., Warren, R., Wilby, R., & Wood, R.A. (2009) *UK Climate Projections Science Report: Climate change projections*. Met Office Hadley Centre, Exeter.

Nagase, A. & Dunnett, N. (2010) Drought tolerance in different vegetation types for extensive green roofs: Effects of watering and diversity. *Landscape and Urban Planning*, **97**, 318–327.

Nagase, A. & Dunnett, N. (2011) The relationship between percentage of organic matter in substrate and plant growth in extensive green roofs. *Landscape and Urban Planning*, **103**, 230–236.

Nagase, A. & Dunnett, N. (2013a) Amount of water runoff from different vegetation types on extensive green roofs: Effects of plant species, diversity and plant structure. *Landscape and Urban Planning*, **104**, 356–363.

Nagase, A. & Dunnett, N. (2013b) Establishment of an annual meadow on extensive green roofs in the UK. *Landscape and Urban Planning*, **112**, 50–62.

Nagase, A. & Dunnett, N. (2013c) Performance of geophytes on extensive green roofs in the United Kingdom. *Urban Forestry & Urban Greening*, **12**, 509–521.

Nagase, A., Dunnett, N., & Choi, M.S. (2010) Investigation of Dynamic Cycles of Semi-Extensive Green Roof. In *Acta Horticulturae II^{nd} International Conference on Landscape and Urban Horticulture* (ed. by G.P. Prosdocimi Gianquinto & F. Orsini), Vol. 881, pp. 653–660. International Society for Horticultural Science, Leuven.

Nagase, A., Dunnett, N., & Choi, M.S. (2013) Investigation of weed phenology in an establishing semi-extensive green roof. *Ecological Engineering*, **58**, 156–164.

Nagendra, H. & Gopal, D. (2010) Street trees in Bangalore: density, diversity, composition and distribution. *Urban Forestry & Urban Greening*, **9**, 129–137.

National Urban Forestry Unit (2001) *Funding urban forestry.* National Urban Forestry Unit, Wolverhampton.

NBN (2012) *NBN gateway.* http://data.nbn.org.uk/.

NE (2009a) *Green Infrastructure Guidance.* Natural England, Sheffield.

NE (2009b) *Our Natural Health Service: The Role of the Natural Environment in Maintaining Healthy Lives.* Natural England, Sheffield.

Neilan, C. (2010) *CAVAT (Capital Asset Value for Amenity Trees) Full Method: User's Guide.* London Tree Officers Association, London.

Neilan, C. (undated) *CAVAT Tree Value: Approaches to Tree Value.* Available at: www.cavattv.org/intro duction.html

NENW (undated-a) *Assessing the potential for green infrastructure developments within projects.* Interim Report Natural Economy Northwest.

NENW (undated-b) *Developing an outline strategy for linking green and grey infrastructure.* Natural Economy Northwest.

NENW (undated-c) *The economic benefits of green infrastructure: developing key tests for evaluating the benefits of green infrastructure.* Natural Economy Northwest.

NENW (undated-d) *The economic value of green infrastructure.* Natural Economy Northwest.

NENW (undated-e) *How to deliver, measure and demonstrate the economic contribution of the natural environment at a project level: a guide for project managers.* DRAFT Natural Economy Northwest.

Newman, A.P., Coupe, S.J., Smith, H.G., Puehmeir, T., & Bond, P.C. (2006) The microbiology of permeable pavements. *8th International Conference on Concrete Block Paving,* November 6–8, San Francisco, California, USA, pp. 181–191.

Newman, A.P., Pratt, C.J., Coupe, S.J., & Cresswell, N. (2002) Oil bio-degradation in permeable pavements by microbial communities. *Water Science & Technology*, **45**, 51–56.

Newman, A.P., Puehmeir, T., Kwok, V., Lam, M., Coupe, S.J., Shuttleworth, A., & Pratt, C.J. (2004) Protecting groundwater with oil-retaining pervious pavements: historical perspectives, limitations and recent developments. *Quarterly Journal of Engineering Geology and Hydrology*, **37**, 283–291.

Newman, L.F., Cosgrove, W., Kates, R.W., Matthews, R., & Millman, S., eds. (1990) *Hunger in History: Food Shortage, Poverty and Deprivation.* Basil Blackwell, Oxford.

Niachou, A., Papakonstantinou, K., Santamouris, M., Tsangrassoulis, A., & Mihalakakou, G. (2001) Analysis of the green roof thermal properties and investigation of its energy performance. *Energy and Buildings*, **33**, 719–729.

NICE (2008a) *Physical activity and the environment: costing report.* National Institute for Health and Clinical Excellence, London.

NICE (2008b) *Promoting and creating built or natural environments that encourage and support physical activity.* NICE Public Health Guidance 8. National Institute for Clinical Excellence, London.

Niemelä, J., Saarela, S.R., Soderman, T., Kopperoinen, L., Yli-Pelkonen, V., Vare, S., & Kotze, D.J. (2010) Using the ecosystem services approach for better planning and conservation of urban green spaces: a Finland case study. *Biodiversity and Conservation*, **19**, 3225–3243.

Nieuwenhuis, M., Knight, C., Postmes, T., & Haslam, S.A. (2014) The relative benefits of green versus lean office space: Three field experiments. *Journal of Experimental Psychology. Applied*, **20**, 199–214.

Nitis, T., Klaic, Z.B., & Moussiopoulos, N. (2005) Effects of topography on urban heat island. In *10th International Conference on Harmonisation within Atmospheric Dispersion Modelling for Regulatory Purposes*, Crete, p. 5.

Nowak, D.J. (1994a) Air pollution removal by Chicago's urban forest. In *Chicago's Urban Forest Ecosystem: Results of the Chicago Urban Forest Climate Project* (ed. by E.G. McPherson, D.J. Nowak & R.A. Rowntree), pp. 63–79. USDA Forest Service, Radnor, Pennsylvania.

Nowak, D.J. (1994b) Atmospheric carbon dioxide reduction by Chicago's urban forest. In *Chicago's Urban Forest Ecosystem: Results of the Chicago Urban Forest Climate Project* (ed. by E.G. McPherson, D.J. Nowak & R.A. Rowntree), pp. 83–94. USDA Forest Service, Radnor, Pennsylvania.

Nowak, D.J. (1994c) Urban forest structure: the state of Chicago's urban forest. In *Chicago's Urban Forest Ecosystem: Results of the Chicago Urban Forest Climate Project* (ed. by E.G. McPherson, D.J. Nowak & R.A. Rowntree), pp. 3–18. USDA Forest Service, Radnor, Pennsylvania.

Nowak, D.J. (1999) Impact of urban forest management on air pollution and greenhouse gasses. *Proceedings of the Society of American Foresters*, pp. 143–148.

Nowak, D.J. (2008) *Species Selector (Beta) Utility.* USDA Forest Service Syracuse, NY. www.itreetools. org/species/resources/SpeciesSelectorMethod.pdf.

Nowak, D.J. & Crane, D.E. (2002) Carbon storage and sequestration by urban trees in the USA. *Environmental Pollution*, **116**, 381–389.

Nowak, D.J., Crane, D.E., & Stevens, J.C. (2006) Air pollution removal by urban trees and shrubs in the United States. *Urban Forestry and Urban Greening*, **4**, 115–123.

Nowak, D.J. & Greenfield, E.J. (2012) Tree and impervious cover change in U.S. cities. *Urban Forestry & Urban Greening*, **11**, 21–30.

Nowak, D.J., Greenfield, E.J., Hoehn, R.E., & Lapoint, E. (2013) Carbon storage and sequestration by trees in urban and community areas of the United States. *Environmental Pollution*, **178**, 229–236.

Nowak, D.J. & Heisler, G.M. (2010) *Air Quality Effects of Urban Trees and Parks. Executive Summary.* National Recreation and Parks Association, Ashburn, Va.

Nowak, D.J., McPherson, E.G., & Rowntree, R.A. (1994) Executive Summary. Chicago's Urban Forest Ecosystem: Results of the Chicago Urban Forest Climate Project. In *Chicago's Urban Forest Ecosystem: Results of the Chicago Urban Forest Climate Project* (ed. by E.G. McPherson, D.J. Nowak & R.A. Rowntree), pp. iii–vi. USDA Forest Service, Radnor, Pennsylvania.

NULBC (2000) *Local Agenda 21: Working together to preserve our planet.* Newcastle under Lyme Borough Council, Newcastle under Lyme.

NWGITT (2007) *North West green infrastructure guide.* North West Green Infrastructure Think Tank, Warrington.

O'Neill, M.S., Breton, C.V., Devlin, R.B., & Utell, M.J. (2012) Air pollution and health: emerging information on susceptible populations. *Air Quality Atmosphere and Health*, **5**, 189–201.

O'Neill, M.S., Carter, R., Kish, J.K., Gronlund, C.J., White-Newsome, J.L., Manarolla, X., Zanobetti, A., & Schwartz, J.D. (2009) Preventing heat-related morbidity and mortality: New approaches in a changing climate. *Maturitas*, **64**, 98–103.

Oberndorfer, E., Lundholm, J., Bass, B., Coffman, R.R., Doshi, H., Dunnett, N., Gaffin, S., Köhler, M., Liu, K.K.Y., & Rowe, B. (2007) Green Roofs as Urban Ecosystems: Ecological Structures, Functions, and Services. *BioScience*, **57**, 823–833.

Ockenden, W.A., Steinnes, E., Parker, C., & Jones, K.C. (1998) Observations on persistent organic pollutants in plants: implications for their use as passive air samplers and POP cycling. *Environmental Science and Technology*, **32**, 2721–2726.

Ode, A.K. & Fry, G.L.A. (2002) Visual aspects in urban woodland management. *Urban Forestry & Urban Greening*, **1**, 15–24.

Ogilvie, D., Bull, F., Powell, J., Cooper, A.R., Brand, C., Mutrie, N., Preston, J., & Rutter, H. (2011) An Applied Ecological Framework for Evaluating Infrastructure to Promote Walking and Cycling: The iConnect Study. *American Journal of Public Health*, **101**, 473–481.

Ogilvie, D., Egan, M., Hamilton, V., & Petticrew, M. (2004) Promoting walking and cycling as an alternative to using cars: systematic review. *British Medical Journal*, **329**, 763–766B.

Ohlsson, T. (2004) *Birds and insects in Augustenborg Ekostad*. Department of Ecology, Lund University, Lund.

Oke, T.R. (1973) City size and the urban heat island. *Atmospheric Environment*, **7**, 769–779.

Oke, T.R. (1979) *Review of urban climatology 1973–1976*. World Meterological Organisation, Geneva.

Oke, T.R. (1981) Canyon geometry and the nocturnal urban heat-island – comparison of scale model and field observations. *Journal of Climatology*, **1**, 237–254.

Oke, T.R., Spronken-Smith, R.A., Jáuregui, E. & Grimmond, C.S.B. (1999) The energy balance of central Mexico City during the dry season. *Atmospheric Environment*, **33**, 3919–3930.

Oliver, I. & Beattie, A.J. (1996) Invertebrate morphospecies as surrogates for species: a case study. *Conservation Biology*, **10**, 99–109.

Olly, L.M., Bates, A.J., Sadler, J.P., & Mackay, R. (2011) An initial experimental assessment of the influence of substrate depth on floral assemblage for extensive green roofs. *Urban Forestry & Urban Greening*, **10**, 311–316.

Onishi, A., Cao, X., Ito, T., Shi, F., & Imura, H. (2010) Evaluating the potential for urban heat-island mitigation by greening parking lots. *Urban Forestry & Urban Greening*, **9**, 323–332.

Optigrün (undated) *Data Sheet: Hydroseeding Process*. Optigrün international AG Krauchenwies-Göggingen, Germany.

Orru, H., Maasikmets, M., Tamm, T., Kaasik, M., Kimmel, V., Orru, K., Merisalu, E., & Forsberg, B. (2011) Health impacts of particulate matter in five major Estonian towns: main sources of exposure and local differences. *Air Quality, Atmosphere & Health*, **4**, 247–258.

Orwa, C., Mutua, A., Kindt, R., Jamnadass, R., & Anthony, S. (2009) *Agroforestry tree database:* Sterculia foetida. *Agroforestree Database: a tree reference and selection guide version 4.0*. International Center for Research in Agroforestry, Nairobi, Kenya, www.worldagroforestrycentre.org/sea/Products/AFDbases/af/asp/SpeciesInfo.asp?SpID=98.

Orwell, R.L., Wood, R.A., Burchett, M.D., Tarran, J., & Torpy, F. (2006) The potted-plant microcosm substantially reduces indoor air VOC pollution: II. Laboratory study. *Water, Air and Soil Pollution*, **177**, 59–80.

Orwell, R.L., Wood, R.L., Tarran, J., Torpy, F., & Burchett, M.D. (2004) Removal of benzene by the indoor plant/substrate microcosm and implications for air quality. *Water, Air and Soil Pollution*, **157**, 193–207.

Östberg, J., Martinsson, M., Stal, O., & Fransson, A.-M. (2012) Risk of root intrusion by tree and shrub species into sewer pipes in Swedish urban areas. *Urban Forestry & Urban Greening*, **11**, 65–71.

Ostendorf, M., Retzlaff, W., Thompson, K., Woolbright, M., Morgan, S., & Celik, S. (2011) Stormwater runoff from green retaining wall systems. In *Cities Alive: 9th Annual Green Roof and Wall Conference*, Philadelphia, p. 15.

Ottelé, M. (2011) *The Green Building Environment: Vertical Greening*. Technical University Delft, Delft.

Ottelé, M., Koleva, D.A., van Breugel, K., Haas, E.M., Fraaij, A.L.A., & van Bohemen, H.D. (2010a) Concrete as a multifunctional ecological building material: a new approach to green our environment. *Ecology & Safety*, **4**, 223–234.

Ottelé, M., Perini, K., Fraaij, A.L.A., Haas, E.M., & Raiteri, R. (2011) Comparative life cycle analysis for green façades and living wall systems. *Energy and Buildings*, **43**, 3419–3429.

Ottelé, M., van Bohemen, H.D., & Fraaij, A.L.A. (2010b) Quantifying the deposition of particulate matter on climber vegetation on living walls. *Ecological Engineering*, **36**, 154–162.

Ould-Dada, Z. & Baghini, N.M. (2001) Resuspension of small particles from tree surfaces. *Atmospheric Environment*, **35**, 3799–3809.

Overstrom, N. (2010) Potsdammer Platz. In *Landscape Architecture Study Tour* (ed. by J. Ahern). Amhurst, MA, Department of Landscape Architecture and Regional Planning, University of Massachusetts.

Owen, J. (2010) *Wildlife of a Garden: A thirty-five year study.* The Royal Horticultural Society, Peterborough.

Oyabu, T., Sawada, A., Onodera, T., Takenaka, K., & Wolverton, B. (2003) Characteristics of potted plants for removing offensive odours. *Sensors and Actuators B,* **89,** 131–136.

Padilla, B. (2011) *Gastronomy – Seville oranges.* Andalucia Com S.L., Malaga. www.andalucia.com/gastronomy/oranges/sevilla.htm.

Pandit, R., Polyakov, M., Tapsuwan, S., & Moran, T. (2013) The effect of street trees on property value in Perth, Western Australia. *Landscape and Urban Planning,* **110,** 134–142.

Panter, J., Griffin, S., & Ogilvie, D. (2014) Active commuting and perceptions of the route environment: A longitudinal analysis. *Preventive Medicine,* **67,** 134–140.

Paoletti, E., Ferrara, A.M., Calatayud, V., Cervero, J., Giannetti, F., Sanz, M.J., & Manning, W.J. (2009) Deciduous shrubs for ozone bioindication: *Hibiscus syriacus* as an example. *Environmental Pollution,* **157,** 865–870.

Paoletti, L., Falchi, M., Viviano, G., Ziemacki, G., Batisti, D., & Pisani, D. (1989) Features of airborne breathable particulate in a remote rural and in an urban area. *Water, Air and Soil Pollution,* **43,** 85–94.

Papworth, S.K., Rist, J., Coad, L., & Milner-Gulland, E.J. (2009) Evidence for shifting baseline syndrome in conservation. *Conservation Letters,* **2,** 93–100.

Parisi, A. & Kimlin, M.G. (1999) Comparison of the spectral biologically effective solar ultraviolet in adjacent tree shade and sun. *Physics in Medicine and Biology,* **44,** 2071–2080.

Parisi, A., Kimlin, M.G., & Turnbull, D.J. (2001) Spectral shade ratios on horizontal and sun normal surfaces for single trees and relatively cloud free sky. *Journal of Photochemistry and Photobiology B: Biology,* **65,** 151–156.

Parisi, A., Kimlin, M.G., Wong, J.C.F., & Wilson, M. (2000) Diffuse component of solar ultraviolet radiation in tree shade. *Journal of Photochemistry and Photobiology B: Biology,* **54,** 116–120.

Park, S.-H. & Mattson, R.H. (2009) Therapeutic influences of plants in hospital rooms on surgical recovery. *Hortscience,* **44,** 102–105.

Parker, D.B., Malone, G.W., & Walter, W.D. (2012) Vegetative environmental buffers and exhaust fan deflectors for reducing downwind odor and VOCs from tunnel-ventilated swine barns. *Transactions of the Asabe,* **55,** 227–240.

Parker, J.H. (1983) Landscaping to reduce the energy used in cooling buildings. *Journal of Forestry,* **81,** 82–84 + 105.

Patel, N. (2006) *Biodiversity & Barclays.* Sheffield University, Sheffield. www.green-roof.group.shef.ac.uk/pdf/nitapatel.pdf.

Pathak, V., Tripathi, B.D., & Mishra, V.K. (2011) Evaluation of Anticipated Performance Index of some tree species for green belt development to mitigate traffic generated noise. *Urban Forestry & Urban Greening,* **10,** 61–66.

Pauleit, S. (2003) Urban street tree plantings: indentifying the key requirements. *Proceedings of the Institution of Civil Engineers-Municipal Engineer,* **156,** 43–50.

Payne, R.M. (2000) *The flora of roofs.* Privately published, Kings Lynn.

Peck, S.W. (2003) Towards an integrated green roof infrastructure evaluation for Toronto. *The Green Roof Infrastructure Monitor* www.greenroofs.ca, **5,** 4–5.

Peck, S.W., Clallaghan, C., Kuhn, M.E., & Bass, B. (1999) *Greenbacks from Green Roofs: Forging a new industry in Canada. Status report on benefits, barriers and opportunities for green roof and vertical garden technology diffusion.* Canada Mortgage and Housing Corporation, Ottawa.

Peper, P.J., McPherson, E.G., Suimpson, J.R., Vargas, K.E., & Xiao, Q. (2008) *City of Indianapolis, Indiana Municipal Forest Resource Analysis.* Centre for Urban Forest Research, Davis, CA.

Perez, G., Rincon, L., Vila, A., Gonzalez, J.M., & Cabeza, L.F. (2011a) Behaviour of green façades in Mediterranean Continental climate. *Energy Conversion and Management,* **52,** 1861–1867.

Perez, G., Rincon, L., Vila, A., Gonzalez, J.M., & Cabeza, L.F. (2011b) Green vertical systems for buildings as passive systems for energy savings. *Applied Energy,* **88,** 4854–4859.

Perini, K. & Magliocco, A. (2014) Effects of vegetation, urban density, building height, and atmospheric conditions on local temperatures and thermal comfort. *Urban Forestry & Urban Greening*, **13**, 495–506.

Perini, K., Ottele, M., Fraaij, A.L.A., Haas, E.M., & Raiteri, R. (2011a) Vertical greening systems and the effect on air flow and temperature on the building envelope. *Building and Environment*, **46**, 2287–2294.

Perini, K., Ottele, M., Haas, E.M., & Raiteri, R. (2011b) Greening the building envelope, façade greening and living wall systems. *Open Journal of Ecology*, **1**, 1–8.

Perini, K. & Rosasco, P. (2013) Cost-benefit analysis for green façades and living wall systems. *Building and Environment*, **70**, 110–121.

Perre, P., Loyola, R.D., Lewinsohn, T.M., & Almeida-Neto, M. (2011) Insects on urban plants: contrasting the flower head feeding assemblages on native and exotic hosts. *Urban Ecosystems*, **14**, 711–722.

Perry, T. & Nawaz, R. (2008) An investigation into the extent and impacts of hard surfacing of domestic gardens in an area of Leeds, United Kingdom. *Landscape and Urban Planning*, **86**, 1–13.

Peters, E.B. & McFadden, J.P. (2010) Influence of seasonality and vegetation type on suburban microclimates. *Urban Ecosystems*, **13**, 443–460.

Petroff, A., Mailliat, A., Amielh, M., & Anselmet, F. (2008) Aerosol dry deposition on vegetative canopies. Part I: Review of present knowledge. *Atmospheric Environment*, **42**, 3625–3653.

Pham, T.-T.-H., Apparicio, P., Landry, S., Séguin, A.-M., & Gagnon, M. (2013) Predictors of the distribution of street and backyard vegetation in Montreal, Canada. *Urban Forestry & Urban Greening*, **12**, 18–27.

Phoothiwut, S. & Junyapoon, S. (2013) Size distribution of atmospheric particulates and particulate-bound polycyclic aromatic hydrocarbons and characteristics of PAHs during haze period in Lampang Province, Northern Thailand. *Air Quality, Atmosphere and Health*, **6**, 397–405.

Phytokinetic (2013) *Iberflora 2012 Award for Best Sustainable Initiative*. Phytokinetic, www.phytokinetic.net/blog/2013/03/15/premi-iberflora-12-a-la-millor-iniciativa-sostenible/.

Pitt, M. (2008) *Learning Lessons from the 2007 Floods*. The Cabinet Office, London.

Pocock, M.J.O., Chapman, D.S., Sheppard, l.J., & Roy, H.E. (2014a) *Choosing and Using Citizen Science: a guide to when and how to use citizen science to monitor biodiversity and the environment*. Centre for Ecology & Hydrology, Wallingford.

Pocock, M.J.O., Chapman, D.S., Sheppard, l.J., & Roy, H.E. (2014b) *A Strategic Framework to Support the Implementation of Citizen Science for Environmental Monitoring*. Centre for Ecology & Hydrology, Wallingford.

Pollard, E., Hooper, M.D., & Moore, N.W. (1974) *Hedges*. Collins, London.

Pomerantz, M., Pon, B., Akbari, H., & Chang, S.-C. (2000) *The Effect of Pavements' Temperatures On Air Temperatures in Large Cities*. Lawrence Berkeley National Laboratory, Berkeley.

Pope III, C.A., Brook, R.D., Burnett, R.T., & Dockery, D.W. (2011) How is cardiovascular disease mortality risk affected by duration and intensity of fine particulate matter exposure? An integration of the epidemiologic evidence. *Air Quality, Atmosphere and Health*, **4**, 5–14.

Posudin, Y. (2010) *Volatile organic compounds in indoor air: scientific, medical and instrument aspects*. National University of Life and Environmental Sciences of Ukraine, Kiev, www.ekmair.ukma.kiev.ua/bitstream/123456789/885/1/Posudin_Volatile%20Organic%20Compounds%20in%20Indoor%20Air.pdf.

Praag, R.v. (2011) *Groen Moet je Doen: Groen moet je doen; de effecten van groen gebruiken in de bouw om het leefklimaat van de mens te verbeteren*. Delft University of Technology, Delft.

Prajapati, S.K. & Tripathi, B.D. (2008) Seasonal variation of leaf dust accumulation and pigment content in plant species exposed to urban particulates pollution. *Journal of Environmental Quality*, **37**, 865–870.

Pratt, C.J. (2004) *Sustainable drainage: a review of published material on the performance of various SUDS components prepared for the Environment Agency*. Coventry University, Coventry.

Pratt, C.J., Newman, A.P., & Bond, P.C. (1999) Mineral oil bio-degradation within a permeable pavement: long term observations. *Water Science & Technology*, **39**, 103–109.

Pretty, J., Griffin, M., Peacock, J., Hine, R., Sellens, M., & South, S. (2005a) *A countryside for health and wellbeing. The physical and mental health benefits of green exercise.* Countryside Recreation Network, Sheffield.

Pretty, J., Peacock, J., Sellens, M., & Griffin, M. (2005b) The mental and physical health outcomes of green exercise. *International Journal of Environmental Health Research,* **15**, 319–337.

Prokop, G., Jobstmann, H., & Schönbauer, A. (2011) *Overview of best practices for limiting soil sealing or mitigating its effects in EU-27.* European Commission DG Environment, doi: 10.2779/15146.

Prusty, B.A.K., Mishra, P.C., & Azeez, P.A. (2005) Dust accumulation and leaf pigment content in vegetation near the national highway at Sambalpur, Orissa, India. *Ecotoxicology and Environmental Safety,* **60**, 228–235.

Pucher, J., Buehler, R., Bassett, D.R., & Dannenberg, A.L. (2010) Walking and Cycling to Health: A Comparative Analysis of City, State, and International Data. *American Journal of Public Health,* **100**, 1986–1992.

Pucher, J. & Dijkstra, L. (2000) Making walking and cycling safer: Lessons from Europe. *Transportation Quarterly,* **54**, 25–50.

Pugh, T.A.M., MacKenzie, A.R., Whyatt, J.D., & Hewitt, C.N. (2012) Effectiveness of green infrastructure for improvement of air quality in urban street canyons. *Environmental Science & Technology,* **46**, 7692–7699.

Qin, J., Zhou, X., Sun, C., Leng, H., & Lian, Z. (2013) Influence of green spaces on environmental satisfaction and physiological status of urban residents. *Urban Forestry & Urban Greening,* **12**, 490–497.

Quine, C.P. & Humphrey, J.W. (2010) Plantations of exotic tree species in Britain: irrelevant for biodiversity or novel habitat for native species? *Biodiversity and Conservation,* **19**, 1503–1512.

Raaschou-Nielsen, O., Andersen, Z.J., Beelen, R., Samoli, E., Stafoggia, M., Weinmayr, G., Hoffmann, B., Fischer, P., Nieuwenhuijsen, M.J., Brunekreef, B., Xun, W.W., Katsouyanni, K., Dimakopoulou, K., Sommar, J., Forsberg, B., Modig, L., Oudin, A., Oftedal, B., Schwarze, P.E., Nafstad, P., De Faire, U., Pedersen, N.L., Östenson, C.-G., Fratiglioni, L., Penell, J., Korek, M., Pershagen, G., Eriksen, K.T., Sørensen, M., Tjønneland, A., Ellermann, T., Eeftens, M., Peeters, P.H., Meliefste, K., Wang, M., Bueno-de-Mesquita, B., Key, T.J., de Hoogh, K., Concin, H., Nagel, G., Vilier, A., Grioni, S., Krogh, V., Tsai, M.-Y., Ricceri, F., Sacerdote, C., Galassi, C., Migliore, E., Ranzi, A., Cesaroni, G., Badaloni, C., Forastiere, F., Tamayo, I., Amiano, P., Dorronsoro, M., Trichopoulou, A., Bamia, C., Vineis, P., & Hoek, G. (2013a) Air pollution and lung cancer incidence in 17 European cohorts: prospective analyses from the European Study of Cohorts for Air Pollution Effects (ESCAPE). *The Lancet Oncology,* **14**, 813–822.

Raaschou-Nielsen, O., Sorensen, M., Ketzel, M., Hertel, O., Loft, S., Tjonneland, A., Overvad, K., & Andersen, Z.J. (2013b) Long-term exposure to traffic-related air pollution and diabetes-associated mortality: a cohort study. *Diabetologia,* **56**, 36–46.

Radford, T. (2004) Do trees pollute the atmosphere? *The Guardian.* www.guardian.co.uk/science/2004/may/13/thisweekssciencequestions3.

Rai, A. & Kulshreshtha, K. (2006) Effect of particulates generated from automobile emission on some common plants. *Journal of Food, Agriculture and Environment,* **4**, 253–259.

Ranasinghe, T.T. (2002) Manual of Law/No-space Agriculture-cum-Family Business Gardens. RUAF Foundation, Leusden, Netherlands.

Randrup, T.B., McPherson, E.G., & Costello, L.R. (2001) Tree root intrusion in sewer systems: review of the extent and costs. *Journal of Infrastructure Systems,* **7**, 26–31.

Rath, J. & Kiessl, K. (1989) Auswirkungen von Fassadenbegrunungen auf die Warme- und Feuchtehaushalt von Aussenwanden ubd schadenstisiko. *Fraunhofer-Institut fur Bauphysik, IBP-Bericht Ftb-4/1989.*

RationalWiki (2012) *Trees cause pollution.* RationalWiki. http://rationalwiki.org/wiki/Trees_cause_pollution.

RCEP (2009) *Artificial Light in the Environment.* The Royal Commission on Environmental Pollution, London.

Reich, P.B., Oleksyn, J., Modrzynski, J., & Tjoelker, M.G. (1996) Evidence that longer needle retention of spruce and pine populations at high elevations and high latitudes is largely a phenotypic response. *Tree Physiology*, **16**, 643–647.

Reid, L. & Hunter, C. (2011) *BeWEL: behaviour for well-being, environment & life. State of understanding report 1: Personal well being and interactions with nature.* University of Aberdeen, Aberdeen.

Renninger, H.J., Wadhwa, S., Gallagher, F.J., Vanderklein, D., & Schafer, K.V.R. (2013) Allometry and photosynthetic capacity of poplar (*Populus deltoides*) along a metal contamination gradient in an urban brownfield. *Urban Ecosystems*, **16**, 247–263.

Rentao, L., Runcheng, B., & Halin, Z. (2008) Dust removal property of major afforested plants in and around an urban area, North China. *Ecology & Environment*, **17**, 1879–1886.

Rey, G., Fouillet, A., Bessemoulin, P., Frayssinet, P., Dufour, A., Jougla, E., & Hemon, D. (2009) Heat exposure and socio-economic vulnerability as synergistic factors in heat-wave-related mortality. *European Journal of Epidemiology*, **24**, 495–502.

Rey, G., Fouillet, A., Jougla, E., & Hemon, D. (2007) Heat waves, ordinary temperature fluctuations and mortality in France since 1971. *Population*, **62**, 533–563.

Rhine, J.B. (1924) Clogging of stomata of conifers in relation to smoke injury and distribution. *Botanical Gazette*, **78**, 226–232.

RHS (2011) *Tulipa saxatilis* (Bakeri Group) 'Lilac Wonder' (15). AGM. Royal Horticultural Society, Wisley. http://apps.rhs.org.uk/plantselector/plant?plantid=4249.

RHS (undated) *Front Gardens: Are We Parking on Our Gardens? Do Driveways Cause Flooding?* Royal Horticultural Society, London.

Rich, C. (2012) *Urban Farms.* Abrams, New York.

Richards, N.A. (1983) Diversity and stability in a street tree population. *Urban Ecology*, **7**, 159–171.

Riddell-Black, D. (1994) Heavy metal uptake by fast growing willow species. In *Willow Vegetation Gilters for Municipal Wasewaters and Sludges: A Biological Purification System* (ed. by P. Aronsson & K. Perttu), pp. 145–150. Sveriges Lantbruksuniversitet, Uppsala.

Roman, L.A. & Scatena, F.N. (2011) Street tree survival rates: Meta-analysis of previous studies and application to a field survey in Philadelphia, PA, USA. *Urban Forestry & Urban Greening*, **10**, 269–274.

Rosenthal, J.K., Crauderueff, R., & Carter, M. (2008) *Urban heat island mitigation can improve New York City's environment: research on the impacts of mitigation strategies.* Sustainable South Bronx, New York.

Rosenzweig, C., Iglesias, A., Yang, X.B., Epstein, P.R., & Chivian, E. (2001) Climate change and extreme weather events; implications for food production, plant diseases, and pests. *Global Change and Human Health*, **2**, 90–104.

Rowe, D.B. (2011) Green roofs as a means of pollution abatement. *Environmental Pollution*, **159**, 2100–2110.

Rowe, D.B., Getter, K.L., & Durhman, A.K. (2012) Effect of green roof media depth on Crassulacean plant succession over seven years. *Landscape and Urban Planning*, **104**, 310–319.

Rudel, R.A., Camann, D.E., Spengler, J.D., Korn, L.R., & Brody, J.G. (2003) Phthalates, alkylphenols, pesticides, polybrominated diphenyl ethers, and other endocrine-disrupting compounds in indoor air and dust. *Environmental Science & Technology*, **37**, 4543–4553.

Rushton, B.T. (2001) Low-impact parking lot design reduces runoff and pollutant loads. *Journal of Water Resources Planning and Management*, **127**, 172–179.

Rustin, S. (2012) Are urban chicken-keepers doing more harm than good? *The Guardian*, 23 November. www.theguardian.com/lifeandstyle/2012/nov/23/urban-chicken-keepers-doing-harm.

Rutgers, R. (2012) *Living façades: a study on the sustainable features of vegetated façade cladding.* MSc, Delft University of Technology, Delft.

Sæbø, A., Benedikz, T., & Randrup, T.B. (2003) Selection of trees for urban forestry in the Nordic countries. *Urban Forestry & Urban Greening*, **2**, 101–114.

Sæbø, A., Borzan, Ž., Ducatillion, C., Hatzistathis, A., Lagerström, T., Supuka, J., García-Valdecantos, J.L., Regio, F., & Van Slycken, J. (2005) The selection of plant materials for street trees, park trees and urban woodland. In *Urban Forests and Trees: A Reference Book* (ed. by C.C. Konijnendijk, K. Nilsson, T.B. Randrup & J. Schipperijn), pp. 257–280. Springer-Verlag, Berlin.

Sæbø, A., Popek, R., Nawrot, B., Hanslin, H.M., Gawronska, H., & Gawronski, S.W. (2012) Plant species differences in particulate matter accumulation on leaf surfaces. *Science of the Total Environment*, **427–428**, 347–354.

Saelensminde, K. (2004) Cost-benefit analyses of walking and cycling track networks taking into account insecurity, health effects and external costs of motorized traffic. *Transportation Research Part a-Policy and Practice*, **38**, 593–606.

Sandhu, A., Halverson, L.J., & Beattie, G.A. (2007) Bacterial degradation of airborne phenol in the phyllosphere. *Environmental Microbiology*, **9**, 383–392.

Sarajevs, V. (2011) *Health Benefits of Street Trees*. Forest Research, Alice Holt.

SCC (2011) *Sheffield Development Framework: Climate Change and Design. Supplementary Planning Document and Practice Guide*. Sheffield City Council, Sheffield.

Schaffner, A., Messner, B., Langebartels, C., & Sandermann, H. (2002) Genes and enzymes for in-planta phytoremediation of air, water and soil. *Acta Biotechnologica*, **22**, 141–151.

Schmidt, M. (2009) Rainwater harvesting for mitigating local and global warming. In *Fifth Urban Research Symposium 2009: Cities and Climate Change: Responding to an Urgent Agenda*, Marseille, France, pp. 1–15.

Scholtz, M. & Grabowiecki, P. (2007) Review of permeable pavement systems. *Building and Environment*, **42**, 3830–3836.

Scholtz, M. & Grabowiecki, P. (2009) Combined permeable pavement and ground source heat pump systems to treat urban runoff. *Journal of Chemical Technology and Biotechnology*, **84**, 405–413.

Schroeder, H., Flannigan, J., & Coles, R. (2006) Resident's attitudes toward street trees in the UK and US communities. *Arboriculture and Urban Forestry*, **32**, 236–246.

Schuler, T.R. (1994) The importance of imperviousness. *Watershed Protection Technology*, **1**, 100–111.

Scotscape (undated) *The ANSystem Modular Living Wall*. Scotscape Ltd, Mitcham Junction.

Scott, K.I., McPherson, E.G., & Simpson, J.R. (1998) Air pollutant uptake by Sacramento's urban forest. *Journal of Arboriculture*, **24**, 224–234.

Scott, K.I., Simpson, J.R., & McPherson, E.G. (1999) Effects of tree cover on parking lot microclimate and vehicle emissions. *Journal of Arboriculture*, **25**, 129–142.

Seamans, G.S. (2013) Mainstreaming the environmental benefits of street trees. *Urban Forestry & Urban Greening*, **12**, 2–11.

SedumDirect (undated) *SD Sedum blankets*. Sedum Direct, Ysselsteyn, Limburg, Netherlands, www.sedumdirect.nl/gb/products/sedum_blankets.

Segschneider, H.J. (1995) Effects of atmospheric nitrogen-oxides ($N_{ox}$) on plant-metabolism – a review. *Angewandte Botanik*, **69**, 60–85.

Sehmel, G.A. (1980) Particle and gas deposition – a review. *Atmospheric Environment*, **14**, 983–1011.

Selosse, M.-A., Richard, F., He, X., & Simard, S.W. (2006) Mycorrhizal networks: des liaisons dangereuses? *Trends in Ecology & Evolution*, **21**, 621–628.

Sendo, T., Kanechi, M., Uno, Y., & Inagaki, N. (2010) Evaluation of growth and green coverage of ten ornamental species for planting as urban rooftop greening. *Journal of the Japanese Society for Horticultural Science*, **79**, 69–76.

SG (2003) *Water Environment and Water Services (Scotland) Act 2003*. The Scottish Government, Edinburgh.

SG (2005) *The Water Environment (Controlled Activities) (Scotland) Regulations 2005*. The Scottish Government, Edinburgh.

SG (2009) *The Flood Risk Management (Scotland) Act 2009*. The Scottish Government, Edinburgh.

Sharkey, T.D., Wiberley, A.E., & Donohue, A.R. (2008) Isoprene emission from plants: why and how. *Annals of Botany*, **101**, 5–18.

Shepherd, J.M., Pierce, H., & Negri, A.J. (2002) Rainfall modification by major urban areas: Observations from spaceborne rain radar on the TRMM satellite. *Journal of Applied Meteorology*, **41**, 689–701.

Sheweka, S. & Magdy, N. (2011) The living walls as an approach for a healthy urban environment. *Energy Procedia*, **6**, 592–599.

Shibata, S. & Suzuki, N. (2002) Effects of the foliage plant on task performance and mood. *Journal of Environmental Psychology*, **22**, 265–272.

Shirley-Smith, C. (2002) Integrated water management as a tool for sustainable urban regeneration. In *The Sustainable City II. Urban Regeneration and Sustainability* (ed. by C.A. Brebbia, J.F. Martin-Duque & L.C. Wadhwa), pp. 117–131. WIT Press, Southampton.

Shuttleworth, C.M., Lurz, P.W.W., Geddes, N., & Browne, J. (2012) Integrating red squirrel (*Sciurus vulgaris*) habitat requirements with the management of pathogenic tree disease in commercial forests in the UK. *Forest Ecology and Management*, **279**, 167–175.

Sieghardt, M., Mursch-Radlgruber, E., Paoletti, E., Couenberg, E., Dinmitrakopoulus, A., Rego, F., Hatzistathis, A., & Randrup, T.B. (2005) The abiotic urban environment: impact of urban growing conditions on urban vegetation. In *Urban Forests and Trees: A Reference Book* (ed. by C.C. Konijnendijk, K. Nilsson, T.B. Randrup & J. Schipperijn). Springer-Verlag, Berlin.

Sillmann, J. & Roeckner, E. (2008) Indices for extreme events in projections of anthropogenic climate change. *Climatic Change*, **86**, 83–104.

Simberloff, D. & Cox, J. (1987) Consequences and costs of conservation corridors. *Conservation Biology*, **1**, 63–71.

Simmons, M.T., Gardiner, B., Windhager, S., & Tinsley, J. (2008) Green roofs are not created equal: the hydrologic and thermal performance of six different extensive green roofs and reflective and non-reflective roofs in a sub-tropical climate. *Urban Ecosystems*, **11**, 339–348.

Simon, E., Braun, M., Vidic, A., Bogyo, D., Fabian, I., & Tothmeresz, B. (2011) Air pollution assessment based on elemental concentration of leaves tissue and foliage dust along an urbanization gradient in Vienna. *Environmental Pollution*, **159**, 1229–1233.

Simpson, J.R. (1998) Urban forest impacts on regional cooling and heating energy use: Sacramento County case study. *Journal of Arboriculture*, **24**, 201–214.

Simpson, J.R. & McPherson, E.G. (2011) The tree BVOC index. *Environmental Pollution*, **159**, 2088–2093.

Sjöman, H. (2012) *Trees for tough urban sites: learning from nature.* PhD, Swedish University of Agricultural Sciences, Alnarp.

Sjöman, H., Gunnarsson, A., Pauleit, S., & Bothmer, R. (2012a) Selection approach of urban trees for inner-city environments: learning from nature. *Arboriculture & Urban Forestry*, **38**, 194–204.

Sjöman, H. & Nielsen, A.B. (2010) Selecting trees for urban paved sites in Scandinavia – A review of information on stress tolerance and its relation to the requirements of tree planters. *Urban Forestry & Urban Greening*, **9**, 281–293.

Sjöman, H., Nielsen, A.B., & Oprea, A. (2012b) Trees for urban environments in northern parts of Central Europe – a dendroecological study in north-east Romania and Republic of Moldavia. *Urban Ecosystems*, **15**, 267–281.

Sjöman, H., Nielsen, A.B., Pauleit, S., & Olsson, M. (2010) Habitat studies identifying potential trees for urban paved environments: a case study from Qinling Mt., China. *Arboriculture & Urban Forestry*, **36**, 261–271.

Sjöman, H., Ostberg, J., & Buhler, O. (2012c) Diversity and distribution of the urban tree population in ten major Nordic cities. *Urban Forestry & Urban Greening*, **11**, 31–39.

SKLCC (2014) *KLCC Park.* Suria KLCC Sdn Bhd, www.suriaklcc.com.my/attractions/klcc-park/, Kuala Lumpur.

Sliney, D.H. (1986) Physical factors in cataractogenesis: ambient ultraviolet radiation and temperature. *Investigative Opthalmology and Visual Science*, **27**, 781–790.

Smiley, E.J. (2008) Root pruning and stability of young willow oak. *Arboriculture & Urban Forestry*, **34**, 123–128.

Smiley, E.T., Calfee, L., Fraedrich, B.R., & Smiley, E.J. (2006) Comparison of structural and noncompacted soils for trees surrounded by pavement. *Arboriculture & Urban Forestry*, **34**, 164–169.

Smit, J., Ratta, A., & Nasr, J. (1996) *Urban Agriculture: Food, Jobs and Sustainable Cities*. UNDP Habitat II Series Vol 1 UNDP, New York.

Smith, A. & Pitt, M. (2011) Healthy workplaces: plantscaping for indoor environmental quality. *Facilities*, **29**, 169–187.

Smith, A.D., Tucker, M., & Pitt, M. (2011a) Healthy, productive workplaces: towards a case for interior plantscaping. *Facilities*, **29**, 209–223.

Smith, C., Dawson, D., Archer, J., Davies, M., Frith, M., Hughes, E., & Massini, P. (2011b) *From Green to Grey; Observed Changes in Garden Vegetation Structure in London 1998–2008*. London Wildlife Trust, London.

Smith, J.P. (1999) Healthy bodies and thick wallets: the dual relation between health and economic status. *The Journal of Economic Perspectives*, **13**, 145–166.

Smith, K.E.C. & Jones, K.C. (2000) Particles and vegetation: implications for the transfer of particle-bound organic contaminants to vegetation. *The Science of the Total Environment*, **246**, 207–236.

Smith, S. (1899) Vegetation a remedy for the summer heat of cities. *Appleton's Popular Science Monthly*, **54**, 433–450.

Smith, W.H. (1977) Removal of atmospheric particulates by urban vegetation: implications for human and vegetative health. *The Yale Journal of Botany and Medicine*, **50**, 185–197.

Snodgrass, E.C. & McIntyre, L. (2010) *The Green Roof Manual: A Professional Guide to Design, Installation and Maintenance*. Timber Press, Portland.

Snodgrass, E.C. & Snodgrass, L.L. (2006) *Green Roof Plants: A Resources and Planting Guide*. Timber Press, Portland.

Solecki, W.D., Rosenzweig, C., Oarshall, L., Pope, G., Clark, M., Cox, J., & Wiencke, M. (2005) Mitigation of the heat island effect in urban New Jersey. *Environmental Hazards*, **6**, 39–49.

Solomon, P.A., Gehr, P., Bennett, D.H., Phalen, R.F., Mendez, L.B., Rothen-Rutishauser, B., Clift, M., Brandenberger, C., & Muhlfeld, C. (2012) Macroscopic to microscopic scales of particle dosimetry: from source to fate in the body. *Air Quality, Atmosphere and Health*, **5**, 169–187.

Solomon, S. (2004) The hole truth. *Nature*, **427**, 289–291.

Solomon, S., Qin, D., Manning, M., Chen, Z., Marquis, M., Averyt, K.B., Tignor, M., & Miller, H.L., eds. (2007) *IPCC, 2007: Summary for policymakers*. Cambridge University Press, Cambridge.

Sonkin, B., Edwards, P., Roberts, I., & Green, J. (2006) Walking, cycling and transport safety: an analysis of child road deaths. *Journal of the Royal Society of Medicine*, **99**, 402–405.

Sorensen, M., Andersen, Z.J., Nordsborg, R.B., Becker, T., Tjonneland, A., Overvad, K., & Raaschou-Nielsen, O. (2013) Long-Term Exposure to Road Traffic Noise and Incident Diabetes: A Cohort Study. *Environmental Health Perspectives*, **121**, 217–222.

Southwood, T.R.E. (1961) The number of species of insect associated with various trees. *Journal of Animal Ecology*, **30**, 1–8.

Southwood, T.R.E., Moran, V.C., & Kennedy, C.E.J. (1982) The Richness, Abundance and Biomass of the Arthropod Communities on Trees. *Journal of Animal Ecology*, **51**, 635–649.

Speak, A.F., Rothwell, J.J., Lindley, S.J., & Smith, C.L. (2012) Urban particulate pollution reduction by four species of green roof vegetation in a UK city. *Atmospheric Environment*, **61**, 283–293.

Specht, K., Siebert, R., Hartmann, I., Freisinger, U.B., Sawicka, M., Werner, A., Thomaier, S., Henckel, D., Walk, H., & Dierich, A. (2014) Urban agriculture of the future: an overview of sustainability aspects of food production in and on buildings. *Agriculture and Human Values*, **31**, 33–51.

Speirs, L.J. (2003) Sustainable planning: the value of green space. In *First international conference on sustainable planning and development* (ed. by E. Beriatos, C.A. Brebbia, H. Coccossis & A. Kungolos), pp. 337–346. WIT Press, Skiathos, Greece.

Spellerberg, I. (1992) *Evaluation and Assessment for Conservation: Ecological Guidelines for Determining Priorities for Nature Conservation*. Chapman & Hall, London.

Spellerberg, I. & Green, D.G. (2008) Trees in Urban and City Environments: a review of the selection criteria with particular reference to nature conservation in New Zealand Cities. *Landscape Review*, **12**, 19–31.

Spellerberg, I.F. & Gaywood, M. (1989) *Linear features: linear habitats & wildlife corridors*. English Nature, Peterborough.

Spellerberg, I.F., Eriksson, N.E., & Crump, V.S.A. (2006) Silver birch (*Betula pendula*) pollen and human health: problems for an exotic tree in New Zealand. *Arboriculture & Urban Forestry*, **32**, 133–137.

Spolek, G. (2008) Performance monitoring of three ecoroofs in Portland, Oregon. *Urban Ecosystems*, **11**, 349–359.

Sram, R.J., Binkova, B., Beskid, O., Milcova, A., Rossner, P., Rossner, P.J., Rossnerova, A., Solansky, I., & Topinka, J. (2011) Biomarkers of exposure and effect—interpretation in human risk assessment. *Air Quality, Atmosphere and Health*, **4**, 161–167.

Šrámek, F. & Dubský, M. (2007) Effect of slow release fertilizers on container-grown woody plants. *Horticultural Science (Prague)*, **34**, 35–41.

Stanton, W.R., Janda, M., Baade, P.D., & Anderson, P. (2004) Primary prevention of skin cancer: a review of sun protection in Australia and internationally. *Health Promotion International*, **19**, 369–378.

Starke, P., Gobel, P., & Coldewey, W.G. (2010) Urban evaporation rates for water-permeable pavements. *Water Science and Technology*, **62**, 1161–1169.

Steffan, C., Schmidt, M., Köhler, M., Hűbner, I., & Reichmann, B. (2010) *Rainwater Management Concepts: Greening Buildings, Cooling Buildings*. Berlin Senate for Urban Development Communications, Berlin.

Steinbach, R., Green, J., Datta, J., & Edwards, P. (2011) Cycling and the city: A case study of how gendered, ethnic and class identities can shape healthy transport choices. *Social Science & Medicine*, **72**, 1123–1130.

Steinbrecher, T., Danninger, E., Harder, D., Speck, T., Kraft, O., & Schwaiger, R. (2010) Quantifying the attachment strength of climbing plants: A new approach. *Acta Biomaterialia*, **6**, 1497–1504.

Sternberg, T., ed. (2010) *Ivy on Walls*. English Heritage, London.

Sternberg, T., Viles, H., & Cathersides, A. (2011a) Evaluating the role of ivy (*Hedera helix*) in moderating wall surface microclimates and contributing to the bioprotection of historic buildings. *Building and Environment*, **46**, 293–297.

Sternberg, T., Viles, H., Cathersides, A., & Edwards, M. (2010) Dust particulate absorption by ivy (*Hedera helix* L) on historic walls in urban environments. *Science of The Total Environment*, **409**, 162–168.

Sternberg, T., Viles, H., & Edwards, M. (2011b) Absorption of Airborne Particulates and Pollutants by Ivy (*Hedera helix* L) in Oxford, UK. *5th International Conference on Bioinformatics and Biomedical Engineering*, p. 4.

Stewart, H.E., Owen, S., Donovan, R.G., MacKenzie, Hewitt, N., Skiba, U., & Fowler, D. (undated-a) *Trees and Sustainable Air Quality: Using Trees to improve Air Quality in Cities*. Dept. of Environmental Science, Lancaster University, Lancaster.

Stewart, H.E., Street, R.A., Scholefield, P.A., & R. Hewitt, C.N. (undated-b) *Isoprene and monoterpene-emitting species survey*. Available at: www.es/lancs.ac.uk/chgroup/iso-emissions.pdf.

Stewart, R.A. & Hytiris, N. (2008) The role of Sustainable Urban Drainage Systems in reducing the flood risk associated with infrastructure. A93 Craighall Gorge to Middle Mause Farm – road re-alignment, Perth, Scotland–Case Study. *11th International Conference on Urban Drainage*. Edinburgh, Scotland, p. 15.

Stifter, R. (1997) Greenery on the roof: a futuristic, ecological building method. In *For a More Human Architecture in Harmony with Nature, Hundertwasser Architecture* (ed. by F. Hundertwasser), pp. 156–158. Benedikt Tashen Verlag GmbH, Cologne.

Stone, B. (2004) Paving over paradise: how land use regulations promote residential imperviousness. *Landscape and Urban Planning*, **69**, 101–113.

Stovin, V. (2010) The potential of green roofs to manage Urban Stormwater. *Water and Environment Journal*, **24**, 192–199.

Strohbach, M.W., Haase, D., & Kabisch, N. (2009) Birds and the city: urban biodiversity, land use and socioeconomics. *Ecology & Society*, **14**, 31. www.ecologyandsociety.org/vol14/iss2/art31/.

Stülpnagel, A.v. (1987) *Klimatische Veränderungen in Ballungsgebieten unter besonderer Berücksichtigung der Ausgeleichswirkung von Grünflächen, dargestellt am Beispiel von Berlin (West)*. Technical University Berlin, Berlin.

Stülpnagel, A.v., Horbert, M., & Sukopp, H. (1990) The importance of vegetation for the urban climate. In *Urban Ecology* (ed. by H.E.A. Sukopp), pp. 175–193. SPB Academic Publishing, The Hague.

SUDSWP (2010) *SUDS for Roads*. Scottish Environment Protection Agency, Stirling.

Sukopp, H. (1990) Urban ecology and its application in Europe. In *Urban Ecology* (ed. by H. Sukopp), pp. 1–22. SPB Academic, The Hague.

Sunyer, J., Schwartz, J., Tobias, A., Macfarlane, D., Garcia, J., & Antó, J.M. (2000) Patients with chronic obstructive pulmonary disease are at increased risk of death associated with urban particle air pollution: a case crossover analysis. *American Journal of Epidemiology*, **151**, 50–56.

Susca, T., Gaffin, S.R., & Dell'Osso, G.R. (2011) Positive effects of vegetation: Urban heat island and green roofs. *Environmental Pollution*, **159**, 2119–2126.

SW (2011) *Sewers for Scotland 2nd Edition: Consultation Report*. Scottish Water, Edinburgh.

Szyszkowicz, M., Porada, E., Kaplan, G.G., & Grafstein, E. (2012a) Ambient ozone as a risk factor for ED visits for cellulitis. *Environment & Pollution*, **1**, 105–111.

Szyszkowicz, M., Porada, E., Kaplan, G.G., & Rowe, B.H. (2010) Ambient ozone and emergency department visits for cellulitis. *International Journal of Environmental Research and Public Health*, **7**, 4078–4088.

Szyszkowicz, M., Porada, E., Searles, G., & Rowe, B.H. (2012b) Ambient ozone and emergency department visits for skin conditions. *Air Quality, Atmosphere and Health*, **5**, 303–309.

Taha, H. (1997) Urban climates and heat islands: Albedo, evapotranspiration, and anthropogenic heat. *Energy and Buildings*, **25**, 99–103.

Takahashi, M., Konaka, D., Sakamoto, A., & Morikawa, H. (2005) Nocturnal uptake and assimilation of nitrogen dioxide by C3 and CAM plants. *Zeitschrift Fur Naturforschung C-a Journal of Biosciences*, **60**, 279–284.

Takano, T., Nakamura, K., & Watanabe, M. (2002) Urban residential environments and senior citizens' longevity in megacity areas: the importance of walkable green spaces. *Journal of Epidemiology and Community Health*, **56**, 913–918.

Tan, H., Barret, M., Rice, O., Dowling, D.N., Burke, J., Morrissey, J.P., & O'Gara, F. (2012) Long-term agrichemical use leads to alterations in bacterial community diversity. *Plant Soil and Environment*, **58**, 452–458.

Tan, P.Y. & Sia, A. (2005) A pilot green roof research project in Singapore. *Proceedings of the 3rd North American Green Roof Conference: Greening Rooftops for Sustainable Communities*, 4–6 May 2005, Washington DC, pp. 13. The Cardinal Group, Toronto.

Tani, A. & Hewitt, C.N. (2009) Uptake of aldehydes and ketones at typical indoor concentrations by houseplants. *Environmental Science & Technology*, **43**, 8338–8343.

Tarran, J., Torpy, F., & Burchett, M. (2007) Use of indoor plants to cleanse indoor air – research review. *Proceedings of Sixth International Conference on Indoor Air Quality, Ventilation & Energy Conservation in Buildings – Sustainable Built Environment*, Vol. 3, pp. 249–256, Sendai, Japan.

Taylor, A.F., Kuo, F.E., & Sullivan, W.C. (2002) Views of nature and self-discipline: evidence from inner city children. *Journal of Environmental Psychology*, **22**, 49–63.

TCF (2011) *What is Green Infrastructure?* The Conservation Fund, Chapel Hill, NC. Available at: www.greeninfrastructure.net/content/definition-green-infrastructure.

TCPA (2004) *Biodiversity by design. A guide for sustainable communities.* Town and Country Planning Association, London.

Tello, M.-L., Tomalek, M., Siwecki, R., Gáper, J., Motta, E., & Mateo-Sagasta, E. (2005) Biotic urban growing conditions – threats, pests and diseases. In *Urban Forests and Trees: A Reference Book* (ed. by C.C. Konijnendijk, K. Nilsson, T.B. Randrup & J. Schipperijn), pp. 325–365. Springer-Verlag, Berlin.

ten Brink, P., Berghofer, A., Schroter-Schlaack, C., Sukhdev, P., Vakrou, A., White, S., & Wittmer, H.

(2009) *TEEB – The Economics of Ecosystems and Biodiversity for National and International Policy Makers 2009. Summary: responding to the value of nature.* UNEP, Kenya.

Thoennessen, M. (2002) Elementdynamik in fassadenbegrűnendm Wilden Wein. *Kölner Geograph. Arbeiten Heft,* **78,** 1–110.

Thomas, J.A. (1983) The Ecology and Conservation of *Lysandra bellargus* (Lepidoptera, Lycaenidae) in Britain. *Journal of Applied Ecology,* **20,** 59–83.

Thönnessen, M. & Werner, W. (1996) Die fassadenbegrünende Dreispitzige Jungfernrebe als Akkumulationsindikator. *Gefahrstoffe – Reinhaltung der Luft,* **56,** 351–357.

Tiwary, A., Morvan, H.P., & Colls, J.J. (2005) Modelling the size-dependent collection efficiency of hedgerows for ambient aerosols. *Journal of Aerosol Science,* **37,** 990–1015.

Tiwary, A., Reff, A., & Colls, J.J. (2008) Collection of ambient particulate matter by porous vegetation barriers: Sampling and characterization methods. *Journal of Aerosol Science,* **39,** 40–47.

Tiwary, A., Sinnett, D., Peachey, C., Chalabi, Z., Vardoulakis, S., Fletcher, T., Leonardi, G., Grundy, C., Azapagic, A., & Hutchings, T.R. (2009) An integrated tool to assess the role of new planting in PM10 capture and the human health benefits: A case study in London. *Environmental Pollution,* **157,** 2645–2653.

Tobermore (2014) *Turfstone Block Paving,* Vol. 2014. Tobermore, Northern Ireland.

Todorova, A., Asakawa, S., & Aikoh, T. (2004) Preferences for and attitudes towards street flowers and trees in Sapporo, Japan. *Landscape and Urban Planning,* **69,** 403–416.

Tolkien, J.R.R. (1937) *The Hobbit.* George Allen & Unwin, London.

Tonietto, R., Fant, J., Ascher, J., Ellis, K., & Larkin, D. (2011) A comparison of bee communities of Chicago green roofs, parks and prairies. *Landscape and Urban Planning,* **103,** 102–108.

Topp, E., Scheunert, I., Attar, A., & Korte, F. (1986) Factors affecting the uptake of $^{14}C$ labelled organic chemicals by plants from soil. *Ecotoxicology and Environmental Safety,* **11,** 219–228.

Torpy, F.R., Irga, P.J., & Burchett, M. (2014) Profiling indoor plants for the amelioration of high $CO_2$ concentrations. *Urban Forestry & Urban Greening,* **13,** 227–233.

Torrance, S., Bass, B., MacIvor, S., & McGlade, T. (2013) *Design Guidelines for Biodiverse Green Roofs.* City of Toronto, Toronto.

Tota-Maharaj, K. & Scholz, M. (2010) Efficiency of permeable pavement systems for the removal of urban runoff pollutants under varying environmental conditions. *Environmental Progress and Sustainable Energy,* **29,** 358–369.

Tota-Maharaj, K. & Scholz, M. (2013) Combined permeable pavement and photocatalytic titanium dioxide oxidation system for urban run-off treatment and disinfection. *Water and Environment Journal,* **27,** 338–347.

Tota-Maharaj, K., Scholz, M., Ahmed, T., French, C., & Pagaling, E. (2010) The synergy of permeable pavements and geothermal heat pumps for stormwater treatment and reuse. *Environmental Technology,* **31,** 1517–1531.

Treeconomics (2011) *Torbay's Urban Forest: Assessing Urban Forest Effects and Values. A Report on the Findings from the UK i-Tree Eco Pilot Project.* Treeconomics, Exeter.

Tubby, K. (2006) *Final report on the condition survey of non-woodland amenity trees.* Forest Research, Alice Holt.

Turnbull, D.J. & Parisi, A. (2006) Effective shade structures. *The Medical Journal of Australia,* **184,** 13–15.

Tyndall, J. & Colletti, J. (2007) Mitigating swine odor with strategically designed shelterbelt systems: a review. *Agroforestry Systems,* **69,** 45–65.

Tyndall, J.C. & Larsen, G.L. (undated) *Factsheet vegetated environmental buffers for odour mitigation.* Pork Information Gateway.

Tyrväinen, L., Gustavsson, R., Konijnendijk, C., & Ode, A. (2006) Visualization and landscape laboratories in planning, design and management of urban woodlands. *Forest Policy and Economics,* **8,** 811–823.

Tyrväinen, L. & Miettinen, A. (2000) Property prices and urban forest amenities. *Journal of Environmental Economics and Management,* **39,** 205–223.

Tyrväinen, L., Pauleit, S., Seeland, K., & de Vries, S. (2005) Benefits and uses of urban forests and trees. In *Urban Forests and Trees* (ed. by C.C. Konijnendijk, K. Nilsson, T.B. Randrup & J. Schipperijn), pp. 81–114. Springer, Dordrecht.

Tzoulas, K., Korpela, K., Venn, S., Yli-Pelkonen, V., Kazmierczak, A., Niemela, J., & James, P. (2007) Promoting ecosystem and human health in urban areas using Green Infrastructure: A literature review. *Landscape and Urban Planning*, **81**, 167–178.

UEL (2011) *Urban Orchard*. University of East London, London.

Ugolini, F., Tognetti, R., Raschi, A., & Bacci, L. (2013) *Quercus ilex* L. as bioaccumulator for heavy metals in urban areas: Effectiveness of leaf washing with distilled water and considerations on the trees distance from traffic. *Urban Forestry & Urban Greening*, **12**, 576–584.

UKGK (2014) UK Go Karting. UK Go Karting, Tenterden, Kent. www.uk-go-karting.com/contact.

Ulrich, R.S. (1983) Aesthetic and affective response to natural environment. In *Human Behaviour and Environment: Advances in Theory and Research* (ed. by I. Altman & J.F. Wohlwill), Volume 6: Behaviour and the Natural Environment, pp. 85–60. Plenum Press, New York.

Ulrich, R.S. (1984) View through a window may influence recovery from surgery. *Science,* **224**, 420–421.

UNI-Group (2014) *UNI Eco-stone®*. UNI-Group, U.S.A., Palm Beach Gardens, Florida, www. uni-groupusa.org/eco-stone.htm.

Unsworth, M.H. & Wilshaw, J.C. (1989) Wet, occult and dry deposition of pollutants on forests. *Agricultural and Forest Meteorology*, **47**, 221–238.

UoC (undated) *Brownian Motion and Diffusion*. University of California, San Diego.

USDAFS (2014a) *i-Tree Streets User's Manual v5.0*. USDA Forest Service. www.itreetools.org/resources/manuals/Streets_Manual_v5.pdf.

USDAFS (2014b) *What is i-Tree?* USDA Forest Service. www.itreetools.org/.

USGBC (2014) *LEED*. US Green Building Council, Washington, DC. www.usgbc.org/leed/leed_main.asp.

Van Bohemen, H.D., Fraaij, A.L.A., & Ottele, M. (2008) Ecological engineering, green roofs and the greening of vertical walls of buildings in urban areas. *Ecocity World Summit 2008 Proceedings*, pp. 1–10.

van den Berg, L.M. (2002) Urban agriculture between allotment and market gardening: contributions to the sustainability in African and Asian cities. In C.A. Brebbia, J.F. Martin-Duque & L.C. Wadhwa (eds) *The Sustainable City II: Urban Regeneration and Sustainability*, 945–959. WIT Press, Southampton.

van den Berg, A.E. & Custers, M.H.G. (2011) Gardening promotes neuroendocrine and effective restoration from stress. *Journal of Health Psychology*, **16**, 3–11.

van den Berg, A.E., van Winsum-Westra, M., de Vries, S. & van Dillen, S.M.E. (2010) Allotment gardening and health: a comparative survey among allotment gardens and their neighbors without an allotment. *Environmental Health*, **9**, 74.

van Dillen, S.M.E., De Vries, S., Groenewegen, P.P., & Spreeuwenberg, P. (2012) Greenspace in urban neighbourhoods and residents' health: adding quality to quantity. *Journal of Epidemiology and Community Health*, **66:e9**, 5.

van Loon, L.C., Bakker, P.A.H.M., & Pieterse, C.M.J. (1998) Systemic resistance induced by rhizosphere bacteria. *Annual Review of Phytopathology*, **36**, 453–483.

Van Renterghem, T. & Botteldooren, D. (2008) Numerical evaluation of sound propagating over green roofs. *Journal of Sound and Vibration*, **317**, 781–799.

Van Renterghem, T. & Botteldooren, D. (2011) In-situ measurements of sound propagating over extensive green roofs. *Building and Environment*, **46**, 729–738.

Van Renterghem, T., Hornikx, M., Forssen, J., & Botteldooren, D. (2013) The potential of building envelope greening to achieve quietness. *Building and Environment*, **61**, 34–44.

van Setten, B.A.A., Makkee, M., & Moulijn, J.A. (2001) Science and technology of catalytic diesel particulate filters. *Catalysis Reviews*, **43**, 489–564.

VanWoert, N.D., Rowe, D.B., Andresen, J.A., Rugh, C.L., Fernandez, R.T., & Xiao, L. (2005) Green roof stormwater retention: Effects of roof surface, slope, and media depth. *Journal of Environmental Quality*, **34**, 1036–1044.

Varshney, C.K. & Mitra, I. (1993) Importance of hedges in improving urban air-quality. *Landscape and Urban Planning*, **25**, 75–83.

Velazquez, L.S. (2004) *SkyGardens – travels in landscape architecture*. Greenroofs.com. www.greenroofs.com/archives/sg_jan-apr04.htm.

Velikova, V., Tsonev, T., Loreto, F., & Centritto, M. (2011) Changes in photosynthesis, mesophyll conductance to $CO_2$, and isoprenoid emissions in *Populus nigra* plants exposed to excess nickel. *Environmental Pollution*, **159**, 1058–1066.

Vergnes, A., Le Viol, I., & Clergeau, P. (2012) Green corridors in urban landscapes affect the arthropod communities of domestic gardens. *Biological Conservation*, **145**, 171–178.

Vergnes, A., Kerbiriou, C., & Clergeau, P. (2013) Ecological corridors also operate in an urban matrix: A test case with garden shrews. *Urban Ecosystems*, **16**, 511–525.

Vertigarden (2011) *VertiGarden frame and irrigation*. Vertigarden. www.vertigarden.co.uk.

VHG (2014) *The Roof Gardens*. Virgin Hotels Group Ltd. www.roofgardens.virgin.com/.

Vidrih, B. & Medved, S. (2013) Multiparametric model of urban park cooling island. *Urban Forestry & Urban Greening*, **12**, 220–229.

Viles, H., Sternberg, T., & Cathersides, A. (2011) Is ivy good or bad for historic walls? *Journal of Architectural Conservation*, **17**, 25–41.

Viles, H.A. & Wood, C. (2007) Green walls?: integrated laboratory and field testing of the effectiveness of soft wall capping in conserving ruins. In *Building Stone Decay: from diagnosis to conservation* (ed. by P. R. & B.J. Smith). Geological Society, London.

Viljoen, A. & Bohn, K. (2014) *Second Nature Urban Agriculture: Designing Productive Cities*. Routledge, Abingdon.

Villiger, J. (1986) 'Green roofs and walls' – a necessity in the city. *Anthos,* **1/86**, 1–3.

Viswanathan, B., Volder, A., Watson, W.T., & Aitkenhead-Peterson, J.A. (2011) Impervious and pervious pavements increase soil $CO_2$ concentrations and reduce root production of American sweetgum (Liquidambar styraciflua). *Urban Forestry & Urban Greening*, **10**, 133–139.

Vlachokostas, C., Achillas, C., Moussiopoulos, N., Kalogeropoulos, K., Sigalas, G., Kalognomou, E.A., & Banias, G. (2012) Health effects and social costs of particulate and photochemical urban air pollution: a case study for Thessaloniki, Greece. *Air Quality Atmosphere and Health*, **5**, 325–334.

Vollmer, J.J. & Gordon, S.A. (1975) Chemical communication: II between plants and insects and among social insects. *Chemistry*, **48**, 6–11.

Wang, H., Shi, H., & Li, Y. (2011) Leaf dust capturing capacity of urban greening plant species in relation to leaf micromorphology. *International Symposium on Water Resource and Environmental Protection (ISWREP)*, pp. 2198–2201.

Wang, S., Ang, H.M., & Tade, M.O. (2007) Volatile organic compounds in indoor environment and photocatalytic oxidation: State of the art. *Environment International*, **33**, 694–705.

Wang, Z.Q., Wu, L.H., Animesh, S., & Zhu, Y.H. (2009) Phytoremediation of rocky slope surfaces: selection and growth of pioneer climbing plants. *Pedosphere*, **19**, 541–544.

Wani, B.A. & Khan, A. (2010) Effect of cement dust pollution on the vascular cambium of *Juglans regia* (L.). *Journal of Ecology and the Natural Environment*, **2**, 225–229.

Warscheid, T. & Braams, J. (2000) Biodeterioration of stone: a review. *International Biodeterioration & Biodegradation*, **46**, 343–368.

Watermatic (2011) *Green wall irrigation systems*. Watermatic. www.watermaticltd.co.uk/green-wall.html.

Watt, J. & Ball, D.J. (2009) *Trees and the Risk of Harm*. National Tree Safety Group, Edinburgh.

WBOC (2013) *Berlin's new 'rain tax' sticks atlantic general with large bill*. WBOC16. www.wboc.com/story/24311245/berlins-rain-tax-sticks-atlantic-general-with-large-bill.

Wen, L.M. & Rissel, C. (2008) Inverse associations between cycling to work, public transport, and overweight and obesity: Findings from a population based study in Australia. *Preventive Medicine*, **46**, 29–32.

White, M.P., Alcock, I., Wheeler, B.W., & Depledge, M.H. (2013) Would you be happier living in a greener urban area? A fixed-effects analysis of panel data. *Psychological Science*, **24**, 920–928.

Whitehead, P.G., Wilby, R.L., Battarbee, R.W., Kernan, M., & Wade, A.J. (2009) A review of the potential impacts of climate change on surface water quality. *Hydrological Sciences Journal-Journal Des Sciences Hydrologiques*, **54**, 101–123.

Whitford, V., Ennos, A.R., & Handley, J.F. (2001) "City form and natural process" – indicators for the ecological performance of urban areas and their application to Merseyside, UK. *Landscape and Urban Planning*, **57**, 91–103.

Whittaker, R.H. & Feeny, P.P. (1971) Allelochemicals: chemical interactions between species. *Science*, **171**, 757–770.

Whittinghill, L. (2011) *Cities Alive: Green Roofs for Healthy Cities*. Michigan State University, East Lansing, MI.

Whittinghill, L.J. & Rowe, D.B. (2011) Salt tolerance of common green roof and green wall plants. *Urban Ecosystems*, **14**, 783–794.

Whittinghill, L.J., Rowe, D.B., & Cregg, B.M. (2013) Evaluation of vegetable production on extensive green roofs. *Agroecology and Sustainable Food Systems*, **37**, 465–484.

Wickham, S. & Hallam, A. (2006) Structural and Construction Issues. *Green Roof Conference 2006*, Sheffield University, Sheffield, pp. 26.

Wilby, R.L. (2003) Past and projected trends in London's urban heat island. *Weather*, **58**, 251–260.

Wild, E., Dent, J., Thomas, G.O., & Jones, K.C. (2006) Visualizing the air-to-leaf transfer and within-leaf movement and distribution of phenanthrene: Further studies utilizing two-photon excitation microscopy. *Environmental Science & Technology*, **40**, 907–916.

Williams, R., ed. (2000) *Orchards & Wildlife*. Common Ground, London.

Willis, S.G., Hill, J.K., Thomas, C.D., Roy, D.B., Fox, R., Blakeley, D.S., & Huntley, B. (2009) Assisted colonization in a changing climate: a test-study using two U.K. butterflies. *Conservation Letters*, **2**, 45–51.

Wilmers, F. (1988) Green for melioration of urban climate. *Energy and Buildings*, **11**, 289–299.

Wilmers, F. (1991) Effects of vegetation on urban climate buildings. *Energy and Buildings*, **15**, 507–514.

Wilson, E.O. (1984) *Biophilia – the human bond with other species*. Harvard University Press, Cambridge, Massachusetts.

Witherspoon, J.P. & Taylor, F.G.J. (1969) Retention of a fallout simulant containing 134Cs by pine and oak trees. *Health Physics*, **17**, 825–829.

Witter, G. (1986) A campaign for climbing plants in the city. *Anthos*, **1**, 29–34.

Wolf, D. & Lundholm, J.T. (2008) Water uptake in green roof microcosms: Effects of plant species and water availability. *Ecological Engineering*, **33**, 179–186.

Wolf, K.L. (2005) Business district streetscapes, trees, and consumer response. *Journal of Forestry*, **103**, 396–400.

Wolkoff, P. (2003) Trends in Europe to reduce the indoor air pollution of VOCs. *Indoor Air*, **13**, 5–11.

Wolverton, B.C. (1997) *How to Grow Fresh Air: 50 Houseplants That Purify Your Home or Office*. Penguin Books, London.

Wolverton, B.C., Douglas, W.L., & Bounds, K. (1989a) *A study of interior landscape plants for indoor air pollution abatement: an interim report*. NASA, Stennis Space Center, MS.

Wolverton, B.C., Johnson, A., & Bounds, K. (1989b) *Interior landscape plants for indoor pollution abatement – final report*. NASA, Stennis Space Center, MS.

Wolverton, B.C., McDonald, R.C., & Mesick, H.H. (1985) Foliage plants for indoor removal of the primary combustion gasses carbon monoxide and nitrogen dioxide. *Journal of the Mississippi Academy of Sciences*, **30**, 1–8.

Wolverton, B.C., McDonald, R.C., & Watkins, E.A.J. (1984) Foliage plants for removing indoor air pollutants from energy-efficient homes. *Economic Botany*, **32**, 224–228.

Wolverton, B.C. & Wolverton, J.D. (1993) Plants and soil microorganisms: removal of formaldehyde, xylene, and ammonia from the indoor environment. *Journal of the Mississippi Academy of Sciences*, **38**, 11–15.

Wolverton, B.C. & Wolverton, J.D. (1996) Interior plants: their influence on airborne microbes inside energy efficient buildings. *Journal of the Mississippi Academy of Sciences*, **41**, 99–105.

Wong, N.H., Tan, A.Y.K., Chen, Y., Sekar, K., Tan, P.Y., Chan, D., Chiang, K., & Wong, N.C. (2010a) Thermal evaluation of vertical greenery systems for building walls. *Building and Environment*, **45**, 663–672.

Wong, N.H., Tan, A.Y.K., Chiang, K., & Wong, N.C. (2010b) Acoustics evaluation of vertical greenery systems for building walls. *Building and Environment*, **45**, 411–420.

Wong, N.H., Tan, A.Y.K., Tan, P.Y., & Wong, N.C. (2009) Energy simulation of vertical greenery systems. *Energy and Buildings*, **41**, 1401–1408.

Wood, R.A., Burchett, M.D., Alquezar, R., Orwell, R.L., Tarran, J., & Torpy, F. (2006) The potted-plant microcosm substantially reduces indoor air VOC pollution I. Office field-study. *Water, Air and Soil Pollution*, **175**, 163–180.

Wood, R.A., Orwell, R.L., Tarran, J., Torpy, F., & Burchett, M. (2002) Potted-plant/growth media interactions and capacities for removal of volatiles from indoor air. *Journal of Horticultural Science & Biotechnology*, **77**, 120–129.

Woodell, S. (1979) The flora of walls and pavings. In *Nature in Cities* (ed. by I.C. Laurie), pp. 135–157. John Wiley and Sons, Chichester.

WT (2014) *The nature's calendar survey*. The Woodland Trust, Grantham. www.naturescalendar.org.uk/.

Xia, L., Lenaghan, S.C., Zhang, M., Wu, Y., Ahao, X., Burris, J.N., & Stewart, C.N.J. (2011) Characterization of English ivy (*Hedera helix*) adhesion force and imaging using atomic force microscopy. *Journal of Nanoparticle Research*, **13**, 1029–1037.

Xu, Z.J., Qin, N., Wang, J.G., & Tong, H. (2010) Formaldehyde biofiltration as affected by spider plant. *Bioresource Technology*, **101**, 6930–6934.

Xu, Z.J., Wang, L., & Hou, H.P. (2011) Formaldehyde removal by potted plant-soil systems. *Journal of Hazardous Materials*, **192**, 314–318.

Yang, D.S., Pennisi, S.V., Son, K.-C., & Kays, S.J. (2009) Screening indoor plants for volatile organic pollutant removal efficiency. *HortScience*, **44**, 1377–1381.

Yang, J., McBride, J., Zhou, J., & Sun, A. (2005) The urban forest in Beijing and its role in air pollution reduction. *Urban Forestry & Urban Greening*, **3**, 65–78.

Yang, J., Yu, Q., & Gong, P. (2008) Quantifying air pollution removal by green roofs in Chicago. *Atmospheric Environment*, **42**, 7266–7273.

Yoo, M.H., Kwon, Y.J., Son, K.-C., & Kays, S.J. (2006) Efficacy of indoor plants for the removal of single and mixed volatile organic pollutants and physiological effects of the volatiles on the plants. *Journal of the American Society for Horticultural Science*, **131**, 452–458.

Yoshimura, H., Zhu, H., Wu, Y.Y., & Ma, R.J. (2010) Spectral properties of plant leaves pertaining to urban landscape design of broad-spectrum solar ultraviolet radiation reduction. *International Journal of Biometeorology*, **54**, 179–191.

Yu, X.D., Zhou, P., Zhou, X., & Liu, Y.H. (2005) Cyanide removal by Chinese vegetation–quantification of the Michaelis-Menten kinetics. *Environmental Science and Pollution Research International*, **12**, 221–226.

Zakaria, N.A., Ghani, A.A., Abdullah, R., Sidek, L.M., & Ainan, A. (2003) Bio-ecological drainage system (BIOECODS) for water quantity and quality control. *International Journal of River Basin Management*, **1**, 1–15.

Zhang, B., Xie, G., Zhang, C., & Zhang, J. (2012) The economic benefits of rainwater-runoff reduction by urban green spaces: A case study in Beijing, China. *Journal of Environmental Management*, **100**, 65–71.

Zheng, Y., Lyons, T., Ollerenshaw, J.H., & Barnes, J.D. (2000) Ascorbate in the leaf apoplast is a factor mediating ozone resistance in *Plantago major*. *Plant Physiology and Biochemistry*, **38**, 403–411.

Zhou, X. & Kim, J. (2013) Social disparities in tree canopy and park accessibility: A case study of six cities in Illinois using GIS and remote sensing. *Urban Forestry & Urban Greening*, **12**, 88–97.

Zhou, Y. & Shepherd, J.M. (2010) Atlanta's urban heat island under extreme heat conditions and potential mitigation strategies. *Natural Hazards*, **52**, 639–668.

Zhu, Y., Hinds, W.C., Kim, S., Shen, S., & Sioutas, C. (2002a) Study of ultrafine particles near a major highway with heavy-duty diesel traffic. *Atmospheric Environment*, **36**, 4323–4335.

Zhu, Y., Hinds, W.C., Kim, S., & Sioutas, C. (2002b) Concentration and size distribution of ultrafine particles near a major highway. *Journal of the Air and Waste Management Association*, **52**, 1032–1042.

Zulfacar, A. (1975) Vegetation and urban environment. *Journal of the Urban Planning and Development Division of the American Society of Civil Engineers*, **101**, 21–33.

# INDEX